Water in Kentucky

Natural History, Communities, and Conservation

Edited by Brian D. Lee, Daniel I. Carey, and Alice L. Jones

UNIVERSITY PRESS OF KENTUCKY

Scholarly publisher for the Commonwealth,
serving Bellarmine University, Berea College, Centre
College of Kentucky, Eastern Kentucky University,
The Filson Historical Society, Georgetown College,
Kentucky Historical Society, Kentucky State University,
Morehead State University, Murray State University,
Northern Kentucky University, Transylvania University,
University of Kentucky, University of Louisville,
and Western Kentucky University.

Editorial and Sales Offices: The University Press of Kentucky
663 South Limestone Street, Lexington, Kentucky 40508-4008
www.kentuckypress.com

Cataloging-in-Publication data available from the Library
of Congress

ISBN 978-0-8131-6868-5 (hardcover : alk. paper)
ISBN 978-0-8131-6870-8 (pdf)
ISBN 978-0-8131-6869-2 (epub)

This book is printed on acid-free paper meeting
the requirements of the American National Standard
for Permanence in Paper for Printed Library Materials.
∞

Manufactured in the United States of America.

Member of the Association of American University Presses

Special thanks to our sponsors

Sustainable Streams

College of Agriculture, Food, and Environment
Department of Landscape Architecture
and
University of Kentucky

Contents

Introduction

Water and the processes driven by water are often essential in defining how we identify different regions, as well as issues across the Commonwealth of Kentucky. *Water in Kentucky: Natural History, Communities, and Conservation* is about the past, present, and some aspects of the future of water in the commonwealth. The volume's overall objective is to explore the variety of ways water shaped and continues to shape life in Kentucky through telling both the biophysical and the historical and contemporary stories of water's impact. By reading this book, you should gain an improved understanding of the role water had and continues to play in each of our daily lives.

The volume is also intended as an information source and an inspiration to make a positive difference in our communities. We hope that readers will be challenged to help solve local water-quality problems, reaffirm what they are already doing and learn about new ideas, or do something different in order to be more successful.

The book is aimed at readers who have an interest in water and water issues but are not necessarily experts. The audiences we have in mind include teachers, researchers, practitioners, community watershed organization members, government agency personnel, and students. We see this volume potentially being used in a variety of undergraduate fields of study, such as geography, history, environmental science, forestry, hydrology, watershed management, and landscape architecture, to name a few.

As editors, we spent a great deal of time discussing from our perspective some of the essential and untold topical stories across the commonwealth. We sought to include a wide range of stories told by a variety of people who have professional and often deeply personal experiences with water in the commonwealth. Lead contributors were tasked with a minimal topical story charge to develop stories either alone or with selected cocontributors.

During the writing and reviewing of contributors' drafts, the book has gone through several organizational iterations. Undoubtedly, you may see other ways to organize sections or the entire book, as we have as well. As editors, though, a key theme we sought to express is one of hope. We could have focused only on stories of despair surrounding water. In fact, some of the chapters will start at a point of desperation and move on to what was learned and could be applied to prevent a tragic event from occurring or what could be done when such an event does occur.

We encouraged contributors to keep in mind Kentucky's storytelling culture and to weave a compelling story of people, landscapes, and events that are grounded in and supported by formal literature. The chapters use known human, social, biological, and hydrologic science sources and link to real on-the-ground Kentucky examples to illustrate historical or anticipated situations. Although the chapters are bound as a book, each chapter should tell a story that can stand on its own outside the context of the edited volume. Each chapter includes an ending that states the two or three major implications, actions, or other takeaway points that the reader could apply to other situations. Because of space limitations, in many cases contributors have also identified additional resources for further reading, such as books, popular-press articles, academic literature, or research reports beyond the literature cited.

Some readers may be familiar with the genre of how-to books on watershed management. The volume before you is not such a book. We see *Water in Kentucky* as an excellent companion to community watershed-planning guidebooks, but we specifically did not intend to cover the same territory as a modern watershed-management book. We also do not see this book as a prescription for what should be done in the future, although forward-looking contributions are included.

This book was written over several years, and dozens of people took part in telling the stories of water in Kentucky. There have been other published books about Kentucky's water resources. The reason that this book is needed is that it fills a void in the literature by examining how water has shaped Kentucky's landscapes, people, and communities over time and space. At twenty-three chapters, the book is not all-inclusive, and we knew that we could not have every story told in a single volume, but we pushed the limits of what a single volume could hold. We also sought to have the volume read from beginning to end, from any starting point, and by individual chapters. Accomplishing this goal with perfect efficiency is a challenge.

This book is long overdue for reasons that could fill a book themselves. Bringing this book to publication has been a very sinuous journey. The contributors have given tirelessly and freely to this effort with a great deal of patience. We are indebted to each and every contributor who has steadfastly put up with a production process whose greatly protracted timeline has felt at times almost frozen or at least stuck in molasses. We are also grateful to the University Press of Kentucky for seeing the promise in this project and supporting it throughout. The outside reviewers have contributed to helping this book take its final form through broad and detailed comments. It was clear that the reviewers fully supported this book and wanted it to be the best it could be. As editors, we are grateful to each other for the perspectives, styles, and vision we brought to the project. We each have grown and respect the contributions that have been brought to make this endeavor a reality.

Chapter One

The Waters of Kentucky

Daniel I. Carey

More than 32 trillion (32 million million) gallons of water normally fall from the skies over Kentucky each year. Where does this precipitation come from? Most of Kentucky's water has been lifted from the western Gulf of Mexico by the sun (evaporation) and carried to the state by southerly winds. The evaporation process leaves the sea salts behind and yields fresh water. As the air cools, the water vapor condenses into droplets and falls as precipitation.

Kentucky receives an average of 46 inches of precipitation per year, mostly in the form of rain (Krieger, Cushman, and Thomas 1969). South-central and southeastern Kentucky average over 50 inches, and northern Kentucky 40 inches. The precipitation falls fairly uniformly throughout the year, about 4 inches each month, slightly less in September and October.

There have been large variations from the average precipitation. Lexington received less than 25 inches in 1930, the worst drought on record, and Caneyville in Grayson County received more than 7 feet of rain in 1979. Communities across Kentucky received no rainfall in October 1963 (Kentucky Climate Center 2013). Many areas received record rainfalls in January 1937, which caused the worst floods in Kentucky on record. The skies dumped nearly 23 inches of rain on Earlington in Muhlenberg County that January. Despite variations, precipitation in Kentucky is greater and more nearly the same in all months than in most areas of the world.

About 20 trillion gallons (63 percent) of Kentucky's annual gift of water is taken back to the air (and on to neighboring states) by evaporation or transpiration (similar to sweating) by plants. Most of this occurs in late summer and early fall, when temperatures are high and plant growth is at its peak. Nearly 9 trillion gallons (28 percent)

run off directly to streams, and 3 trillion gallons (9 percent) percolate into bedrock to become groundwater, primarily during winter and early spring, when vegetation is dormant, temperatures are low, and the ground is often saturated. Most floods occur at this time of year.

Groundwater subsequently provides water for springs, wells, and streams and maintains stream flow (also known as baseflow, drought flow, groundwater recession flow, low flow, or sustained flow) during droughts. The 7-day, 10-year low flow (7Q10) is often used as a measure of the minimum flow necessary to maintain water quality. During the drought of 1930, the flow at Lock 10 on the Kentucky River near Winchester was less than 42 cubic feet per second (cfs), the 7Q10, for more than 120 days. This flow was sustained by nearly 120 billion gallons of groundwater. Normal flows are about 1,500 cfs.

Surface Water: Rivers, Streams, and Creeks

The surface of Kentucky tilts to the north and west, the general flow direction of its major streams. The Cumberland River system, starting in southeastern Kentucky, takes a detour into Tennessee before returning to western Kentucky.

All but 3 percent of the runoff from Kentucky's seven major river basins goes into the Ohio River; the remainder flows directly into the Mississippi River below its confluence with the Ohio (figure 1.1). Effectively all of Kentucky's runoff eventually returns to the western Gulf of Mexico via the Mississippi, completing the hydrologic cycle.

More than 90,000 miles of streams have been mapped in Kentucky at a scale of 1:24,000. Streams generally flow

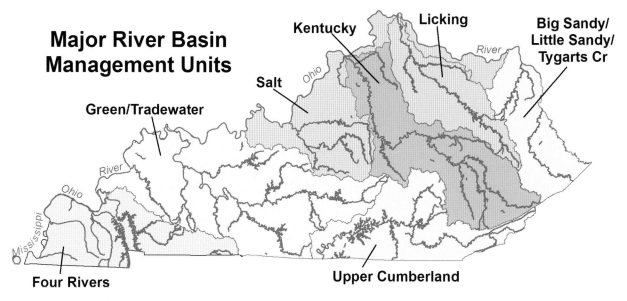

1.1. The major river basins in Kentucky.

in a pattern resembling tree branches (dendritic) within a drainage basin or watershed (figure 1.2).

A perennial stream has continuous flow in parts of its bed year-round during years of normal rainfall. Intermittent streams normally cease flowing for weeks or months each year, and ephemeral channels flow only for hours or days after rainfall or snowmelt. One way streams are characterized by hydrologists (those who study the movement, distribution, and quality of water throughout Earth) is by the term "order." Initial stream segments in a watershed are designated as first order. Two first-order streams combine to form a second-order stream, and so forth.

Kentucky is blanketed by a web of streams (figure 1.3). On average, about 2¼ miles of streams flow across each of Kentucky's 40,405 square miles of area.

Stream density, the length of streams per square mile, varies by region (figure 1.4). In the Sinkhole Plain of the Pennyroyal region, where many streams flow into sinkholes to underground conduits (flow channels), the surface stream density is half that of the rest of the state. On the impermeable (resistant to water infiltration) shale of hilly areas of the Bluegrass region, there are nearly 3 miles of surface streams per square mile.

Less than half of the 90,000 stream miles in the state have been given names. The amount of naming also var-

ies by region. Physiographic regions, shaped by geology, have distinctive topography, resources, and, to some extent, cultures. In the mountains of eastern Kentucky, where alluvial valleys (overbank stream deposits built up over geologic time) provide living space for homes and communities, the people are more intimately connected with the streams, and more than 75 percent of the stream miles are named. In the Inner and Outer Bluegrass, where streams lie in the background, less than one-third of the stream miles have been named. Forms and guidelines to officially name a stream can be obtained from the U.S. Geological Survey (USGS) through the U.S. Board on Geographic Names.

Surface Water: Wetlands, Ponds, Lakes, and Reservoirs

About 215,000 wetlands, ponds, lakes, and reservoirs in Kentucky slow the flow of water back to the Gulf. Ninety percent of these have an area of 1 acre or less. There are 1,429 lakes and reservoirs with an area greater than 10 acres, 136 greater than 100 acres, and 20 greater than 1,000 acres. Kentucky Lake has the largest surface area, 250 square miles, of any man-made lake east of the Mississippi, and Lake Cumberland can store more water,

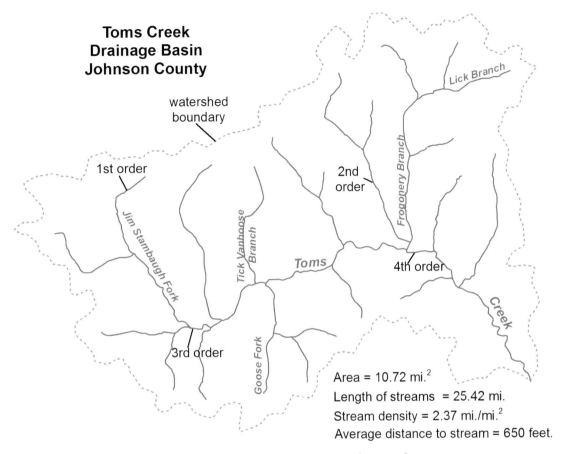

**Toms Creek
Drainage Basin
Johnson County**

watershed
boundary

1st order

2nd
order

Jim Stambaugh Fork

Tick Vanhoose Branch

Frogonery Branch

Lick Branch

Toms

4th order

Creek

3rd order

Goose Fork

Area = 10.72 mi.^{2}
Length of streams = 25.42 mi.
Stream density = 2.37 mi./mi.^{2}
Average distance to stream = 650 feet.

1.2. Toms Creek Drainage Basin in Johnson County, depicting a stream network nomenclature.

nearly 2 trillion gallons, than any lake east of the Mississippi and south of the Great Lakes.

Larger lakes and reservoirs, nearly all man-made, provide water for communities, industry, and recreation. They are often used to generate electric power or to provide flood protection. Lock and dam structures on the Kentucky, Green, and Barren Rivers—originally built for transportation—now provide pools for recreation and water supply.

Thousands of Kentuckians and visitors fish, boat, and take part in water sports on Kentucky's lakes. The U.S. Army Corps of Engineers operates 21 lakes in Kentucky with a total of 205,000 water acres and 4,250 miles of shoreline. These facilities are estimated to generate $480 million annually from 22 million person-trips per year (U.S. Army Corps of Engineers 2006). For example, the

fishing is generally good below the outlet spillway of Lake Malone in the Sandstone Hills region of Muhlenberg County. The lake is surrounded by 50-foot Caseyville Formation sandstone bluffs and hardwood forests with hiking trails. Constructed in the late 1950s, the lake covers 825 acres in Muhlenberg, Todd, and Logan Counties. The lake was made possible by local game and fishing clubs. Spillways and outlet works allow controlled releases that support aquatic life in streams below dams during times of low flow.

The National Wetlands Inventory Program of the U.S. Fish and Wildlife Service has mapped the nation's wetlands (U.S. Fish and Wildlife Service 2013). Wetlands serve a variety of functions: reducing flooding by providing temporary flood storage; recharging groundwater; improving water quality by filtering out impurities;

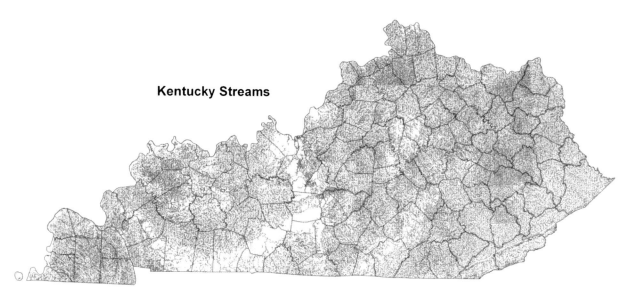

Kentucky Streams

1.3. The blue lines indicate streams, with a notable absence of surface streams in the karst-dominated landscapes.

recycling nutrients; providing habitat for fish, wildlife, and plants; and providing wildlife and recreation opportunities. Kentucky has over 440,000 acres of palustrine (marsh, swamp, pond) wetlands.

Groundwater

In addition to Kentucky's surface-water storage facilities, there are many reservoirs beneath the ground in Kentucky. These reservoirs are recharged through percolation, primarily during the winter and early spring, and discharge back to the surface through springs, seeps, and streams throughout the year. Groundwater discharge during summer and fall sustains stream flow.

A good underground reservoir, called an aquifer, is made of sand, gravel, or rock that provides a large volume of spaces for water, and those spaces are connected so that the water may move relatively freely. The availability of groundwater varies across the state, depending on the geology.

The unconsolidated (loosely arranged) sand and gravel beneath the ground in the Jackson Purchase region and in the Ohio River alluvium (see figure 1.4) have a large volume of well-connected pore space for water storage.

Wells in these areas yield hundreds of gallons of water per minute. The Louisville Water Company's Riverbank Filtration Project uses the natural filtration of the alluvial sand and gravel along the Ohio River to reduce turbidity and protect the water from pathogens, herbicides, and pesticides. The water is still chlorinated, but riverbank filtration reduces treatment requirements and is considered a green alternative to traditional methods.

Since unconsolidated sand and gravel make a good aquifer, we might suspect that sandstone and conglomerate would also. Sandstone and conglomerate are rocks that have been formed by cementing sand and gravel in the process of lithification (rock making). The cementing material is mostly where the grains touch, leaving the spaces between the grains open. Therefore, the rock is generally porous and well connected, so it can store and transmit significant volumes of water. Most wells drilled into the sandstone, siltstone, and conglomerate of the Eastern and Western Coal Fields will yield enough water for home use, and some will yield enough for public and industrial water supplies.

Shale (lithified clay) impedes the flow of groundwater and restricts its movement. A plot of 48,000 water wells (Kentucky Groundwater Repository, Kentucky Geologi-

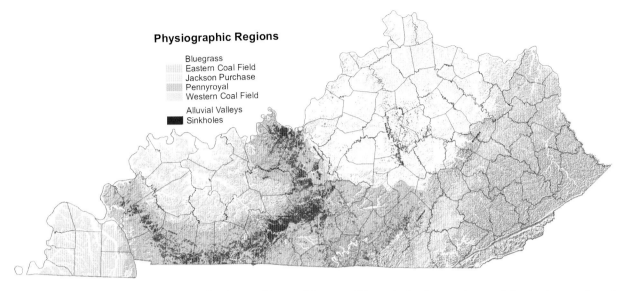

Physiographic Regions

Bluegrass
Eastern Coal Field
Jackson Purchase
Pennyroyal
Western Coal Field

Alluvial Valleys
Sinkholes

1.4. The physiographic regions that are a fundamental basis of the rise of Kentucky landscapes that result in shaping people and communities.

cal Survey [KGS]) reveals areas where the lack of wells indicates that shale in the bedrock may be a factor, such as the Bluegrass Hills region (figure 1.5). Groundwater in these areas is not a reliable water source. Of particular note is the groundwater environment in karst regions, which is different from that of the rest of the state because of the geologic composition and structure. Carbon dioxide, which forms very weak carbonic acid (comparable to vinegar) when it dissolves in water, is present in large amounts in rain and soil water. In karst regions, this acidic water percolates through limestone cracks and joints, dissolves the limestone, and enlarges the flow paths. Given enough time, conduits and large caves flowing with underground streams are formed. The storage and transmission of groundwater in limestone may vary significantly within a short distance depending on the jointing and dissolution.

Karst underground flow basins may not follow the topography of the surface. Water that flows underground through a sinkhole or sinking stream in one watershed may emerge from a spring in a different surficial watershed. Underground flow conditions may also change depending on the size of the springshed (the area that a spring drains, similar to a watershed), and some conduits may flow only during large storms or flooding.

In order to better understand the nature of underground flow in karst areas, hydrogeologists inject nontoxic dyes into the water where it enters the ground and place dye detectors at a number of downstream springs to determine where the water reemerges. In this way, they can estimate the location of underground flow paths and create groundwater basin maps that can be used to more effectively manage water quality and stormwater runoff.

Flooding

Kentuckians are blessed, on average, with all the water they need or want, but sometimes they get too little—periodic droughts—and sometimes they get too much. The National Weather Service estimates flood damages in Kentucky in the tens of millions of dollars each year.

U.S. Department of Agriculture (USDA) soil-survey data for Kentucky indicate that 828,751 acres are frequently flooded and 1,841,269 acres are occasionally

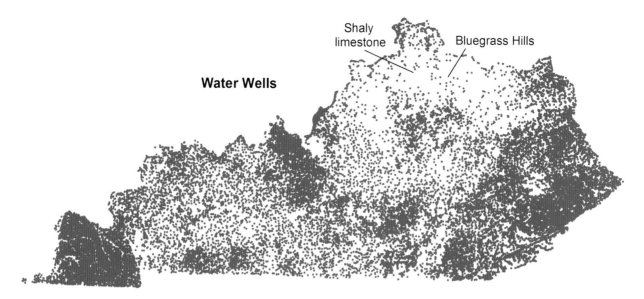

1.5. The blue dots represent the tens of thousands of domestic water wells that are related to the geologic formations, human populations, and water-supply capacity/need for domestic use.

flooded. Therefore, 2,670,000 acres (4,170 square miles), or roughly 10 percent of the state, are subject to flooding.

For example, some areas of eastern Kentucky have been devastated by stream flooding. For many years, we humans felt that we could control nature and eliminate flood damage by constructing dams, levees, and other flood-control works. We now recognize that a more useful way to reduce flood damage is to avoid construction in the floodplain (flat or nearly flat land adjacent to a stream or river that experiences occasional or periodic flooding) and use our floodplains for recreation and other purposes only marginally affected by flooding. The Paul Hunt Thompson Golf Course and the adjacent Stumbo Park near Allen in Floyd County are floodplain uses that reduce damage due to flooding.

The lack of surface streams in karst regions does not mean that flooding is not an issue. Some flood-prone areas are miles from the nearest surface stream or floodplain, and property owners may not realize that they are at risk until a flood occurs (American Geosciences Institute). Sinkhole flooding causes more than a million dollars in damage each year in Kentucky (KGS 2010b). For example, sinkhole flooding can be observed in highly developed karst of Fayette County in the Bluegrass region.

Water Usage

When our ancestors crawled onto the land, they brought the sea with them; we are mostly water. A 200-pound man contains 120 pounds, about 15 gallons, of water—in muscles and kidneys (79 percent water), brain and heart (73 percent water), lungs (83 percent water), bones (31 percent water), and skin (64 percent water) (USGS 2013b). Water carries nutrients in our bodies and hauls away the waste. We must continually replenish the water in our bodies and take in a minimum of slightly over 3 quarts of water each day (Gleick 1996).

Road names give an indication of the importance of water power in early Kentucky—Parkers Mill Road, Fords Mill Road, Craigs Mill Road, Clays Mill Road, and others. All the early mills were powered by streams. The iconic Weisenberger Mill has been continuously operated by six generations of Weisenbergers since 1865 in Midway, Scott County, and is one of the few remaining operating examples. South Elkhorn Creek provides hydropower to the mill's turbines.

Kentuckians use about 4.5 billion gallons of water each day (Downs and Caldwell 2007, updated). Power generation uses about four times as much water as all

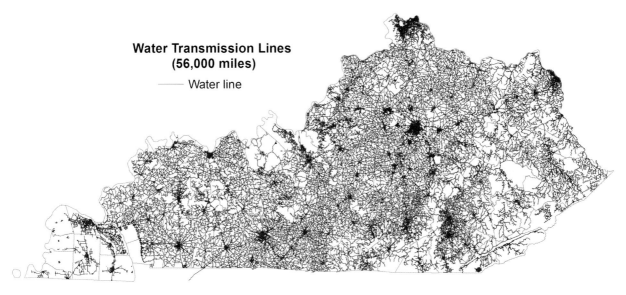

**Water Transmission Lines
(56,000 miles)**

—— Water line

1.6. A web of water lines across the commonwealth supplies running water to the vast majority of Kentucky households.

other uses combined, over 3.5 billion gallons each day. All other uses total an estimated 950 million gallons per day (mgd): public and domestic, 615 mgd (65 percent); industrial, 190 mgd (20 percent); agriculture, 65 mgd (7 percent); mining, 38 mgd (4 percent); and commercial, 34 mgd (4 percent).

Most uses of water are nonconsumptive; that is, the water is not used up but is borrowed from the ground or a stream and is later returned to the hydrologic cycle. Kentuckians on community water systems use about 150 gallons per person per day in homes, stores, and factories and for fire protection, swimming pools, street cleaning, and other uses.

Over 400 public water systems provide drinking water to 1,750,000 Kentucky households (97 percent) through 56,500 miles of water transmission lines (Kentucky Division of Water [KYDOW] 2010) (figure 1.6). The remaining Kentucky homes rely on wells, springs, cisterns, hauled water, and other sources.

About two-thirds of the water systems rely on surface-water sources (streams, lakes, reservoirs) and one-third on groundwater (figure 1.7). In 2005, 489 mgd came from surface sources, and 69 mgd came from groundwater sources (Downs and Caldwell 2007).

There are more than four times as many miles of public water lines as public sewer lines (13,250 miles) in Kentucky. Fourteen percent (1,885 miles) of the public sewer lines serve as combined storm-sewer overflow (CSO) (Kentucky Infrastructure Authority [KIA] 2013) (figure 1.8). CSOs carry storm runoff, domestic sewage, and industrial wastewater. During large storms, they overflow, causing wastewater to bypass the treatment plant and flow directly to streams, rivers, or other water bodies. In many urban areas, they are a major water-pollution concern (U.S. Environmental Protection Agency).

Most of the water used in homes is returned to streams through sewers and treatment plants or to the ground through on-site treatment systems. Although 97 out of 100 Kentuckians are on public water, only 52 out of 100 are on public sewers, and 48 percent—2 million Kentuckians—rely on private package plants, septic systems, artificial wetlands, other systems, or no treatment (KIA 2010) (figure 1.9).

In the Bluegrass region, it is generally known that a percentage of on-site systems are straight pipes. In other regions, the percentage is probably higher. Conventional on-site septic systems (not to mention straight pipes) cause surface and groundwater pollution affecting public health

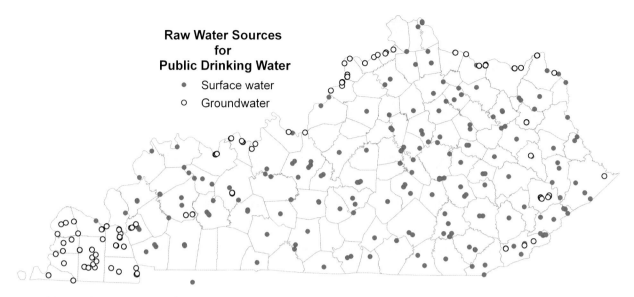

Raw Water Sources
for
Public Drinking Water
- ● Surface water
- ○ Groundwater

1.7. Public drinking water comes from a mixture of groundwater and surface-water sources.

in areas with unsuitable soil (USDA 2010) (figure 1.10) and geologic (KGS 2002–7) (figure 1.11) conditions. Typical limitations include poor permeability, high water table, depth to bedrock, slope, flooding, and karst.

Water Quality

Water clings to itself and everything it touches (*surface tension:* the attractive force exerted on the surface molecules of a liquid by the molecules beneath that tends to draw the surface molecules into the bulk of the liquid and makes the liquid assume the shape having the least surface area; *cohesion:* molecular attraction by which the particles of a body are united throughout the mass; *adhesion:* the molecular attraction exerted between surfaces of bodies in contact). Water is the near-perfect universal solvent; given enough time, it can dissolve practically anything. For this reason, by the time the nearly pure water that was lifted from the Gulf of Mexico at the beginning of the hydrologic cycle returns from Kentucky on its thousand-mile overland journey to the Gulf, it carries bits of nearly everything it has touched along the way.

Water quality affects animals, plants, and aquatic life. We describe the quality of water by the dissolved chemicals, gases, minerals, sediment, and living plants and animals it contains at any particular stage of its journey from sky to sea. All human activities—urban development, resource extraction, agriculture, industry, power generation—affect the quality of water and the life that it supports.

Polluted water can come from point or nonpoint sources. Point-source pollutants include pathogens from sewage-treatment plants, chemicals from industry, and mercury from power plants (indirectly from air to water). Nonpoint-source pollutants include failing (or no) septic fields for homes; agricultural runoff carrying animal wastes, pesticides, nutrients, and sediments; and urban stormwater runoff containing oil, grease, road salts, lawn fertilizer, heavy metals (lead), bacteria, and polychlorinated biphenyls (PCBs).

The leading sources of pollution of Kentucky's surface water and groundwater are agriculture, resource extraction (coal mining and petroleum operations), sewage-treatment plants, on-site sewage systems, and urban runoff (KYDOW, Water Quality Reports). Sediment/siltation, dissolved solids, nutrient eutrophication, pathogens, PCBs, and mercury are the primary pollutants. Eutrophication is an increase in nutrient concentration leading to algal bloom and depletion of oxygen in the water with resulting damage to aquatic life. For example, excessive

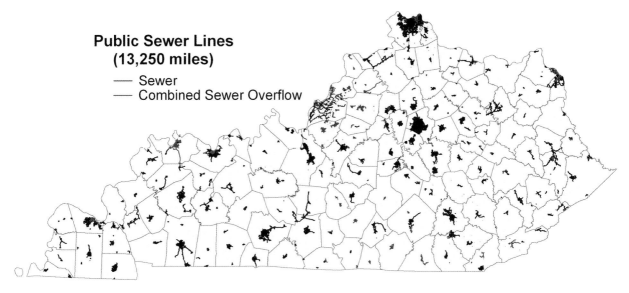

1.8. Sanitary sewer lines are used to move wastewater to treament facilities. Combined storm and sewer lines were once used and now present special design, policy, economic, and legal challenges for people and communities.

nutrients in runoff from nearby farmland can create algal bloom in ponds, which is unfortunately a common sight in Kentucky. Algal bloom can create taste and odor problems in drinking water but is typically not a dire health issue.

Pathogens are disease-causing microorganisms that can cause a variety of gastrointestinal problems, primarily diarrhea, that can sometimes be severe. PCBs are compounds of chlorine atoms attached to a molecule of benzene rings and are used in transformers and capacitors and as coolants. PCBs were classified as a persistent organic pollutant, and PCB production was banned in the United States in 1979, but PCBs are still present in Kentucky's waters. PCBs have been linked to hormone disruption and reproductive problems in aquatic invertebrates, fish, birds, and mammals. Mercury can injure the brain and kidneys. The heavy metal bioaccumulates as it moves up the food chain. No method of cooking or cleaning contaminated fish will reduce the amount of mercury in them. In April 2000, a statewide fish-consumption advisory for sensitive populations (children six years of age and younger and women of childbearing age) was issued over concerns about mercury found in fish tissue statewide, with a recommendation that no more than 8 ounces of local fish per week be consumed by any individual.

Surface-Water Quality

For the general public, the Kentucky Division of Water (DOW) monitors and assesses Kentucky's streams, lakes, and reservoirs and provides a biennial water-quality inventory report to the U.S. Congress under Sections 305(b) and 303(d) of the Clean Water Act. These assessments determine how well those waters support aquatic life, swimming (primary-contact recreation), boating, fishing, and wading (secondary contact recreation), drinking water, and suitability for fish consumption to protect human health.

The DOW has designated five basin management units (BMUs): Kentucky River; Salt-Licking Rivers; Upper Cumberland and 4-Rivers; Green-Tradewater; and Big Sandy/Little Sandy/Tygarts Creek. Each BMU is given an in-depth assessment of water quality on a five-year rotation. The 2010 assessment was based primarily on monitoring results from April 2007 to March 2009 on the Big Sandy/Little Sandy/Tygarts Creek (1,157 stream miles assessed) and Kentucky River (2,239 stream miles assessed)

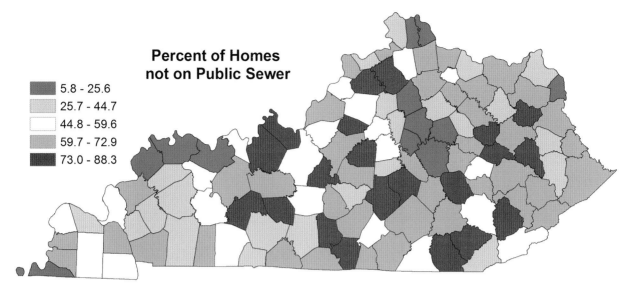

1.9. Most Kentuckians are able to use public water, but the majority of residents are not able to use similarly operated wastewater-treatment systems for their homes.

BMUs. The 2010 report also provided a statewide update of monitoring results.

The 2010 report showed that 62 percent of assessed streams (754 miles) fully supported fish consumption and 38 percent (456 miles) did not, about the same as previous assessments. In the Tygarts Creek basin, only 5 out of 51 miles assessed in 2010 fully supported fish consumption, a decrease of 75 percent from the 2004 assessment. Pollutants were mercury and PCBs. On the Kentucky River, the miles of assessed waters fully supporting fish consumption decreased from 74 percent (of 326 miles assessed) in 2006 to 50 percent (of 308 miles assessed) in 2010 because of mercury.

Nearly 10,000 miles (11 percent) of streams were assessed for aquatic life; 52 percent fully supported the designated use, and 48 percent did not. Over 60 percent of assessed streams in the Salt River, Tygarts Creek, and Upper Cumberland basins fully supported aquatic life. Thirty percent or less of assessed streams in the Tradewater River, Mississippi River, and Big Sandy River basins fully supported aquatic life. Statewide, the percentage of streams fully supporting aquatic life declined from 35 percent in 2004 to 27 percent in 2010. Biological surveys in streams of the Big Sandy/Little Sandy/Tygarts

Creek BMU found that the percentage of assessed streams fully supporting aquatic life fell from 12.5 percent in 2002 to 6 percent in 2007. Pollutants were sedimentation/siltation, nutrient eutrophication, and dissolved solids (such as chlorides, nitrates, sulfates, phosphates, sodium, magnesium, calcium, and iron).

Statewide, nearly 4,800 miles of streams were assessed for primary-contact recreation (swimming); 31 percent fully supported swimming and 69 percent did not, little changed from previous assessments. In the Kentucky River basin, 23 percent of assessed stream miles fully supported swimming in 2010, compared with 54 percent in 2006. The most frequent cause of nonsupport for swimming was pathogens.

The Kentucky DOW assessed 220,000 acres of publicly owned lakes, ponds, and reservoirs for aquatic life. Of those, 211,448 acres (96 percent) fully supported aquatic-life use, while 8,560 acres did not. The top three pollutants affecting lakes for this use were nutrient eutrophication, pH, and dissolved oxygen. The 2010 results were comparable to those of 2008.

Support for aquatic life in lakes of the Big Sandy/ Little Sandy/Tygarts Creek BMU rose from 49 percent (3,625 acres) in 2004 to 100 percent (7,383 acres) in 2010.

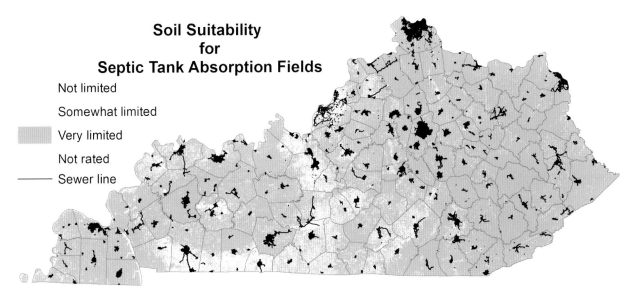

1.10. Soils are an important factor in considering limitations to proper effluent-absorption-field function.

Similar support for assessed lakes in the Kentucky River BMU rose from 36 percent in 2006 to 52 percent (3,113 of 5,965 acres assessed) in 2010.

Statewide, 205,635 acres of public lakes have been assessed for fish consumption: 59 percent (122,247 acres) fully supported fish consumption in 2010, up from 55 percent (of 204,732 assessed acres) in 2008. None of the 3,794 assessed lake acres (51 percent of those in the BMU) in the Big Sandy/Little Sandy/Tygarts Creek BMU fully supported fish consumption in 2010, compared with 100 percent support for 1,089 assessed acres in 2004. Primary pollutants here are mercury and PCBs in fish tissue. Of the more than 7,000 acres of public lakes in the Kentucky River BMU, none of the assessed acres fully supported fish consumption in 2006 (3,724 acres assessed) or 2010 (4,440 acres assessed); the pollutant of concern in this basin is (methyl)mercury in fish tissue.

The Federal Water Pollution Control Act requires the management of nonpoint sources of water pollution. Best management practices (BMPs), both structural and nonstructural, have been developed to prevent or reduce the movement of sediment, nutrients, pesticides, and other pollutants from the land to surface water or groundwater. Often the approach is as simple as keeping livestock away from or out of ponds and waterways. For example, the U.S. Department of Agriculture's Natural Resources Conservation Service provides cost sharing to farmers through the Environmental Quality Incentive Program to help with implementation.

Groundwater Quality

Groundwater quality has been assessed in all BMUs using well samples and data from the Kentucky Groundwater Data Repository (Fisher, B. Davidson, and Goodmann 2007a, 2007b, 2008; Fisher, O. B. Davidson, and Goodmann 2004). The overall quality of the groundwater was deemed generally good. For example, some areas in the Eastern Kentucky Coal Field have high levels of iron or sulfur in the groundwater (red dog), with iron in water flowing from underground precipitating out when the flow encounters oxygen, resulting in iron oxide (rust) red color.

Nitrate-nitrogen concentrations far exceeding natural contributions are common in agricultural areas. Nitrate-nitrogen levels greater than 10 milligrams per liter can present a hazard to babies and pregnant women. Ammonia-nitrogen concentrations were also commonly

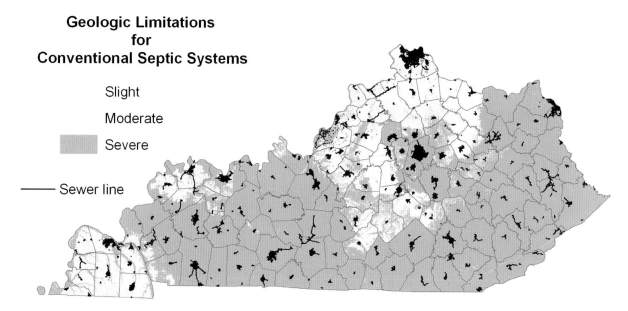

Geologic Limitations for Conventional Septic Systems

Slight

Moderate

Severe

——— Sewer line

1.11. Geologic conditions also influence considerations in operating a conventional septic system. The map indicates where systems will likely have limited function because of the geologic formation in that part of the landscape.

above recommended limits in the coal fields, possibly from coal beds or leaf litter. Available pesticide data were not sufficient to completely characterize any potential problems, and the authors concluded that "pesticides may be a greater health threat at some times of the year than these data suggest." With regard to volatile organic chemicals (VOCs), "The detection of volatile organic chemicals in springs and shallow wells that were previously thought to be free of such compounds suggests that volatile organic chemicals are entering regional groundwater systems." Some VOCs are considered carcinogenic. Finally, "Springs and shallow wells are more likely to have high levels of metals, nutrients, pesticides, and volatile organic chemicals than intermediate or deep wells. The potential contamination of the shallow groundwater system (springs and shallow wells) is cause for concern, as is potential contamination of the intermediate and deeper groundwater system."

The water that falls from the sky over Kentucky touches us in every aspect of our lives—where we live, work, and play. Water shapes the landscape, the people, and the resulting communities. As that water passes by us through the hydrologic cycle, it is our responsibility to leave it as pure as we received it for our neighbors and the plants, animals, and aquatic life that rely on it.

There are a variety of programs, resources, and groups across the state that can help Kentuckians learn about water, get involved in water-quality monitoring and improvement programs, clean up the waterways, adopt a watershed, and, in general, enjoy and help preserve the beauty and water resources of the place where we live: the Kentucky Division of Water Kentucky Water Watch, the Kentucky Waterways Alliance, Watershed Watch organizations, Eastern Kentucky PRIDE, Bluegrass Greensource, Kentucky Riverkeeper, the Appalachian Heritage Alliance, and many others.

Acknowledgments

Thanks to Ashley Osborne, University of Kentucky Agricultural Extension Services; Randy Payne, Kentucky Division of Water, Water Quality Branch; Roger Recktenwald, Kentucky Association of Counties; Jack Stickney, Kentucky Rural Water Association; and Jim Kipp, University of Kentucky Water Resources Research Institute, for their thoughtful and constructive review comments.

References and Additional Resources

American Geosciences Institute (AGI). www.americangeosciences.org (accessed October 5, 2016).

Appalachian Heritage Alliance, P.O. Box 1003, Campton, KY 41301. www.appalachianheritagealliance.org (accessed October 5, 2016).

Bluegrass Greensource. 2013. www.bgpride.org (accessed October 5, 2016).

Carey, D. I. 1992. *Water Quality in the Kentucky River Basin.* Kentucky Geological Survey, series 11, Information Circular 37, 66 p. http://www.uky.edu/KGS/pdf/ic11_37.pdf (accessed October 27, 2015).

Carey, D. I., et al. 1993. *Quality of Private Ground-Water Supplies in Kentucky.* Kentucky Geological Survey, series 11, Information Circular 44, 160 p. http://www.uky.edu/KGS/pdf/ic11_44.pdf (accessed October 5, 2016).

Currens, J. C. 2002. *Kentucky Is Karst Country! What You Should Know about Sinkholes and Springs.* Kentucky Geological Survey, series 12, Information Circular 4, 29 p. http://kgs.uky.edu/kgsweb/olops/pub/kgs/ic04_12.pdf (accessed October 27, 2015).

Downs, Aimee C., and William E. Caldwell. 2007. *Kentucky Public Water-Supply Withdrawals during 1995, 2000, and 2005.* U.S. Geological Survey Open-File Report 2007-1209, 13 p. http://pubs.usgs.gov/of/2007/1209/pdf/OFR20071209.pdf (accessed October 27, 2015).

Eastern Kentucky Pride. 2013. www.kypride.org (accessed October 5, 2016).

Fisher, R. S., B. Davidson, and P. T. Goodmann. 2007a. *Groundwater Quality in Watersheds of the Kentucky River, Salt River, Licking River, Big Sandy River, Little Sandy River, and Tygarts Creek (Kentucky Basin Management Units 1, 2, and 5).* Kentucky Geological Survey, series 12, Report of Investigations 16, 97 p. http://kgs.uky.edu/kgsweb/olops/pub/kgs/water/RI16_12/RI16entire.pdf (accessed October 27, 2015).

———. 2007b. *Regional Groundwater Quality in the Watersheds of the Upper Cumberland, Lower Cumberland, and Lower Tennessee Rivers, and the Jackson Purchase Region (Kentucky Basin Management Unit 3).* Kentucky Geological Survey, series 12, Report of Investigations 15, 105 p. http://kgs.uky.edu/kgsweb/olops/pub/kgs/water/RI15_12/index.html (accessed October 27, 2015).

———. 2008. *Groundwater Quality in Watersheds of the Big Sandy River, Little Sandy River, and Tygarts Creek (Kentucky Basin Management Unit 5).* Kentucky Geological Survey, series 12, Report of Investigations 19, 95 p. http://kgs.uky.edu/kgsweb/olops/pub/kgs/RI19_12.pdf (accessed October 27, 2015).

Fisher, R. S., O. B. Davidson, and P. T. Goodmann. 2004. *Summary and Evaluation of Groundwater Quality in the Upper Cumberland, Lower Cumberland, Green, Tradewater, Tennessee, and Mississippi River Basins.* Kentucky Geological Survey, series 12, Open File Report 04-004, 157 p. http://kgs.uky.edu/kgsweb/olops/pub/kgs/of_04_04.pdf (accessed October 27, 2015).

Gleick, P. H. 1996. Basic Water Requirements for Human Activities: Meeting Basic Needs. *Water International* 21: 83–96. http://pacinst.org/wp-content/uploads/sites/21/2012/10/basic_water_requirements-1996.pdf (accessed October 27, 2015).

Hopkins, H. T. 1966. *The Fresh-Saline Water Interface in Kentucky.* U.S. Geological Survey, Kentucky Geological Survey, 19 p. http://kgs.uky.edu/kgsweb/download/wrs/SALINE.PDF (accessed October 27, 2015).

Kentucky Climate Center. 2013. Western Kentucky University. http://kyclimate.org/.

Kentucky Division of Water. *Water Quality Reports.* http://water.ky.gov/waterquality/Pages/IntegratedReport.aspx (accessed October 27, 2015).

Kentucky Division of Water Kentucky Water Watch. http://water.ky.gov/ww/pages/default.aspx (accessed October 27, 2015).

Kentucky Geological Survey. 2002–7. *Generalized Geologic Data for Land-Use Planning in Kentucky Counties.* http://kgs.uky.edu/kgsweb/download/geology/landuse/lumaps.htm (accessed October 27, 2015).

———. 2010a. *Karst Land in Kentucky.* http://www.uky.edu/KGS/water/general/karst/ (accessed October 27, 2015).

———. 2010b. *Sinkhole Flooding.* http://www.uky.edu/KGS/water/general/karst/sinkhole_flooding.htm (accessed October 27, 2015).

———. 2010c. *Water Data: Data Sources for Groundwater and Surface Water in Kentucky.* http://www.uky.edu/KGS/water/library/ (accessed October 27, 2015).

Kentucky Infrastructure Authority (KIA). 2013. *Water Resource Information System.* http://kia.ky.gov/wris/ (accessed October 27, 2015).

Kentucky Riverkeeper. 2013. http://kyriverkeeper.org/ (accessed October 27, 2015).

Krieger, R. A., Cushman, R. V., and Thomas, N. O., 1969, *Water in Kentucky.* Kentucky Geol. Survey, ser. 10, Spec. Pub. 16, 51 p. http://www.uky.edu/KGS/pdf/sp1.pdf.

Leopold, L. B., and W. B. Langbein. 1960. *A Primer on Water.* U.S. Geological Survey, 50 p. http://pubs.usgs.gov/gip/7000045/report.pdf (accessed October 27, 2015).

Pielke, Jr., R. A., M. W. Downton, and J. Z. Barnard Miller. 2002. Flood Damage in the United States, 1926-2000: A Reanalysis of National Weather Service Estimates. Boulder, CO: UCAR.

Veni, G., et al. 2001. *Living with Karst: A Fragile Foundation.* American Geological Institute, 64 p. http://www.americangeosciences.org/sites/default/files/karst.pdf (accessed October 27, 2015).

Strahler, A. N. 1957. Quantative Analysis of Watershed Geomorphology. *Transactions of the American Geophysical Union* 8 (6): 913–920. http://onlinelibrary.wiley.com/doi /10.1029/TR038i006p00913/abstract (accessed October 27, 2015).

U.S. Army Corps of Engineers. 2006. *Value to the Nation.* www.usace.army.mil/.

U.S. Department of Agriculture, Natural Resources Conservation Service. 2013. *Web Soil Survey.* http://websoilsurvey .sc.egov.usda.gov/App/HomePage.htm (accessed October 27, 2015).

U.S. Environmental Protection Agency. National Pollutant Discharge Elimination System (NPDES) Home. http:// cfpub.epa.gov/npdes/pareas.cfm (accessed October 27, 2015).

U.S. Fish and Wildlife Service. 2013. *National Wetlands Inventory Program: Overview.* https://www.fws.gov/Wetlands /nwi/Overview.html (accessed October 27, 2015).

U.S. Geological Survey. 2013a. Executive Secretary, U.S. Board of Geographic Names, 523 National Center, Reston, VA 20192. http://geonames.usgs.gov/domestic/index.html (accessed October 27, 2015).

———. 2013b. *Water Science for Schools: The Water in You.* http://ga.water.usgs.gov/edu/propertyyou.html (accessed October 27, 2015).

Waller, R. M. 1994. *Ground Water and the Rural Homeowner.* U. S. Geological Survey, 37 p. http://pubs.usgs.gov/gip/gw _ruralhomeowner/ (accessed October 27, 2015).

Watershed Watch in Kentucky. 2013. https://sites.google.com /site/watershedwatch/ (accessed October 27, 2015).

Chapter Two

A Cartographic View of Kentucky's Watershed and Landscapes

The Early Years and Modern Accomplishments

Daniel I. Carey

Maps let us see the spatial relationships among things on, under, and above the ground and enable us to find the answers to questions that help us live in harmony with the land and water. For example, maps can be made to show different land areas of Kentucky and their potential for agricultural uses or to answer such questions as the following: Where is the unstable shale in relation to proposed urban development? Where are the wetlands in my county, and how big are they? What is the likely source of the human fecal coliform I see in the creek? Where are the soils and bedrock suitable for on-site septic systems, and where are they not? On the basis of topography and geology, what would the best alignment be for this new highway? Where are the valuable natural resources? Where are the water resources that might be affected by the chemical spill? Maps help us see the important locational relationship between us and elements of the landscape and better understand the place on the land where we live, work, and play.

Early Maps

Stream valleys were often the primary routes used by American Indians, explorers, and early settlers. When the first settlers moved across Kentucky, they built a mental image of the land: the rivers were here, the hills were there, and the rolling lands were over there; it took this long to get from there to here, so it must be about this far; this is the route of an Indian trail; there is a salt

lick at this point. Early surveyors and cartographers took those mental pictures and put them on paper with pen and ink. Maps of Kentucky were a significant improvement, as can be seen in a 1795 version by surveyor and cartographer Elihu Barker.

Map Scale

Maps represent a bird's-eye view of the land at a given scale. The scale of a map is a reflection of the level of detail we can see on a map of a given size. One inch on the map represents so many units (inches, feet, meters, or miles) on the earth. It is hard to imagine a usable map larger than a room, and most maps are no larger than 3 or 4 feet on a side. A large desk-sized map of Kentucky might have a scale of 1:633,600: 1 inch on the map equals 633,600 inches, or 52,800 feet at 12 inches per foot and 10 miles at 5,280 feet per mile, across the land as the crow flies. On the same-sized sheet, Allen County might be represented at 1:57,024, or 1 inch equals 0.9 mile, and the city of Campbellsville at 1:6,000, or 1 inch equals 500 feet, with increasing levels of detail.

Ways of Seeing the Land

During the Civil War, the Corps of Topographical Engineers used hot-air balloons to get a bird's-eye view of the land and make maps. Now we can go up in an airplane and take aerial photos showing roads, streams,

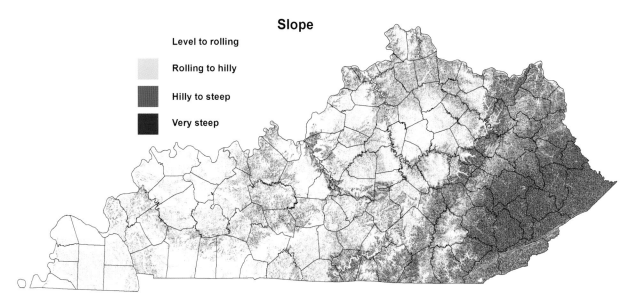

2.1. Today, digitally encoded elevation data and computers are used to make a statewide slope map in a matter of minutes.

farmland, residential areas, and other topographic features. One example is the U.S. Department of Agriculture Farm Services Agency's National Agriculture Imagery Program, which has collected growing-season aerial imagery in Kentucky, and whose maps are available online for 2004, 2006, 2008, 2010, 2012, and 2014.

Topographic maps show us the shape of the land: where it is steep or flat. Surveyors measured the hills and valleys of Kentucky from 1949 to 1956 to map the topography on 783 maps at a scale of 1:24,000. Each map covers an area of 7.5 minutes latitude by 7.5 minutes longitude. The completion of this large-scale topographic mapping provided an essential foundation for additional mapping of soils, geology, watersheds, and many other features. These maps are still very valuable today because they provide a snapshot of what the landscape was like just decades ago.

The key feature of a topographic map is the contour lines because they illustrate the shape of the land. Contour lines indicate the elevation of the land above sea level. Along a contour line, the elevation remains the same. When contour lines are close together, on hillsides, the land is steeper. In stream valleys and on hilltops, the land is flatter, and contour lines are farther apart. Different

maps have different contour intervals. For example, a 20-foot contour interval is common for a map. Contour intervals range from 10 feet in relatively flat western Kentucky to 40 feet in the rougher eastern Kentucky landscape. The accuracy of the elevations is +/−½ contour interval. A key capability that a topographic map affords is the derivation of a topographic slope analysis (figure 2.1), as well as the ability to determine watershed boundaries to better manage water resources.

The U.S. Geological Survey and the U.S. Department of Agriculture's Natural Resources Conservation Service determined watershed and subunit drainage boundaries based on topographic divides. These units are characterized by hydrologic unit codes (HUCs), with basins and subunits ranging from large, HUC-6, to small, HUC-14. There are 13 HUC-6 basins that make up the Commonwealth of Kentucky. The largest is the Green River Basin at 8,814 square miles. There are 616 HUC-11 units with an average area of 65.59 square miles and 9,109 HUC-14 units with an average area of 4.44 square miles (figure 2.2).

Topographic mapping of the state allowed geologists to map sinkholes (topographic depressions; figure 2.3). Knowing the locations of sinkholes, which characterize

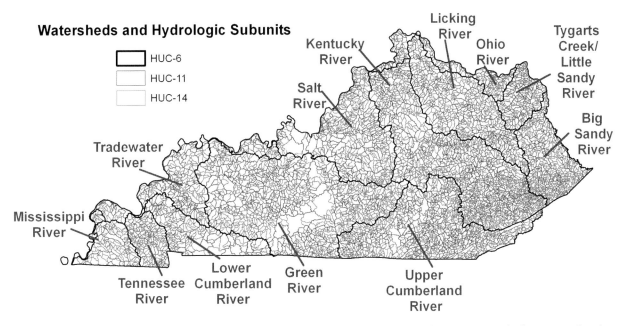

2.2. Hydrologic unit codes (HUCs). HUC-14s, the smallest watersheds, are nested within HUC-11s, which are nested within HUC-6s as a unified system to designate all land areas within the commonwealth and across the United States.

Kentucky's karst regions, is necessary to protect water resources from a variety of land uses.

In order to better understand the magnitude of the mapping effort, consider the following example. Using the topographic maps as a base, more than two hundred geologists spent a total of 661 working years from 1960 to 1978 doing field studies and mapping the geology of Kentucky at a scale of 1:24,000 on 7.5-minute quadrangle maps. Other than Rhode Island, Kentucky is the only state with geology mapped at this level of detail.

Geologic maps show rock formations beneath the surface, faults, mines, and mineral deposits (figure 2.4). Geologic maps are used in the design and construction of highways, dams, and reservoirs, in the development of resources such as coal, oil and gas, and minerals, and in land-use planning, environmental protection, and, essentially, any activity that may be affected by the underlying bedrock in that part of the watershed or landscape.

Because of its topography and the nature of the underlying rocks, Kentucky is subject to a variety of geologic hazards, such as sinkholes, shrinking and swelling shales,

slope stability, landslides, earthquakes, and floods. Maps are used to illustrate the location and hazard type(s). (figure 2.5).

Mapping and understanding geologic formations helps people develop, use, and protect surface-water and groundwater resources. For example, understanding geologic-formation limitations helps people and communities plan for on-site residential septic systems as described in Chapter 1 (figure 1.11) as well as shown in figure 2.6.

Kentucky has had generalized geologic maps for land-use planning for every county since 2010 (http://kgs.uky.edu/kgsweb/download/geology/mapstoteachers.htm). The map series, produced by the Kentucky Geological Survey, shows the bedrock limitations for various development types, discusses and illustrates local potential hazards, and uses photos and diagrams to illustrate county-specific conditions.

Building on the geologic maps, geologists have also mapped Kentucky soil types statewide. These soil maps are an invaluable data source for a variety of land uses, including helping preserve prime farmland and guide urbanization, as depicted for Anderson County in figure 2.7.

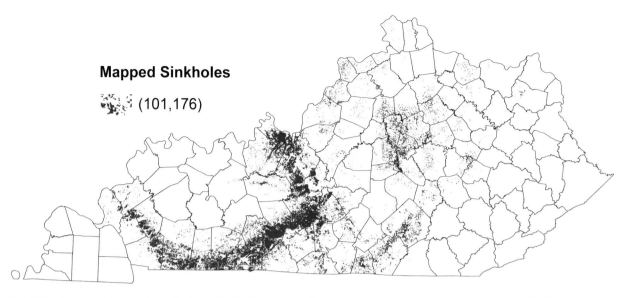

Mapped Sinkholes

(101,176)

2.3. Using topographic data, karst hydrogeologists have mapped and continue to map more than 100,000 sinkholes, ranging in size from less than forty square feet to more than twelve square miles.

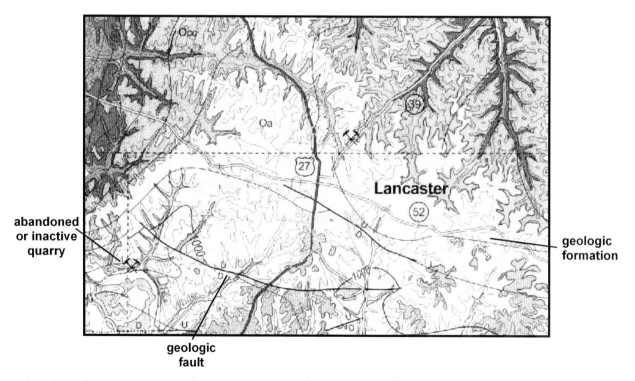

2.4. Geology of the Lancaster area in Garrard County. Each color represents a different geologic formation.

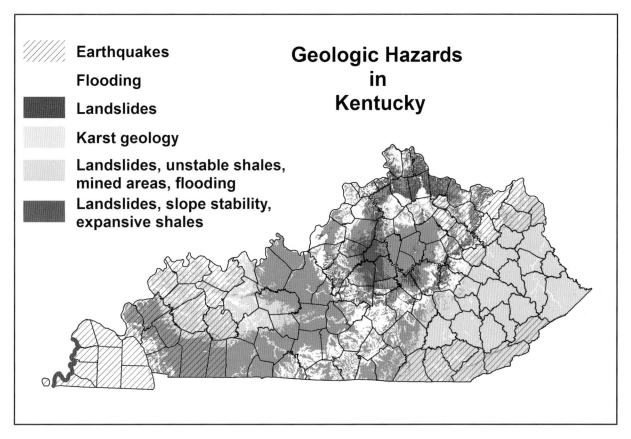

2.5. There are a variety of geologic hazards that contribute to the shaping of communities across the commonwealth.

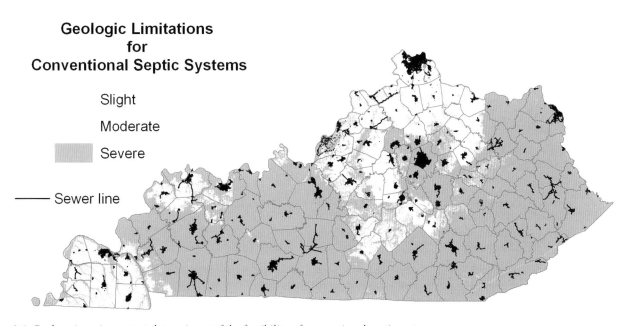

2.6. Geology is an important determinant of the feasibility of conventional septic systems.

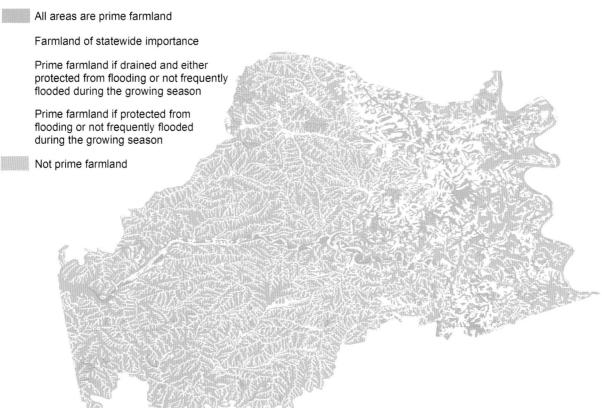

Anderson County Farmland

All areas are prime farmland

Farmland of statewide importance

Prime farmland if drained and either protected from flooding or not frequently flooded during the growing season

Prime farmland if protected from flooding or not frequently flooded during the growing season

Not prime farmland

2.7. Prime farmland and farmland of statewide importance based on soils in Anderson County, Kentucky.

The Kentucky Environmental and Public Protection Cabinet, under contract to the Kentucky Department of Fish and Wildlife and in cooperation with the U.S. Fish and Wildlife Service, created digital data for Kentucky wetlands from the final 1:24,000-scale National Wetlands Inventory map sheets (2002). Nearly 233,000 palustrine wetlands (marshes, swamps, ponds) with an average size of 1.9 acres were delineated, digitized, and made freely available via the Kentucky Geography Network (figure 2.8).

Maps in Schools

In addition to helping homeowners, planners, local officials, environmentalists, and resource developers, maps are used inside and outside the classroom to help students of all ages better understand the landscape in which they live (http://kgs.uky.edu/kgsweb/download /geology/mapstoteachers.htm).

Recent Developments in Mapping

Before the 1980s, mapmaking was a relatively slow and tedious process requiring significant manual labor. The advent of computer technology, aerial and satellite positioning, range finding, and the development of geographic information systems (GISs) made it possible to generate a variety of more precise, accurate, and detailed maps more efficiently and at relatively lower cost. Maps that formerly took months to produce can now be made in days or less through integration of a number of technologies. Since

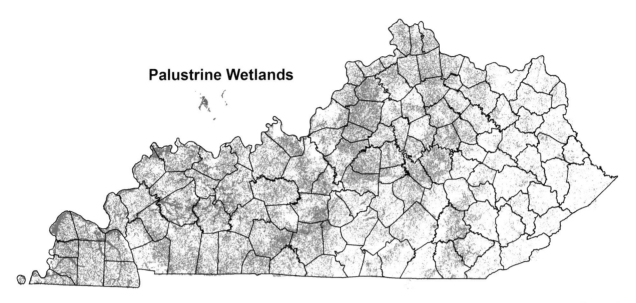

Palustrine Wetlands

2.8. Nearly half a million acres of palustrine (marsh, swamp, or pond) wetlands in Kentucky provide habitat for wildlife and water for agriculture, reduce flooding, and improve water quality.

the data and maps are now typically stored digitally on a computer, they can be easily corrected, updated, and used in a variety of ways. People use a GIS to efficiently manage the array of currently available digital data to produce maps or view them on any number of digital devices.

A GIS is any system capable of capturing, storing, analyzing, and displaying geographically referenced information; that is, data identified by location. Your brain working with a map is a GIS. Computer hardware and GIS software are part of a GIS that provides tools for visualizing, exploring, querying, editing, and analyzing information linked to geographic locations.

The power of a GIS comes from the ability to relate different information in a spatial context and to reach a conclusion about this relationship. For example, where are the sinkholes in relation to a proposed highway location? Where is the unstable shale in relation to a proposed subdivision or commercial development? Where are the water sources that might be affected by a chemical spill or a coal-slurry spill? This is accomplished by using a location reference system, such as latitude and longitude, to understand the spatial relationships in our environment. A GIS also allows us to perform spatial analyses and make maps that show all relevant features in an area of

interest. A GIS can reveal and communicate important information that leads to a better understanding of the place where people live and lets people make better decisions on how to live in the landscape. For example, students can learn about earth science and the community in which they live using maps created with a GIS system.

Hundreds of people in Kentucky have worked together to produce one of the best GIS systems of any state in the United States. The nationally recognized and invaluable Kentucky Geography Network developed and maintained by the Kentucky Division of Geographic Information currently provides access to nearly 600 GIS data sets and products—GIS data layers, aerial and satellite imagery, digital elevation models, maps, and related documents—twenty-four hours a day and seven days a week. Available Kentucky GIS data include information on the following:

- Cities, towns, counties, property parcels, schools, and libraries
- Highways, roads, streets, railroads, and airports
- Soils, geology, and sinkholes
- Streams, water bodies, wetlands, flood areas, water quality, watersheds, watershed-protection areas, groundwater data, and groundwater dye traces

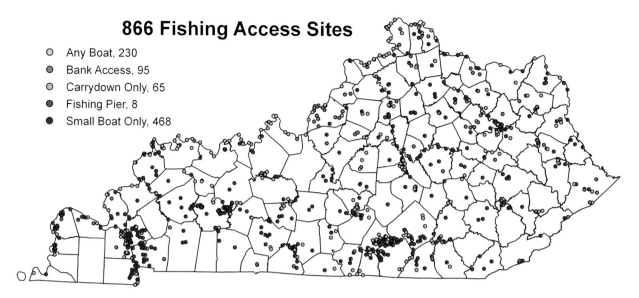

866 Fishing Access Sites

- ○ Any Boat, 230
- ◑ Bank Access, 95
- ○ Carrydown Only, 65
- ● Fishing Pier, 8
- ● Small Boat Only, 468

2.9. Boat-launching points and fishing piers as compiled by the Kentucky Department of Fish and Wildlife Resources (2010).

- Waste-disposal sites, feedlots, industrial sites, and mined areas
- Water, sewer, and electrical utilities
- Oil and gas, coal, and mineral resources
- Ecoregions, wildlife-management areas, wild and aquatic life, and conservation areas
- State and national parks, trails, fishing, boating, hunting, and recreation areas
- Census data
- And much more

With a GIS, a person can answer questions about where he or she lives. How many miles of roads are in my county? How many square miles is my county? Where has coal been mined in my county? Where are the four-wheeler trails? How many miles of streams are there? Where are the lakes, and how big are they? Where are the boat ramps I can use when I go fishing (figure 2.9)?

GIS on the Internet

With GIS software, a computer, and Kentucky's data resources, anyone can become a mapmaker. Internet GIS map services provide information on communities, development planning, transportation, water resources,

energy, geology, mineral resources, and recreation. With a GIS, a person can create a map of the watershed that his or her school is in and look at other features surrounding that school (figure 2.10).

GIS Internet mapping software has been used to create a number of interactive maps showing geology, soils, water and sewer infrastructure, zoning and property information, flood information, highway development, energy resources, and many more features. For example, hunting and fishing sites can be located in a map service/viewer, as can parks and riding trails for recreation.

The making of maps has come a long way since the first European settlers moved across Kentucky. We now have the ability to map virtually every physical feature of the commonwealth at a relatively high level of detail and accuracy. We can combine data to see and understand in new and different ways the spatial relationships of the landscapes where we live. We can be almost anywhere and answer a variety of questions; for example, if I build my house in this part of the county, what are the soils and underlying rocks? Will they provide a suitable foundation for my house? Can I build a pond that will not leak? Is the area on public water? Are the soils and geology suitable for a septic system if there is no public sewer at the site? Are there geologic hazards I should be aware of, such as regu-

Mill Creek Watershed

Two-meter aerial photos from the U.S. Department of Agriculture, Farm Services Administration, National Agricultural Imagery Program, 2006.

For more information about the water and rocks of Kentucky, go to www.uky.edu/kgs

Mill Creek

Mill Creek Elementary School

1 inch equals 225 feet

0   220   440   880   1,320   1,760
                                                               Feet

Map compiled by Dan Carey, Kentucky Geological Survey

2.10. A watershed map for Millcreek Elementary School in Fayette County made with GIS in a few hours (by the author) using the U.S. Department of Agriculture, Farm Services Administration, National Agricultural Imagery Program (2006).

lar seasonal flooding or unstable slopes? Where is the best location for the new county school? The new industrial park? The new commercial district? The new bypass? Or one can determine from existing data that some types of urbanization are not even needed. We now have mapping tools that can provide answers to these kinds of questions. Using these tools and acting on the knowledge derived from them will help us live in harmony with the land.

Additional Resources

Cressman, E. R., and M. C. Noger. 1981. *Geologic Mapping of Kentucky—A History and Evaluation of the Kentucky Geological Survey–U.S. Geological Survey Mapping Program, 1960–1978.* U.S. Geological Survey Circular 801, 22 p. http://pubs.usgs.gov/circ/1981/0801/report.pdf (accessed October 28, 2015).

Kentucky Division of Geographic Information. 2013. Kentucky Geography Network. http://kygisserver.ky.gov/geoportal/catalog/main/home.page (accessed October 28, 2015).

Kentucky Geological Survey. 2002–7. *Generalized Geologic Data for Land-Use Planning in Kentucky Counties.* http://kgs.uky.edu/kgsweb/download/geology/landuse/lumaps.htm (accessed October 28, 2015).

———. 2008a. *Geologic Hazards in Kentucky.* Map and Chart 185_12. http://kgs.uky.edu/kgsweb/olops/pub/kgs/mc185_12.pdf (accessed October 28, 2015).

———. 2008b. *Kentucky Terrain.* Map and Chart 187_12. http://kgs.uky.edu/kgsweb/olops/pub/kgs/mc187_12.pdf (accessed October 28, 2015).

———. 2009. *River Basin Maps.* http://kgs.uky.edu/kgsweb/download/water/basins.htm (accessed October 28, 2015).

———. 2013a. *Maps and GIS.* http://www.uky.edu/KGS/gis/index.htm (accessed October 28, 2015).

———. 2013b. *Maps-to-Teachers.* http://www.uky.edu/KGS/announce/landuse_teacher.htm (accessed October 28, 2015).

Kentucky Infrastructure Authority. 2013. *Water Resource Information System.* http://kia.ky.gov/wris/ (accessed October 28, 2015).

National Oceanic and Atmospheric Administration (NOAA). 2013. *NOAA Digital Coast Project LIDAR Data Archive.* http://www.ngdc.noaa.gov/mgg/bathymetry/lidar.html (accessed October 28, 2015).

Raitz, K., N. O'Malley, J. Levy, D. Gilbreath, C. Rulo, B. Shearer, N. O. Hammon, T. Conforti, and J. Renner. 2008. *Kentucky's Frontier Trails: Warrior's Path, Boone's Trace, and Wilderness Road.* Kentucky Transportation Cabinet, poster POS_019_KYTC.

Ruthven, C. L., J. D. Kiefer, S. F. Greb, and W. M. Andrews Jr. 2003. *Geologic Maps and Geologic Issues in Kentucky: A Citizen's Guide.* Kentucky Geological Survey, series 12, Special Publication 3, 27 p. http://kgs.uky.edu/kgsweb/olops/pub/kgs/sp03_12.pdf (accessed October 28, 2015).

U.S. Department of Agriculture, Natural Resources Conservation Service. 2013. *SSURGO Soils Data.* http://soildatamart.nrcs.usda.gov/County.aspx?State=KY. (Soil Data Mart has been discontinued and has been replaced by the Web Soil Survey at http://websoilsurvey.sc.egov.usda.gov/App/HomePage.htm, accessed October 28, 2015.)

U.S. Fish and Wildlife Service. 2013. *National Wetlands Inventory: Overview.* http://www.fws.gov/wetlands/ (accessed October 28, 2015).

Chapter Three

Springs and the Settlement of Pioneer Kentucky

Gary A. O'Dell

> It was very common for 4 or 5 families to be settled together by some good spring. It was so here. And so Kentucky was settled.
> —Samuel Matthews, early Kentucky pioneer (Draper 11CC157)

Kentucky is a land of springs. There are literally thousands upon thousands of free-flowing natural springs, large and small, across the length and breadth of the commonwealth. Small seeps and trickles may be found in all parts of Kentucky wherever there is fractured rock, but springs are most numerous and abundantly flowing in those extensive areas of the state where cavernous limestone bedrock is present. This is a type of landscape known as karst, where bedrock fractures have been dissolutionally enlarged by the circulation of groundwater to create complex networks of conduits fed by sinkholes and sinking streams, from which water discharges as springs. Nearly half of the land area of Kentucky is characterized as karst, including most of the Inner Bluegrass region and a vast area encircling the Outer Bluegrass from western Kentucky to the mountains of the east.

During the pioneer era in Kentucky, mineral springs and the most prominent freshwater springs served as landmarks in the wilderness, the focal points of a network of trails created by bison in search of salt and by the Native Americans who traversed the region in search of game and to trade with other tribes. The distribution of springs was in large part responsible for the pattern of settlement in pioneer Kentucky. Throughout human history, springwater has been perceived as superior in quality to that of any other source because it has been thought to issue in pristine form from the depths of the earth. A pure and reliable water supply was one of the most

important criteria for any potential settlement, and the early explorers and settlers were eager to claim land containing a significant spring, which became the site of pioneer stations and communities. The location of many cities and towns in Kentucky today, particularly those of the Bluegrass Region, can be attributed to the presence of a spring during the settlement period thought sufficient to supply the inhabitants.

Exploration of the country west of the Appalachian Mountains began in the 1750s, and the hunters, explorers, and surveyors who traveled through the region brought back reports of a rich land teeming with game: buffalo, bear, deer, elk, geese, and turkeys. Those who penetrated as far as central Kentucky described the country north of the Kentucky River as an earthly garden. The gently rolling landscape of the Inner Bluegrass region was a savanna woodland, deep, fertile soils lightly timbered with wide-spreading oak and ash trees mixed with meadowlands and thick stands of cane. Although many of the chronicles of these early explorers were more detailed and eloquent, surveyor Thomas Hanson succinctly captured the essence of the Bluegrass Country in his journal entry for July 1, 1774, describing the vicinity of Elkhorn Creek: "All the land we passed over today is like a Paradise it is so good & beautiful" (Hanson 1905, 129). Such reports generated considerable excitement in the long-settled lands east of the mountains, where soils had been depleted of their fertility and the game had long since been hunted out. Travel through the western country was dangerous, but this did not deter potential immigrants, who came to Kentucky to carve out a new life in the wilderness (Wharton and Barbour 1991, 19–32; Aron 1996, 6; D. B. Smith 1999, 77–78).

The settlement of Kentucky took place between 1773 and 1792, at first as a trickle of hardy pioneers but soon as a flood of thousands of immigrants. The Appalachians were a formidable barrier that allowed only a few points of access to the western frontier, either by way of the Ohio River or through one of the few mountain passes. During the earliest years of settlement, few pioneers were hardy enough to risk the river route, where the threat of Indian ambush was a constant danger. River travel for large parties, requiring construction or purchase of a flatboat, could also be quite expensive, and so, although the land route was nearly as hazardous and the physical obstacles challenging, most early immigrants chose instead to make the long and difficult overland trek through the Cumberland Gap into the region. After 1783, when the danger had lessened, most settlers bypassed the mountains and came down the Ohio River in canoes and flatboats, putting ashore at the mouth of Limestone Creek (the site of present-day Maysville) or another of the river landings that served as thresholds to the Bluegrass. Either approach set pioneers on a network of trails that had been used for millennia by animals and Native Americans to traverse the region (O'Malley 1994, 19–20; 1999, 58–59).

A profusion of narrow pathways had been developed in the wilderness by deer and Native Americans, but the most prominent trails, called "buffalo traces" by Anglo-American settlers, were wide and well-trampled routes produced by the seasonal migrations of bison herds as they traveled between their foraging grounds and the many mineral springs of the region. Such trails had been in existence for millennia, as evidenced by the bones of prehistoric bison (immense shaggy creatures twice the size of modern bison), mastodons, and other extinct animals found by the early pioneers at locations such as Big Bone Lick and the Blue Licks (Hedeen 2008). Many animals, large and small, were attracted by the salt-encrusted earth on the margins of such saline springs, which were termed "licks" by the settlers of the region. Referring to

the salt springs of Kentucky, John Filson wrote in 1784 (32–33), "The amazing herds of Buffaloes which resort hither, by their size and number, fill the traveler with amazement and terror, especially when he beholds the prodigious roads they have made from all quarters, as if leading to some populous city; the vast spaces of land around these springs defolated [sic] as if by a ravaging enemy, and hills reduced to plains." Another traveler through central Kentucky, Nicholas Cresswell, observed in 1775 (1924, 85) that the buffalo "eat great quantities of a sort of reddish clay found near brackish springs. I have seen amazing large holes dug or rather cut by them in this sort of earth."

The "Buffaloes" encountered by the pioneers were relatively recent arrivals to the Ohio Valley region. Before the sixteenth century, a combination of factors prompted movement of the plains bison into the eastern woodlands, including the Indian practice of setting fires to flush game and provide agricultural land that created, enlarged, and maintained prairie enclaves along the eastern margin of the Mississippi. The bison of the Southeast were never as numerous as those of the Great Plains, seldom gathering in herds of more than five hundred individuals, but their movements re-created the system of trails first developed by Pleistocene megafauna. These trails tended to traverse the ridges because the herds were reluctant to ford watercourses, with diversions from the trails to freshwater springs and canebrakes in the lowlands (Rostlund 1960; Jakle 1968, 1969; Belue 1996, 7–10). "The buffalo seldom visited the licks in the winter," Nathan Boone wrote, referring to the hunting practices of his father, Daniel Boone. "They then would keep near the cane as the best winter's range and lived in summer mainly on grass" (Draper 6S103). In the Ohio Valley, geographer John A. Jakle observed (1968, 302), the main objective of the buffalo was "always a salt lick, for basically, the traces were routes of maximum convenience connecting the larger salt springs."

Eastern Kentucky is barricaded by two long, parallel ridges, each more than a hundred miles in length. The outermost of these ridges is Cumberland Mountain, which reaches elevations of 2,200 to 3,500 feet and is separated from Pine Mountain to the northwest, 1,800 to 2,200 feet in elevation, by little more than a dozen miles. Few significant breaks interrupt these ridges. Immigrants could have crossed Cumberland Mountain at Pennington Gap, but there was no equivalent gap in the vicinity allowing passage over Pine Mountain. Settlers instead traveled forty miles farther south to the Cumberland Gap, where a fortuitous alignment of notches through each of the two mountain ridges allowed access to Kentucky. Here, the movements of large animals and Native Americans had for ages been funneled through the Cumberland Gap and along the Cumberland River through a water gap in Pine Mountain (the site of present-day Pineville), developing a well-marked trail. The Cumberland Gap thus became the primary gateway into Kentucky during the earliest years of exploration and settlement (Jakle 1968, 302).

The buffalo trace forded the Cumberland River at the Pine Mountain gap and followed the river eight miles farther to Flat Lick, a relatively level area with several salt springs that was a junction of several major trails. Native Americans of the region frequented the primary spring to make salt and to hunt buffalo attracted to the lick. The notorious British officer Henry Hamilton, captured at Vincennes, Indiana, by George Rogers Clark's men, was brought through Flat Lick in 1779 on his way to captivity at Williamsburg, Virginia, and noted in his journal that here was "a remarkable Buffaloe salt lick." Hamilton also observed that the trees bore markings and characters made by the Indians to describe their various exploits, with the bark removed and the designs colored with red dye (Barnhart 1951, 198). The little stream in the vicinity was known as Stinking Creek, allegedly from the habit of hunters, Anglo and Indian, of leaving decaying carcasses and offal along the banks (Kincaid 1947, 70, 75).

Two primary trails split off from the vicinity of Flat Lick, the Warriors' Path and the Wilderness Trail. The Warriors' Path, a western branch of the long-established Native American trail, the Great Indian Warpath or Athawominee, that traversed the length of the Great Appalachian Valley from upper New York State to Georgia, was a trade and war route linking the tribes of eastern Tennessee with those of the mid–Ohio Valley. The Kentucky branch of the Warpath followed the bison trace across the gap to Flat Lick and then cut northward to Es-

kippakithiki (Indian Old Fields) in eastern Clark County, an abandoned Shawnee town located in the vicinity of several salt licks. The trail split again here, one fork running northwest to the Upper Blue Licks and on to the Ohio River, the other fork curving northeast and passing by Mud Lick (later renamed Olympian Springs) and Salt Lick in Bath County to the Ohio River opposite the mouth of the Scioto River, where the Indian community known as Lower Shawnee Town was located (figure 3.1) (Jillson 1934, 37–39; Kincaid 1947, 30; Jakle 1968, 302; Raitz, Levy, and Gilbreath 2010). These and other licks led explorers to bestow the name "Great Salt Lick Creek" on the river on which they were located, a designation later altered to Licking River. Thomas Hutchins, who came down the Ohio in 1766, later noted that "Great Salt lick-creek is remarkable for fine land, plenty of buffalo, salt-springs, white clay, and limestone. Small boats may go to the crossing of the war-path without any impediment" (Hutchins 1904, 95). The Warriors' Path mainly followed the river valleys, and some sections were developed along buffalo traces.

The Wilderness Trail also began at Flat Lick, although the entire route through southwestern Virginia and into Kentucky was soon known by this name by Anglo-American settlers, and later as the Wilderness Road. The trace only skirted the Bluegrass; it was not until Daniel Boone blazed a trail to Boonesborough in March 1775, departing from the existing trail in Rockcastle County, that there was a direct route from Cumberland Gap into central Kentucky (Kincaid 1947, 101–5). From Flat Lick, the trail followed hunters' paths and segments of buffalo traces to Knob Lick near present-day Danville and continued westward to the Falls of the Ohio, passing by Bullitt's Lick and Mann's Lick. Salt River, the major watercourse in this vicinity, was so designated because of these and the numerous other licks along its tributaries. The buffalo trace continued on the north side of the Ohio, opposite the falls, across southern Indiana to French Lick and crossed the Wabash River at the site of Vincennes (McDowell 1956; Jakle 1968, 302).

With the near elimination of hostile Indian activity in Kentucky by 1783, the Ohio River became the preferred entry route into Kentucky. The trail known to the Native Americans as Alanant-o-wamiowee or the Buffalo

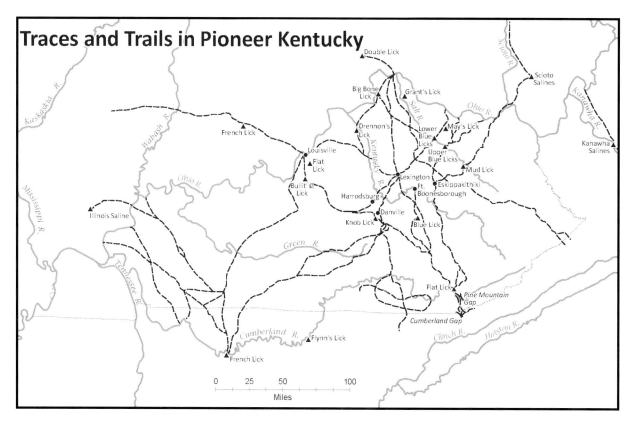

3.1. Map of buffalo traces and trails used by the pioneers during the settlement era. Derived from trail maps in Jakle (1968) and Ulack, Reitz, and Pauer (1998). Cartography by Jeffrey E. Levy, Gyula Pauer Center for Cartography and GIS, University of Kentucky, Lexington.

Path followed a semicircular course through northern Kentucky from lick to lick, connecting two Ohio River landings at the sites of present-day Covington and Maysville (Jakle 1968, 302). This trail began at a ford near the mouth of the Licking River, used by the buffalo as they crossed into Kentucky from their winter range on the prairies of central Ohio. From the Ohio River ford, the trace headed southward to Big Bone Lick (Jillson 1934, 46). George Croghan (1904, 135), in his journal entry for May 31, 1765, described the appearance of the trace as his party of explorers approached Big Bone Lick: "We went to the great lick. . . . On our way we passed through a fine timbered clear wood; we came into a large road which the buffalos have beaten, spacious enough for two wagons to go abreast, and leading straight into the lick." The trail continued south from Big Bone, passing by Drennon's Lick to Leestown on the Kentucky

River just north of Frankfort. Filson (1784, 30) noted that the ford at Leestown was "worthy of admiration; a great road large enough for waggons made by the buffalo, sloping with an easy descent from the top to the bottom of a very large steep hill, at or near the river."

From Leestown, the trace veered east to a salt lick in Scott County, where several trails converged and the herds so trampled the earth in the vicinity that it became known as the Stamping Ground. The Alanant-o-wamiowee continued to the Great Crossing of North Elkhorn Creek and passed through the site of Georgetown and on to Lexington. From downtown Lexington, the trace turned northeast, following the present Limestone Street and Bryan's Station Road, passing the pioneer fort known as Bryan's Station, and ran to the Lower Blue Lick and Mays Lick before ending at the mouth of Limestone Creek, the site of present-day Maysville (Hammon

2000, 129–31). Traveler James Smith (1907, 372) noted in 1795 that the old buffalo road between Lexington and the Lower Blue Lick was "generally 200 feet wide."

There were, of course, many more buffalo traces in Kentucky than the few described here, and each of the major routes also had many branches and alternate courses. These traces guided much of the movement of pioneers during the settlement era and led to the establishment of farms, taverns, blacksmith shops, and other businesses near them, which in turn reinforced continued use of the same routes. Many modern highways and railway routes in Kentucky today follow or parallel trails and roads used during the initial settlement (Raitz, Levy, and Gilbreath 2010).

The saline springs of Kentucky represented valuable natural resources to both Native Americans and Anglo-American settlers. The relatively high population density of the mound-building Mississippian cultures has been attributed, in part, to the numerous saline springs of the region. The Native Americans apparently used salt only as a condiment, as there is no indication that it was employed in preserving meat or fish. Salt could be made by boiling the mineral waters in containers down to crystallization. During prehistoric times, the inhabitants of the region made salt in ceramic pans (figure 3.2), either by placing the vessel on a fire or by simply exposing it to the sun to evaporate the water (Brown 1980). After contact with Anglo settlers, many tribes obtained iron kettles by trade or as plunder during raids, finding them useful not only for cooking but also for more efficient salt making.

Unlike the Native Americans, who smoked rather than salted their meat for preservation, the American colonists were accustomed to a lot of salt in their diet. Salted cod, herring, pork, and beef were important staples in eastern communities (Kurlansky 2002, 217–18) but were not available to the early trans-Appalachian pioneers, who were dependent on the wild game of the region, such as deer and buffalo, salted, dried, or smoked. The explorers of Kentucky thus kept a sharp lookout for salt licks and springs, since the manufacture of salt would be essential for survival of the settlements. As the western country became more settled and the Indian danger subsided, salt manufacture became an important commercial industry.

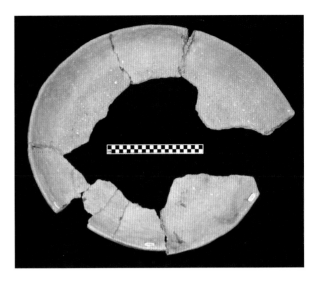

3.2. This ceramic pan, 28 inches (70 cm) in diameter and 3 inches (8 cm) deep, was found during 1984 archaeological excavations of a Fort Ancient Native American site in Bracken County and is similar in type to those used to make salt at various saline springs. The Augusta site appears to have been occupied from about 1400 to 1730 CE. The artifact is curated at the Kentucky Office of State Archaeology (OAS) in Lexington. Photograph by Gary A. O'Dell.

The first commercial saltworks in Kentucky was established at Bullitt's Lick in the fall of 1779, and it remained the most important saltworks in the state, employing hundreds of men at its peak, engaged as woodcutters, kettle tenders, water drawers, and teamsters (McDowell 1956). By 1790, saltworks had been established at most of the larger salt licks throughout the state. Salt making was dangerous work, and scalds and severe burns were common occupational hazards. It was also very destructive of the local environment, since it required cutting of large quantities of timber to feed the fires beneath the kettles. A large saltworks could swiftly create a zone of deforestation in the wilderness that covered hundreds of acres. Production from Kentucky's major salt licks supplied salt to a significant portion of the midwestern region until the emerging industry at the Illinois Saline and, later, the Kanawha salines in western Virginia (near present-day Charleston), the strongest brines yet discovered, came to dominate the market in the early nineteenth century. The relatively weak brines of the Kentucky salines required far more labor to

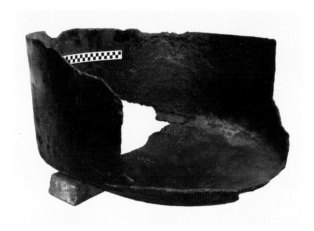

3.3. Drennon's Lick in present-day Henry County was the site of a pioneer saltworks that operated intermittently from 1785 to about 1795. During 1994–5, the Kentucky Department of Highways conducted archaeological excavations and unearthed several iron kettles and furnace remains. Approximately 35 gallons (132 liters) in capacity, this kettle measures 26 inches (67 cm) in diameter across the top and 15 inches (38 cm) deep. The artifact is curated at OAS, Lexington. Photograph by Gary A. O'Dell.

produce a bushel of salt, and so the industry here was ultimately abandoned (Jakle 1969; Stealey 1993). Beginning about 1800, a few of the salt springs, including Olympian Springs (Mud Lick), the Lower Blue Lick, and Drennon's Lick (figure 3.3), achieved new economic significance when they were developed as health resorts, attracting travelers for the alleged curative powers of the mineral waters (Coleman 1955).

During the pioneer era, salt licks were certainly seen as valuable assets and thus were promptly claimed by immigrants but, with the exception of the vicinity of Bullitt's Lick, generally did not attract settlements. Salt was necessary, to be sure, but potable water was far more important to everyday life. If it was the location of the mineral springs that led to the establishment of a network of bison-generated trails, linking lick to lick, that guided pioneer explorations in Kentucky, it was the location of the freshwater springs in the vicinity of these trails that determined where actual settlements were made. Many of the earliest settlers were under the initial impression that the country was poorly watered—Levi Todd recollected that at Boonesborough in 1775, "we

then thought springs of water scarce and that the country would be thinly inhabited"—but they soon learned that clear and free-flowing springs were abundant throughout central Kentucky (Draper 15CC157).

As the most significant freshwater springs were discovered by the Anglo-American pioneers, they became reference points in the wilderness, serving hunters and explorers as navigational markers and locations for rendezvous with other frontiersmen. In the spring of 1779, for example, John Pleakenstalver, Ralph Morgan, and a few others set out from Boonesborough to the vicinity of Elkhorn Creek and along the way "encamped at Todd's spring," where Colonel John Todd had erected a cabin (Collins 1878, 2:179). Like the salt licks, freshwater springs were also good places to wait in concealment to bag a deer or other wild game. Josiah Collins, one of the men who, in April 1779, helped erect the first fort at Lexington, recalled having trouble finding the spring where the settlement was to be located: "The old woodsmen about Boonesborough directed us how to find the camp. Old hunters knew where the spring was we were to build at. In truth they knew just where we were camped, and every big spring through the country, and just the place where this was" (Draper 12CC100).

The Native Americans, who had been hunting in the region for generations and were well acquainted with the springs of Kentucky, used some of the same tactics on the pioneers that had been successful in hunting game for the larder. A raiding party might lie quietly at a well-known spring for hours or even days to ambush some luckless frontiersman as he came to drink (Withers 1831, 144). In the summer of 1780, a year after helping raise the Lexington blockhouse, Josiah Collins and a companion, Thornton Farrow, were approaching Hugh McGary's station on Shawnee Run in present-day Mercer County on an errand from the commander at Lexington. Although Farrow claimed to have seen an Indian on the bank of the Kentucky River at the mouth of Shawnee Run, Collins scoffed, and believing themselves secure, recklessly they "lay down our bellies and drank" from a spring a few hundred yards from the settlement. No harm befell them from this, but Farrow had been correct, and the two men had been followed. On the very next day, a man named Hinton, who had gone out to tend his field, was ambushed and killed by Indians when he drank from the same spring (Draper 12CC68).

Beginning about 1773, the region that would become Kentucky was invaded by a horde of would-be settlers as well as land speculators, who began to choose and survey prime tracts of land. Most of the initial claims were of lands in the central, or Bluegrass, region of the state; other areas were settled later. The land claims were made on the basis of crude surveys, which might constitute no more than corners marked by blazing or girdling trees or carving initials into a tree or on a boundary rock, and through "improvements," such as clearing an acre or two and planting a corn crop or constructing a rough cabin. Surveys of these lands were based on the "metes and bounds" system, which consisted of bearings and distances from one physical landmark to another, and land parcels were further identified by the primary watercourse draining the land. Certificates were awarded by the Virginia government to those settlers who could establish a legally valid claim, and many of the certificates contain references to karst features. There was a well-developed terminology used by the pioneers to describe such features as springs, "sinking springs," "lick springs," "boiling springs," "blue holes," "cave springs," and the like.

Since it was often difficult in the wilderness to locate markers for someone else's survey or even to determine whether the land had been claimed at all, a host of conflicting claims arose. In 1779, the Virginia government sent a commission to Kentucky to judge the land claims and resolve the conflicts, which issued certificates for valid claims. A typical certificate might include a description such as the following, issued by the commission seated at St. Asaph's (Logan's Fort) on April 22, 1780:

Andrew Steele having obtained Certificate for preemption of 1000 Acres of land in the District of Kentucky [in] February 1780 now comes into Court and makes it appear that he is intitled also to a settlement by Virtue of raising a crop of Corn in the year 1776 lying about 2 Miles up from the mouth of Steel run waters of the South fork of Elkhorn to include a large Rocky Spring & his improvements ordered that a certificate issue accordingly. (Kentucky Historical Society, 290)

The uncertainty concerning overlapping claims led to a profusion of litigation that occupied Kentucky courts for decades and enriched a generation of lawyers. An example from the court records is indicative of the nature of depositions in the trials and the importance of springs to settlers claiming land. On April 5, 1801, Samuel Boggs made the following statement in regard to a claim in present-day Scott County, Kentucky. In his deposition, Boggs made clear the competitive nature of land claiming and that settlers were well aware of problems with potentially conflicting claims:

That in year 1776 this deponent in company with William Lindsay deceased and others made an improvement by building a cabbin at the Cave spring where Henry Lindsay now lives which improvement was made for William Lindsay . . . also made several other improvements on the run above and below the Cave spring to keep other people from making improvements that might interfere with those at the Cave spring and this place. (Fayette Circuit Court, *Complete Record Book A,* 381)

The case of the initial settlement of Royal Spring, the site of present-day Georgetown, illustrates both the routes taken by early immigrants into the region and the importance attached to springs as settlement sites (figure 3.4). The spring was first discovered in 1774 by John Floyd, deputy surveyor for Fincastle County, as Virginia's Kentucky lands were then known. In mid-April, William Preston, the official surveyor and sheriff of the new county, sent the young man across the mountains to the west to survey lands under military warrants, one of three such parties he dispatched at about the same time (Cartlidge 1968, 325). As chronicled by Thomas Hanson, a member of the party, while surveying lands in the vicinity of North Elkhorn Creek, Floyd decided to take a short break on July 9 and, with William Nash, "went in search of a spring, which they found." Hanson described the spring as "the largest I have ever seen in the whole country, and forms a creek by itself." During the following two days, the party surveyed 1,000 acres around the spring for John Floyd, breaking camp on July 21 and setting out for a rendezvous at the cabin of James Harrod (Hanson 1905, 129–31).

In late October 1775, Robert Patterson "left the Pittsburg country . . . with John McClelland and family and six other young men for the promised land, Kentucky" (Draper MM, 16). John and Alexander McClelland,

3.4. In 1775, the Royal Spring in Scott County was chosen as the site of McClelland's Station, one of the region's earliest pioneer settlements. Although the station was soon abandoned, the community of Georgetown was founded on the same location, and the spring today still serves as the town's primary water source. The photograph shows the appearance of the spring circa 1888. Source: Kentucky Department of Libraries and Archives.

William McConnell, Francis McConnell Jr., Andrew McConnell, and David Perry had just returned from an exploring trip that had set out the previous April and had spent several months surveying and making improvements in the vicinity of Elkhorn. William McConnell had first explored the Kentucky lands in 1774 and, as the most experienced, was probably the leader of this party (Collins 1878, 2:178). The McClelland party knew exactly where they were going to establish their settlement. John McClelland had very likely discovered Floyd's Spring during his explorations of the Elkhorn region a few months earlier and was unaware that the land around the spring had been surveyed and claimed by John Floyd.

The party of ten men and one woman set off from Pittsburgh, some in canoes, into which they had loaded such goods and supplies as they thought necessary to set up households in a new land, while others kept pace along the bank of the Ohio, driving a few head of live-stock. These were the first stock imported into Kentucky, nine horses and fourteen cattle. In November, after a journey of nearly 400 river miles, the party reached the mouth of Salt Lick Creek, presently the site of the community of Vanceburg, in Lewis County, Kentucky (Draper MM16–21; Fayette Circuit Court, *Complete Record Book A*, 307–13).

Here the party divided, arranging to rendezvous in a few weeks at Leestown, a buffalo ford of the Kentucky River located about a mile north of where Frankfort would later be established. Most of the party, including the McClellands, remained with the canoes and continued down the Ohio River, planning to turn up the Kentucky River and follow it southeastward to Leestown. This would require an additional journey of about 230 miles by water. Patterson, William McConnell, and Stephen Lowrey bade farewell to the McClellands and, led by David Perry, turned their steps away from the Ohio

River, driving the cattle and horses and scouting ahead for the best route for the animals. They headed almost due west, following Salt Lick Creek for a time, crossing Cabin Creek (in present-day western Lewis County), advancing to Stone Lick (now Orangeburg in Mason County, about nine miles southeast of Maysville), and striking the old buffalo trace at Mays Lick (Mason County) that led from the Ohio River to the Lower Blue Licks. They crossed the buffalo road, continued west a few miles until they encountered another trail known as the middle trace, and then turned southward and followed the trace to the Lower Blue Licks. Here Patterson and company ran into Simon Kenton and John Williams, who told them that they knew of no other white person then in the country (Draper MM67–68, 15C25.9–25.10, MM16–21).

Herding the livestock, they crossed the Licking River at this point and traveled westward along another buffalo trace that led to Hinkston's Station (just south of Cynthiana in Harrison County), from there following Townsend Creek southwest to yet another buffalo trail. This trail took them westward to Leestown on the Kentucky River, passing very near Floyd's Spring along the way. The drovers had managed to reach the rendezvous ahead of the McClelland party and there waited for them several days; when the canoes arrived and the separate groups were reunited, they lost no time in striking off to the east, following the same buffalo road back to Floyd's Spring. At this spring, which was renamed Royal Spring, the men established what became known as McClelland's Station, erecting a fortified house (Draper MM17). In May 1776, after John Floyd learned of squatters at his "big spring," he went to see John McClelland, "determined to drive them off, but on seeing his wife & three small children, who must have been distressed, I sold it for £300" (Draper 33S298). Tragically, before the sale could be finalized, John McClelland was killed during an attack on the station at the end of the year.

The majority of pioneer habitations were of the sort termed "stations," similar to that built by the McClellands at Royal Spring. These were more substantial than the primitive pole or log "cabins" constructed as necessary "improvements" to legally secure land claims, which were small, crude, and hastily built. Such improvement cabins were never intended to be occupied on anything but the most temporary basis, if at all. Stations, in con-trast, were in general sturdily built log houses that could be barricaded in event of attack, erected singly or in clusters for common defense for a group of families. Picket stockades consisting of logs placed vertically in the ground were erected at some sites as a further line of defense, often including exterior cabin walls as part of the enclosure. Even the stations, however, were considered only temporary abodes that would be replaced by something more durable when the hazards of the country had been reduced. Large military outposts, or forts, with blockhouses at the corners, were built at Boonesborough, Harrodstown, Lexington, and St. Asaph's (present-day Stanford). These forts differed from stations in that they were sizable complexes intended to house numerous families and military personnel and provide regional administration and services; they also served as points of refuge for outlying stations when activity by hostile Native Americans became particularly severe (O'Malley 1994, 23–29).

Although the land was chosen for its soil quality, sites for station and fort construction were selected on the basis of two important criteria: the defensibility of the location and access to a reliable source of potable water (O'Malley 1999, 61). Every station settlement in central Kentucky was situated near one or more freshwater springs. Sites adjacent to the larger and more prominent springs of the region, particularly those near the trail system, were settled first. For travelers on the frontier, as the region was settled, forts and stations quickly replaced the explorers' springs as navigational waypoints, with the exception of the more prominent licks, such as Big Bone, the Blue Licks, Bullitt's, and others.

In every case, the fort or station was constructed at a short distance from the water supply rather than being situated next to the source or including the spring within the stockade enclosure. Although this practice has long puzzled historians, since persons collecting water would be exposed to attack, there were apparently some very practical reasons for it. If the spring were located in the residential area, traffic by people and livestock would soon transform the area around the water supply into an unpleasant mire, and the presence of mosquitos could be a significant annoyance. Another consideration was that nearly all the larger springs were located in valley bottoms, whereas ridgetops were more defensible and, for that reason, were preferred habitation locations. Valley bottoms, being prone to flooding by surface streams, would also be undesirable residential locations (O'Malley 1994, 36). In a study of the water-supply practices of a modern-day, self-supplied, rural population in a region of southeastern Kentucky lacking a public water system, O'Dell (1996, 88–92) found that accessibility, reliability, and perceived water quality were the most important criteria determining which springs were used for household domestic water needs.

Although this situation might pose a hazard in obtaining water during a time of active hostilities with the Native Americans, it was deemed necessary in most cases to avoid potential fouling of the water by proximity to the population. The risks associated with conveying water from a spring that was not secured within the stockade were somewhat reduced, when possible, by clearing away the intervening brush and trees to provide a clear field of fire from the dwelling place. Separation of source and residential area did not necessarily ensure that the spring and its environs would be kept clean. Before the late nineteenth century, there was no comprehension of the role of microbes in transmitting waterborne diseases, only a vague perception that sickness was somehow connected to "filth," and notions of sanitation were rudimentary at best. On a visit to Harrodsburg in February 1780, where the spring was outside the fort but close to the populated section, Colonel William Fleming noted with considerable distaste the condition of the water supply serving that community:

The spring at this place is below the Fort and fed by ponds above the fort so that the whole dirt and filth of the Fort, putrified flesh, dead dogs, horse, cow, hog excrements and human odour all wash into the spring which with the Ashes and sweepings of filthy Cabbins, the dirtiness of the people, steeping skins to dress and washing every sort of dirty rags and cloths in the spring perfectly poisons the water and makes the most filthy nauseous potation of the water imaginable and will certainly contribute to render the inhabitants of this place sickly. (Fleming 1916, 630)

Even after the country had long been settled, sanitation remained a problem at the springs that supplied communities with water for drinking and other needs. On August 15, 1790, the trustees of Lexington published

an order in the *Kentucky Gazette* that "the public spring on Main Street and the one near the school house, no longer be used as washing places." Apparently, no one paid much attention to this notice since, on July 3, 1795, the trustees threatened to prosecute "anyone doing washing at the public spring" (*Kentucky Gazette*).

As the country grew more populous, the original large tracts claimed by pioneers were divided and divided again and resold to newcomers. Real-estate advertisements of the era were certain to emphasize the presence of a spring, a desirable feature that increased the value of the land. These springs were invariably described as "never-failing," as in the case of an advertisement placed in the *Gazette* on December 7, 1793, by James Dunwiddie, who wished to sell two hundred acres located six miles north of Lexington. On this property were "two never-failing springs near to the buildings, which are a dwelling house twenty by eighteen, and two other cabins, the whole of hewed logs." John Bradford, publisher of the *Gazette* in Lexington, advertised for sale on April 24, 1804, the house and town lot on which he resided, noting the presence of a brick springhouse, from which the flow "never fails in the driest season," and in the backyard of the same lot, another "never-failing SPRING of cold water . . . equal to any in the state." Most springs used as household water supplies, both rural and in towns, were eventually enclosed by springhouses or other structures intended to protect the source and to provide cool storage for produce and dairy products.

As time went by, many of the original forts and stations were abandoned, but others more fortunately situated continued to grow and became the towns and cities of present-day Kentucky. It is a safe assumption that every significant community in central Kentucky, including Lexington, Georgetown, Versailles, Stanford, Lancaster, Paris, and many others, owes its location to the presence of a spring used as a water supply by the original settlers of the region. As the population of these communities grew, a single spring usually proved insufficient to meet the increasing demand for water.

In Lexington, the first effort to increase the flow of the "public spring" by enlarging the opening was made in January 1783; resident Martin Wymore recalled that they "dug that in farther, and more, and it got stronger, as they went farther into the bank" (figure 3.5). This eased the problem only for a short while, and soon the citizens set

3.5. Historical tradition holds that the site of Lexington was named in 1775 when a party of explorers, camped at an attractive spring in the Bluegrass region, first heard news of the opening battle of the American Revolution at Lexington, Massachusetts. Although the town spring that once served Lexington as a water supply was long ago buried by urban development, McConnell's Blue Hole, the explorers' campsite, was preserved. Now a city park, the spring tract serves both to celebrate Lexington's beginnings and as an outdoor education center. Photograph by James R. Rebmann.

about digging at all the wet-weather seeps along the banks of Town Branch. The seeps, said Wymore, "gradually opened, and new springs broke out all along the bank" (Draper 11CC131). Communities were in time forced to develop other water-supply sources; since there were no public distribution systems, individual households were required to dig wells or collect rainwater in cisterns to meet their own needs. By the beginning of the twentieth century, community-based public systems based on artificial reservoirs were being developed to supply water to citizens of the largest towns (O'Dell 1993).

With few exceptions, most of the springs that formerly supplied early settlements that later grew into substantial communities have disappeared as a consequence of urban development (figures 3.6 and 3.7). A few Kentucky communities, such as Georgetown and Elizabethtown, still rely on the same karst springs that determined their original settlement. In most rural areas of the state, however, springs have retained use value and have persisted in the landscape. Most of these springs presently serve to water livestock, although their primary historic function may formerly have been domestic supply. Many rural households today continue to depend on springwater for domestic use where public water-system infrastructure is lacking, even in parts of the relatively affluent and densely populated Bluegrass Region.

References

Manuscripts and Other Unpublished Materials

Draper, L. C. Manuscript Collection. State Historical Society of Wisconsin (Madison).
Fayette Circuit Court. *Complete Record Book A*. Kentucky Department for Libraries and Archives (KDLA), Frankfort.
Land Trials. Microfilm. KDLA.

3.6 and 3.7. Many springs that inspired the founding of pioneer settlements and served as initial water-supply sources were later destroyed or covered by urban development, as in the case of Lexington. For some modern communities, brief glimpses of the flow may be seen, as at Glasgow (upper photo). Other Kentucky towns have endeavored to preserve the settlement spring, as in Versailles, where the town spring, although not used as a water supply, still flows freely from a low cave opening behind the courthouse (lower photo). Photographs by Gary A. O'Dell.

Published Materials

Aron, S. 1996. *How the West Was Lost: The Transformation of Kentucky from Daniel Boone to Henry Clay.* Baltimore: Johns Hopkins University Press.

Barnhart, J. D. 1951. *Henry Hamilton and George Rogers Clark in the American Revolution, with the Unpublished Journal of Lieut. Gov. Henry Hamilton.* Crawfordsville, IN: R. E. Banta.

Belue, T. F. 1996. *The Long Hunt: Death of the Buffalo East of the Mississippi River.* Mechanicsburg, PA: Stackpole Books.

Brown, I. W. 1980. *Salt and the Eastern North American Indian: An Archaeological Study.* Cambridge, MA: Peabody Museum, Harvard University.

Cartlidge, A. M. 1968. Colonel John Floyd: Reluctant Adventurer. *Register of the Kentucky Historical Society* 66 (October): 317–25.

Coleman, J. W., Jr. 1955. *The Springs of Kentucky: An Account of the Famed Watering-Places of the Bluegrass State, 1800–1935.* Lexington, KY: Winburn Press.

Collins, R. H. 1878. *History of Kentucky.* 2 vols. Covington, KY: Collins & Co.

Cresswell, N. 1924. *The Journal of Nicholas Cresswell, 1774–1777.* Carlisle, MA: Dial Press.

Croghan, G. 1904. The Journals of George Croghan, 1750–1765. In *Early Western Journals, 1748–1765,* vol. 1, edited by R. G. Thwaites, 47–176. Cleveland: Arthur H. Clark.

Filson, J. 1784. *The Discovery, Settlement and Present State of Kentucke.* Wilmington, DE: printed by James Adams.

Fleming, W. 1916. Colonel William Fleming's Journal of Travels in Kentucky, 1779–1780. In *Travels in the American Colonies,* edited by N. D. Mereness, 617–55. New York: Macmillan.

Hammon, N. O. 2000. Pioneer Routes in Central Kentucky. *Filson Club History Quarterly* 74 (Spring): 125–43.

Hanson, Thomas. 1905. Hanson's Journal. In *Documentary history of Dunmore's War, 1774,* edited by Reuben G. Thwaites and Louise P. Kellogg, 110–33. Madison: Wisconsin Historical Society.

Hedeen, S. 2008. *Big Bone Lick: The Cradle of American Paleontology.* Lexington: University Press of Kentucky.

Hutchins, T. 1904 [1778]. *A Topographical Description of Virginia, Pennsylvania, Maryland, and North Carolina.* Edited by F. C. Hicks. Cleveland: Burrows Brothers.

Jakle, J. A. 1968. The American Bison and the Human Occupance of the Ohio Valley. *Proceedings of the American Philosophical Society* 112 (August): 299–305.

———. 1969. Salt on the Ohio Valley Frontier, 1770–1820. *Annals of the American Association of Geographers* 59 (December): 687–709.

Jillson, W. R. 1934. *Pioneer Kentucky.* Frankfort, KY: State Journal Company.

Kentucky Historical Society. 1992. The Certificate Book of the Virginia Land Commission 1779-1780. Greenville, SC: Southern Historical Press.

Kincaid, R. L. 1947. *The Wilderness Road.* New York: Bobbs-Merrill.

Kurlansky, M. 2002. *Salt: A World History.* New York: Penguin Books.

McDowell, R. E. 1956. Bullitt's Lick: The Related Saltworks and Settlements. *Filson Club History Quarterly* 30 (July): 241–69.

O'Dell, G. A. 1993. Water Supply and the Early Development of Lexington, Kentucky. *Register of the Kentucky Historical Society* 67 (October): 431–61.

———. 1996. The Search for Water: Self-Supply Strategies in a Rural Appalachian Neighborhood. MA thesis, University of Kentucky.

O'Malley, N. 1994. *"Stockading Up": A Study of Pioneer Stations in the Inner Bluegrass Region of Kentucky.* Archaeological Report 127, revised. Frankfort: Kentucky Heritage Council.

———. 1999. Frontier Defenses and Pioneer Strategies in the Historic Settlement era. In *The Buzzel about Kentuck,* edited by Craig T. Friend, 57–75. Lexington: University Press of Kentucky.

Raitz, K. B., J. E. Levy, and R. A. Gilbreath. 2010. Mapping Kentucky's Frontier Trails through Geographic Information and Cartographic Applications. *Geographical Review* 100 (July): 312–35.

Rostlund, E. 1960. The Geographic Range of the Historic Bison in the Southeast. *Annals of the American Association of Geographers* 50 (December 1960): 395–407.

Smith, D. B. 1999. "This Idea in Heaven": Image and Reality on the Kentucky Frontier. In *The Buzzel about Kentuck,* edited by Craig T. Friend, 77–100. Lexington: University Press of Kentucky.

Smith, J. 1907. Tours in Kentucky and the Northwest: Three Journals by the Rev. James Smith of Powhatan County, Virginia, 1783-1795-1797. Edited by J. Morrow. *Ohio State Archaeological and Historical Quarterly* 16: 348–401.

Stealey, J. E. 1993. *The Antebellum Kanawha Salt Business and Western Markets.* Lexington: University Press of Kentucky.

Ulack, Richard, Karl Raitz, and Gyula Pauer. 1998. *Atlas of Kentucky.* Lexington: University Press of Kentucky.

Wharton, M. E., and R. W. Barbour. 1991. *Bluegrass Land and Life: Land Character, Plants, and Animals of the Inner Bluegrass Region of Kentucky.* Lexington: University Press of Kentucky.

Withers, A. S. 1831. *Chronicles of Border Warfare.* Clarksburg, VA: Joseph Israel.

CHAPTER FOUR

Water and Wastewater Service for the Commonwealth of Kentucky

Roger Recktenwald

Eleven years after the Virginia Colony acquired its western lands in 1763—everything south of the Ohio River, east of the Mississippi River, and north of the westward extension of the parallel separating North Carolina and Virginia—a crew of land surveyors, headed by James Harrod from Pennsylvania, boated down the Ohio River and then canoed south, up the Kentucky River, some ninety-eight miles to the mouth of a small creek called Landings Run. From there they traveled overland westward several miles, stopped at a point near a big cold-water spring, and decided to lay out a town that was called Harrods Town, Harrodstown, or Harrodsburg, Kentucky. Although these hardy folk were not the earliest and definitely not the last white settlers to venture into this territory, they validated their claims by complying with Virginia's law to improve their respective lots (i.e., build structures and grow crops) and were jointly credited as being the first permanent settlement (1774–5) by colonials of English extraction west of the Allegheny Mountains. Hence, because of its Kentucky firstness, I will loosely track the development of water infrastructure at Harrodsburg over time as I trace the development of water and wastewater infrastructure in Kentucky ("infrastructure" here means items constructed to capture, hold, treat, deliver, and remove water, such as reservoirs, treatment plants, pipes, pumps, and tanks).

The small band quickly built a fort, and had they not, I would be citing another community as an example, for the Native Peoples were seriously displeased with their presence. The big spring, located about six-tenths of a mile east-northeast of their new Fort Harrod, was quickly dubbed Town Spring and provided plenty of water for the settlement. But raiding parties visited often during the late 1700s and periodically until after the War of 1812. During attacks, settlers would leave their small cabins and field chores and hastily gather in the fort, where a small spring provided water for drinking, cooking, and washing, although it was known to be tainted by human, animal, and other organic waste in decomposition. All pauses in hostilities were welcome but afforded the fort's occupants way too much time to speculate about what all might be in the little spring's water.

Much of the northern area of the city relied on the larger Town Spring for nearly all of the 1800s and early 1900s for communal drinking water. Even today, the cold water surges to the surface, pools briefly, and then under the name Town Creek begins a meander of several miles, entering the Salt River northwest of the city. However, Harrodsburg residents also relied on many other surface springs scattered throughout the area, and because of the geology, there was a high probability of hitting water virtually anywhere a sober diviner scratched an *x*. Those without easy access to spring or well water built cisterns.

In the early 1800s, Harrodsburg experienced economic growth driven by considerable visitor commerce, generated by attraction to the healing properties or, more accurately, the recreational possibilities of the many local mineral springs. Entrepreneurs built hotels complete with large baths similar in concept to those of ancient India, Greece, and Rome, along with fancy saloons, stocked with some of Kentucky's early liquid products, often safer to drink than the water. Like many other small cities throughout Kentucky and across the nation, Harrodsburg was actively seeking to attract new citizens, and local leaders could easily ensure the availability of water and promise, if not guarantee, a healthy environment.

Unfortunately, incoming settlers to the New World carried with them some very unwelcome pestilences. By the mid-1800s, periodic epidemics of various infectious diseases spread quickly and took many citizens' lives, putting the future of some Kentucky communities in jeopardy. Cholera was deadly and quick. Because its true cause was still unknown, each occurrence arrived like a thief in the night, often accompanied by full-scale community panic. Sadly, unknowingly infected victims in one community would seek refuge with friends and family in adjacent communities, only to spread the disease further. Nancy D. Baird states in the *Kentucky Encyclopedia,* "Of the Asiatic cholera outbreaks that struck nineteenth century America, Kentucky was among the states with the highest fatality rate" (Baird 1992, 184). Outbreaks affecting much of Kentucky occurred in 1832–5, in 1848–54, and again in 1872. Of this last instance, Baird notes that it "claimed thousands of lives in . . . communities that relied on wells and shallow streams for their water supply" (ibid.). But only in hindsight do we know that nearly all drinking-water sources, especially in more densely populated communities of the period across Kentucky, had multiple opportunities to mix with waters that had traveled through scenic hog pens, cattle feedlots, outhouse pits of important people, and other interesting locations on the way to a family's drinking water.

In 1842, back across the Atlantic, Edwin Chadwick developed a report titled "The Sanitary Condition of the Laboring Population of Great Britain" that detailed the dire living and working conditions in industrialized London and across England. He used information gleaned from sanitary surveys of areas featured in Charles Dickens's novels to highlight the correlation between water quality and other environmental factors and the very pathetic state of laborers' personal health.

Soon after, in 1850, the newly created Massachusetts Sanitary Commission, headed by Lemuel Shattuck, published a cumulative report that cited measures taken to promote public health and then set out a plan to conduct a sanitary survey of every community in Massachusetts similar to that done in England. For his efforts, Shattuck has been dubbed the "father of public health," and the commission was imitated across the country in an effort to protect individual and communal health through improved hygiene and protection of water from pollutants. (Today, U.S. states, with the assistance of the U.S. Environmental Protection Agency, conduct sanitary surveys of all public systems.)

In 1854, again back in England, John Snow, a physician battling a resurgent cholera epidemic, drew a map illustrating the location of patients' homes and determined by correlation that they relied on the water from a hand-pumped well on the corner of Cambridge Street and Broad Street in London. He convincingly conveyed his finding to the neighborhood governing body, which then removed the pump handle. The number of new cases of the infection in that area dropped dramatically, and the epidemic soon ended. Shortly afterward, Snow replicated his incident-location study along the Thames River below London during a more serious cholera outbreak. His findings regarding cholera as a waterborne disease were published in *On the Mode of Transmission of Cholera* in 1855.

News of these events traveled fast to the Americas, and although the correlation between contaminated water and disease was still not scientifically understood, it became widely accepted. In Kentucky, suspicion increased regarding the relationship between extensive subsurface and numerous surface waterways, but the confusion between clarity and purity persisted. It was not until the maturity of the industrial period in America (mid-1800s to the mid-1900s), when large but unmeasured volumes of solid, semisolid, and liquid wastes, mostly organic, were increasingly discarded or found their way into streams and rivers, that the impact on a given water body become a matter of serious public concern. But regardless of the volumes of strewn refuse or the obvious degradation of a water's quality, the public reaction was mostly laments of wounded communal aesthetics or the loud, colorful complaints of disgruntled boat captains who had to navigate the sea of flotsam generously contributed to waterways by animal slaughterhouses, tanneries, and seasonal timber-floatation events.

But throughout the 1800s, the ravaging impacts of contagious diseases and other health concerns screamed from local newspapers and were only periodically bumped from the front page to make room for coverage of fire-safety issues. After each tragic fire, community leaders were implored to build a water system for community fire suppression. If local leaders failed to take responsive action by the next community fire, their tenure was short lived. Dealing effectively with the epidemics of the 1800s and the deaths and damage from fires in cities of all sizes required what citizens were demanding, a safe and reliable source of community water. Hence both the leadership and the economic attractiveness of a community came to be evaluated by whether the community was keeping pace with the latest water-supply, treatment, and delivery technology for drinking water and fire suppression.

As is somewhat typical of many issues in Kentucky, this scene was played out loudest and most dramatically in Louisville, where many voiced concern for public health and the need for a reliable, central source of clean water for drinking and fire control. But it was not until 1838 that the Kentucky legislature authorized the incorporation of a private company, the Louisville Gas and Water Company, to drill for and distribute natural gas, and subsequently allowed the company "to develop water for fire suppression . . . use on roads and alleys for cleaning and sprinkling for dust control . . . and for all manufacturing and domestic needs"—a very interesting pecking order. However, the so-called Great Fire of 1840, which literally incinerated to ash a large portion of the commercial center of the city (Third and Fourth Streets between Market Street and the river), again brought a focus on the glaring fact that Louisville had no central water system. So, a mere fourteen years later, in 1854 the Kentucky legislature sprang to the call and again authorized the incorporation of a private company, this time to be known as the Louisville Water Works (later changed to Louisville Water Company), which did incorporate and began operations in 1860.

The company became widely known for its innovative work with coagulation and rapid sand filtration in 1896 and is fairly credited with the early experimentation with and use of chlorination for disinfection; by 1913, it chlorinated and filtered all its water. In 1909, all stock in the company was acquired by the City of Louisville. At present, the Louisville Water Company continues to thrive through regional service-area expansions, extensive collaboration with neighboring water providers, and successful deployment of its applied in-house research on cost-effective and innovative technologies, as evidenced in its recent Riverbank Filtration Project, now in use.

Unfortunately, from the late 1800s through the 1960s, the common practice of freely discharging wastes of all descriptions into the nation's rivers and streams increased in type and volume, with little restriction except for local laws and regulation. As awareness grew regarding the need for more extensive sanitation practices, private and public, the waste stream entering waterways came to include more and more organic wastes, especially from new sewage systems that collected waste, provided minimal treatment, and pumped the "treated" water straight to the adjacent water body. The manufacturing of chemicals from petroleum, minerals, and other raw sources created unwanted by-products that were gladly donated to the overall waste stream in its march to streams, rivers, and oceans, often with very toxic effects. Throughout this entire period, the "treatment" mantra was the infamous gross oversimplification "Dilution is the solution to pollution."

By the 1870s, it seemed that nearly all Kentuckians were under duress, whether from the harsh economic times, the painful aftermath of the Civil War, or as victims or grieving survivors of a bout of cholera, smallpox, or typhoid. State government leaders were constantly challenged to maintain the rural/farm–city/industry constituents' handshake, graphically depicted on the state's flag. Stakes were high, and fixes in communities of every size were expensive but politically unavoidable because of growing citizen demands that government eradicate the epidemics and provide safe drinking water and fire protection.

Two complementary, healthful events, one on the world stage and the other at Frankfort, occurred in 1878. Each event would shape forever a much more positive future for Kentucky than could have been reasonably predicted at the time. First, the long-surmised and much-rumored relationship among microorganisms, health, and water's role as a principal medium was thrust front and center on the world stage. A seminal work of French scientist Louis Pasteur (1822–95) and his colleagues titled "Germ Theory and Its Application to Medicine and

Surgery" was read before the French Academy of Science on April 29, 1878. Their work provided the empirical proof of how the human body is benefited by clean water and can be subjected to disease by polluted water. This work included the development of methods to ensure the safety of water, as well as the means to eradicate the most common forms of epidemic disease that had followed humans since memory. The prospects for radically improved public health via public infrastructure investments to ensure clean, safe drinking water and to provide for proper collection and treatment of human and other organic wastes began immediately to reshape the role of governments in the world and in Kentucky.

Also, in the spring of 1878, a bill authorizing the creation of the State Board of Health was proposed by a candidate in the governor's race, Luke Pryor Blackburn, which passed overwhelmingly in both houses of the Kentucky legislature and was signed by Governor James McCreary prior to Governor Blackburn's inauguration; Kentucky was one of the first states to take this action. Headed by a trained physician, Joseph N. McCormack, the board's legislative mandate was to "do everything necessary to protect the public health." For nearly the next hundred years, the Kentucky State Board of Health, acting through its Bureau of Sanitary Engineering, was the principal state agency charged with ensuring the safety of drinking water and translating germ-theory science into procedures, practices, regulations, and enforcement actions for disease control through inoculation therapies, as well as treatment of wastewater and management of solid waste to protect and improve the health of all Kentuckians.

Increasing Governmental Role in Water and Wastewater Management

The evolution of the various roles of the local, state, and federal governments in management of public water and wastewater in the United States generally kept pace with the broadening scientific understanding of the relationship of water chemistry to biology and vice versa, as well as the unfolding awareness of the natural relationship between humans and the environment shared with other life forms. Tasks were divided such that federal and state legislation established and enforced water-quality standards, and local governments determined how best to achieve those standards.

Role of the Federal Government in Water and Wastewater Management

The attention span of the federal government during its first eighty-four years was consumed by struggles to achieve national independence, followed quickly by the bloodier struggle to ensure national unity. The scope and impact of public health problems confronting the nation from its inception obviously extended beyond any individual state's borders. Despite early, long, loud, and sometimes rabid expressions of allegiance to states' rights from some members of Congress, true patriot-survivalist members eventually succeeded in placing public health on the federal table alongside national defense and regulation of interstate commerce. In 1798, the federal government took its first health-related action, establishing the U.S. Public Health Service within the Department of the Treasury to address chronic health issues of disabled merchant seamen. The agency established marine hospitals in the early 1800s along the East Coast and eventually served inland lake and river boatmen. A marine hospital opened in 1852 in the Portland neighborhood of Louisville, and later another hospital was constructed in Paducah. The service is now directed by the surgeon general under the U.S. Department of Health and Human Services.

The overarching federal role in managing water quality and public sanitation began tangentially, hesitantly, and nearly one hundred years later. The ever-constant wanderlust that seems a dominant national cultural trait, powering early settlement and spurring commerce, became widely evident in the large ridership of the rail passenger trains that crisscrossed the United States in the late 1800s. However, the limited fiscal and legal capacity of localities and states to handle outbreaks of disease across state lines demanded that the federal government step up. Its efforts in true nation building, that is, achieving and maintaining a consistent quality of life for all citizens, which had begun timidly with the U.S. Public Health Service, resumed in the late 1800s with a series of federal acts and programs, many directed to water-quality management. The most important of these before World War II were the following:

- The Interstate Quarantine Act (1893) gave the Public Health Service the authority to control the spread of disease carried by interstate commerce; it mostly affected railroad travelers.
- The Rivers and Harbors Act (1899) (Refuse Act) prohibited discarding materials that obstruct navigation; it was broadened in the 1960s to prohibit discharging any industrial waste into navigable waters.
- Federal Water Quality Standards (1914) were promulgated by the Public Health Service to regulate water used on interstate trains; they were not mandatory for local water systems, but in lieu of no regulations, the standards were widely, if reluctantly, adopted by local public water systems.
- The Public Works Administration (1933), a New Deal agency, publicly designed, bid on, and oversaw construction of 140 drinking-water, 31 sewer, and 18 waste-disposal plants scattered across Kentucky, from Paintsville to Paducah.
- The Civil Works Administration (1933), a New Deal agency, constructed major water and sewer projects, waste incinerators, and outhouses across Kentucky.
- The Works Progress Administration (WPA) (1933), a New Deal agency, was one of the principal components of President Franklin Delano Roosevelt's recovery programs and continued until 1943, when increased private-sector employment related to the World War II effort rendered the program unnecessary.
- The Civilian Conservation Corps (1933), a New Deal agency, focused in Kentucky on planting trees and developing parks. Waterways used for drinking water were benefited by extensive erosion-control and bank-stabilization projects.
- The Rural Electrification Administration (1936), a New Deal agency, was the primary source of funding to extend electricity into rural Kentucky in areas deemed not feasible by the major utilities; it provided a prerequisite for development of water service in rural Kentucky.

For the duration of World War II (1939–45), the entire attention of the United States was understandably focused on the war efforts in Europe and the Pacific. After World War II, the United States sought to sustain the high pace of industrial production by retooling the war-supply industries not only to maintain full employment for current workers but also to accommodate returning soldiers. Many new petrochemicals and synthetic substances benefited manufacturing, but they or their by-products were toxic and joined the long list of pollutants discharged into the air, onto the land, or into waters adjacent to production facilities. Hence, from the late 1930s through the 1970s, the skies over many industrial cities throughout the country became choked to the point of near-zero visibility at high noon, while the nation's waterways and coastal estuaries could not support life, and some even threatened life, both human and aquatic, and discarded materials and refuse rendered large tracts of land unusable. The infamous Kentucky example, the Valley of the Drums, and many others across the nation became the subject of a special federal reclamation effort known as the Superfund, so named because of the cost to eliminate health safety hazards and reclaim natural resources. In this setting, amid cries that it would "slow America down" and "make us less competitive," the U.S. Congress began hearings related to environmental protection. These hearings resulted in a series of congressional acts focused simultaneously on development and environmental quality concerns. These include:

- The Federal Pollution Control Act (1948) was the first major U.S. law to comprehensively address water pollution. Major amendments in the 1960s, 1970, 1972, and 1986 required assessments of every water body in the country; results consistently illustrated that nearly every river, lake, and coastal region of the United States was so severely polluted that it threatened human health and jeopardized aquatic life. This initiated major capital investments at federal, state, and local levels to restore and protect drinking-water supplies.
- The Economic Development Act (1965) provides grant assistance for community water infrastruc-

ture to assist new industries and industry expansions that create jobs.

- The Appalachian Regional Development Act (1965), a partnership between the thirteen Appalachian states and the federal government, supports counties and cities in eastern Kentucky to build water and wastewater facilities. The act, drafted in large part by Kentucky native John Whisman, was reauthorized again in July 2015 for five years and is critical for development in the multistate region.

- The Housing and Urban Development Act (1965) required community planning before requesting federal funding assistance. In each of Kentucky's fifteen substate regions or area development districts, a comprehensive water and sewer plan was prepared by the mid-1970s; 1974 amendments created the Community Development Block Grant Program, providing $10 million to $20 million each year in Kentucky for water and wastewater facilities.

- The Environmental Protection Agency (EPA) (1970) was created to implement and enforce the Federal Pollution Control Act; it is responsible for verification of the underlying science and promulgation of environmental quality standards. This major environmental legislation was signed into law by President Richard Nixon.

- The Consolidated Farm and Rural Development Act (1972) expanded assistance to rural communities to include grants and loans for water and wastewater projects; amendments in 1980 and 1990 increased infrastructure assistance and established the Rural Utilities Service, Kentucky's unparalleled largest and most effective source of funding assistance to rural communities for water and wastewater projects.

- The Federal Water Pollution Control Act (1972) is the nation's most comprehensive step toward clean water; it incorporates all facets of protection and restoration of water quality and mandates continuous planning for water and infrastructure. It was amended in 1977 to create the National Pollutant Discharge Elimination System (NPDES), a comprehensive permit program controlling every discharge or incursion into the "waters of the United States," meaning every body of water. Portions of permit administration and enforcement are delegated to states via award of "primacy" (for Kentucky, the Division of Water); amendments in 1977 increased local water-planning flexibility and established the Clean Water State Revolving Loan Fund to encourage construction of regional facilities.

- The Federal Safe Drinking Water Act (1974), the principal federal law that protects the quality of Americans' drinking water, sets quality standards and oversees state regulators who implement and enforce standards on local systems. Amended significantly in 1986 and in 1996, it requires protection of drinking water at all sources and testing and monitoring of water chemistry at each raw (untreated) water source and at various locations and points during the processes of treatment, storage, transmission, and distribution. A revolving loan fund assists public systems with infrastructure development.

By the mid-1980s, as the smoke actually cleared from the skies across the United States, and other federal environmental protections of water began to be just as effective, people came to realize that these many environmental acts were required for economic sustainability and even national survival. Ironically, in this battle for the environment, we the people had become the enemy within. Habits of excessive resource consumption, coupled with irresponsible discharge of waste and toxins, had become a dangerous extreme of American freedom, but now environmental quality is recognized as a true American achievement. Nonetheless, the balance of power between the federal and state governments in the United States in matters of the environment remains contentious. Balancing critical issues—jobs and water quality—is difficult but is made somewhat easier to accomplish in the civic arena by the availability of information on health statistics and increased longevity rates.

Role of the State Government in Water and Wastewater Management

Throughout the early 1900s, the Kentucky State Board of Health worked with the U.S. Public Health Service in helping the fledgling county health departments across

Kentucky monitor and regulate sanitary conditions. From the outset, the health departments—state and county—had dual charges: individual preventive health via inoculations of vaccines to protect against epidemic diseases that plagued the world and public preventive health via regulation of public food preparation, hotel accommodations, and, even more critical, drinking water and wastewater management. As the science of safe drinking water evolved during the late 1800s and early 1900s and more and more Kentucky cities developed water systems, the role of the Kentucky State Board of Health, Bureau of Sanitary Engineering in Frankfort became increasingly directive, moving beyond monitoring to proscriptive action, including approval of engineering plans for proposed treatment plants and eventually for proposed distribution and collection lines, reservoirs, storage tanks, septic systems, and many other facilities.

As noted earlier, the WPA era of the 1930s was a true watershed for public water and wastewater infrastructure in Kentucky. Because of the matching-funds requirement and shared project-implementation responsibilities among the federal, state, and local governments in the New Deal programs, there was also a required integration of governmental health regulations. The collaboration of local, state, and federal health agencies hammered out in the 1930s and early 1940s continues today.

Kentuckians' reaction to the creation of the EPA was mixed, some greeting it as long overdue and others swearing that it was the manifestation of evil incarnate. Once it was apparent that the EPA was not going away and that participation was rewarded with direct federal funding to communities for planning, design, and construction of wastewater facilities, many came to see the value in change. And what a change it was! Kentucky, like many other states, passed the responsibility for water quality, environmental protection, wastewater, and eventually drinking water from what had become the Kentucky State Department of Health, Division of Sanitary Engineering to a newly created Kentucky Natural Resources and Environmental Protection Cabinet, created by an executive order of Governor Wendell Ford in late 1972 and subsequently legislated. This role continues today through the successor agency, the Energy and Environmental Cabinet. The commonwealth is required to

have environmental legislation that is at least equal to or exceeds that of the federal government. The cabinet establishes and enforces regulations dealing with drinking water, wastewater, air, and solid waste and regulates land impacts of any actions taken or proposed to be taken by private individuals, businesses, and industries, as well as those of local, state, and federal government agencies.

Although the Clean Water Act and related regulations were hard fought in the 1970s, they have generated many positive health and environmental results now taken for granted by a majority of Kentuckians. However, for Kentucky's major extractive-industry sectors (coal, minerals, oil, and natural gas companies) and major manufacturers, together with land developers and governmental agencies alike, the federal and parallel state environmental legislative actions from the late 1940s to this day have effected a quantum shift in what had been normal business. Nearly every action affecting water, land, and air now requires permits and approvals, which have become and remain the new business norm. The permitting process radically changed the way of doing business by extending the preparation time before work can begin, requiring intensive monitoring during work, and ensuring that environmental protections remain as planned after work is completed, all of which increases the costs of doing business.

The Roles of Kentucky Cities and Counties in Water and Wastewater Management

City Water

As a large number of cities came to the "gotta do something" point in the late 1800s and early 1900s regarding development of a community water system, often investors were solicited or volunteered a proposal and were generally welcomed. Such arrangements varied, but investors typically required that the city purchase a percentage of the bonds. Additionally, investors sought exclusive franchise agreements to eliminate competition and best ensure fiscal viability. Typically the city granted easements to cross and run with streets for installation of utility lines, pumping stations, storage tanks, and other necessary equipment.

Almost inevitably, facilities construction was slow to get off or, more accurately, into the ground. Many arrangements unraveled, and the city government would eventually have to assume ownership and proceed to construct, operate, and manage the system. A dramatic exception to this scenario is Harry Reid. Relocating from New York to Versailles, Kentucky, in 1905, he purchased a small electricity-generating company. After much difficulty reaching agreement with the city on franchise and rate approvals, he sold the operation in 1912 and moved to Lexington, where he started the Kentucky Utilities Company. Reid presided over the new company, successfully bringing electricity to many rural communities in Kentucky and western Virginia. Fortuitously, he was poised to assist with automation of the fledgling coal industry in eastern and western Kentucky by supplying electricity to motors for conveyor belts and other mining machinery. He retired in 1925, but the company continued to thrive and branched into other utility services. By 1940, the company also owned waterworks, ice-making and delivery services, bus lines, and other ventures. In 1947, in a strategic move to focus resources, most of the company's non-electric-related services were divested, including the water systems that were purchased by the cities of Mount Sterling, Shelbyville, Somerset, Ferguson, Middlesboro, Greenville, Hickman, Stanford, Wickliffe, Earlington, Clinton, Calhoun, and Georgetown. Many of these cities are now major regional water providers.

The history of the Lexington Hydraulic and Manufacturing Company, incorporated in 1883 and operating in 1885, is also interesting. The firm joined the American Water Works System in 1927 and took the name Kentucky-American Water Company in 1973 as a wholly owned subsidiary of American Water Company, providing water to Lexington–Fayette County and portions of nine other central Kentucky counties. A 2003 vote of the Lexington-Fayette Urban County Government to acquire the system via condemnation was reversed by a vote of a new council in 2005.

Many Kentucky cities got into the water business by taking advantage of the opportunities offered by the New Deal programs to construct or buy their first system or expand an existing system after the crash of 1929. Then, as now, the common-safety-net purpose of government at all levels came into focus. Many communities at that time desperately needed both affordable drinking-water and sewer service, together with the private-sector

jobs these services attract. Consequently, it was quite reasonable for cities to take advantage of federal grants and low-interest loan funds for these major capital utility projects. Deliberations seeking local approval seldom required long city council meetings.

The authority granted to Kentucky cities by the General Assembly has evolved over time but is generally contained in Kentucky Revised Statutes (KRS) Chapter 96; Chapters 65, 74, and 76 allow cities various ways to participate in regional water and wastewater entities; and Chapter 278 exempts cities from jurisdiction of the Public Service Commission in most instances. A Kentucky city may opt to own and operate its system(s) directly as a department of its government or, as thirteen of the larger systems have chosen, to manage its system through a utilities board or commission, which serves as an agency of city government, by delegation. The limits of authority for these commissions are set out in the enabling local ordinance, wherein actual ownership of the facilities, borrowing money, setting of customer rates, and appointment of the management entity's members are all reserved to the city governing body.

A city may develop and provide utility infrastructure services in and beyond its corporate boundary when, how, and as it determines in its own best interest, except that its service facilities must be "contiguous." Of the 425 cities in Kentucky in 2015, 205 owned drinking-water systems with either a treatment plant and distribution facilities or a distribution system only, purchasing water from another city, a water district, or a regional water commission. Each of these cities took on the responsibility at a point in its history in response to the question "What is best for our citizens' health and the future of our community?" Each city's experience owning a water or wastewater system is uniquely different, is rich with lore and colorful characters, and deserves to be told. The evolution of the City of Harrodsburg's water infrastructure is similar to that of many Kentucky cities.

The City of Harrodsburg built its first water-treatment plant in 1893 right next to the Salt River, just off Mackville Road, Kentucky 152, the road to Springfield, about 1.5 miles from downtown. Drawing water from a pool created by a low-water dam, the treatment process was state-of-the-art for the early 1900s, with sedimentation basins and sand filters capable of producing approximately

200,000 gallons per day. Both intake pumps and high-service pumps were initially steam powered and were later converted to electrically powered pumps. Because of seasonal changes in available water in the Salt River at Harrodsburg and increased demand for a greater volume of treated water, the location on the Salt River was maintained until the city's new plant came online in 1956. Basins at the old plant were modified and other facilities were reworked to become the city's first public swimming pool, which was later razed because of safety concerns and was replaced years later with the new pool at the Anderson-Dean Community Park just north of the city off U.S. 127.

The city's second water-treatment plant was constructed on the Kentucky River, approximately 0.5 mile from Shaker Village and some 7 miles northeast of the city. The plant crested the hill just off Kentucky 33. The cast-iron raw-water intake lines for the new plant extend from the Kentucky River near the western pier of High Bridge to a wet well from which high-service turbine pumps bring the water up some four hundred vertical feet through cast-iron lines constructed along Cedar Creek to the plant's settling basins. This new plant's capacity of two million gallons per day (mgd) served well until the mid-1970s, when industrial development and the expanding water-distribution services in the city, as well as those of the North Mercer County Water District, the Lake Village Water Association, and the City of Burgin, called for further expansion.

This plant's first major expansion was completed and brought online in 1981 with a capacity of 4 mgd. Again, increased demand from city service expansions, significant extensions of the Water District and the Water Association, and welcome new industrial growth at Harrodsburg have prompted further expansion of the plant. Completed in 2013, the second major expansion of the facility is now rated as a 6 mgd plant costing approximately $12 million that sports many significant physical modifications, including new flocculation and settling basins, a new lab, a consolidated chemical building, and others, including fully automated system controls with integrated telemetry, giving system operators the capability to change major valve and pump control settings from off-site as may be necessary in emergency situations. A significant creek-bank-stabilization project af-

fecting the raw-water line from the river was completed just before the start of construction of this most recent plant expansion.

The city's water-distribution system dates from the early 1900s, and, as in many older systems, lines were constructed of an array of materials, from wood to cast iron to black iron to galvanized steel to PVC (polyvinyl chloride, a type of plastic), resulting in a number of issues. The wooden lines had a relatively short lifespan and have long since been abandoned, but the multiplicity of other line materials requires that an equally wide inventory of repair couplings, saddle-taps (for setting meters for new customers), valve replacements, and other supplies be available to city crews, as well as the appropriate tools for use with each of these various materials. Consequently, also as in many older systems, challenges arise daily even in routine tasks, such as setting a new tap or responding to a line break where a transition between old and new is required, which is nearly always, with little time for head scratching. Harrodsburg's first elevated storage tank was constructed in 1930 and was taken out of service in 2005. The city presently has two 1-million-gallon tanks and one 750,000-gallon tank, strategically located to maintain adequate service throughout the system. In January 2016, the city had 3,703 residential and commercial water customers, including the wholesale customers cited earlier, each with customers living in Mercer County, including North Mercer County Water (with 3,959 customers), Lake Village Water Association (with 2,087 customers), and the City of Burgin (with 495 customers). Of the 10,098 housing units in Mercer County in 2015, it is estimated from data in the Water Resource Information System that 99 percent of these houses have access to public water service.

City Wastewater

Of the 425 cities in Kentucky in 2015, 213, or 50.1 percent, own wastewater systems with both a collection system and a treatment plant or a collection system only, from which collected waste is transported to another system's treatment plant. Nearly all cities with wastewater collection or treatment systems across Kentucky actively engage in community-based regional infrastructure planning and, as opportunity provides, extend their utilities' services, serving residents both in and out of the city alike, with rates based on the actual cost for providing the service.

Of special note, the Metropolitan Sewer District, serving Jefferson County and portions of adjacent counties with both wastewater and stormwater management, created in accord with KRS 76, is the largest wastewater system in Kentucky. The second largest is Sanitation District 1 of Northern Kentucky, created in 1947 in accord with KRS 220. This system provides wastewater and stormwater management to the citizens of Boone, Kenton, and Campbell Counties and thirty cities across the region. These two older and more weathered public utilities began by inheriting hundreds of small, haphazardly constructed, and failing stand-alone systems and by the processes of reconstruction, elimination, and consolidation now service the largest, most densely populated areas in the commonwealth, including all or portions of five counties and over one hundred cities. Engaging in watershed planning, these utilities continue to blaze a trail through a myriad of difficult environmental, public business, and political issues over time, and many other Kentucky communities have benefited and will benefit from their experiences. Again, the City of Harrodsburg's historical experience establishing a wastewater treatment plant and a collection system and periodically expanding and rebuilding these facilities over time is not unlike the experience of many cities across the commonwealth.

The city's first wastewater-treatment plant was a small facility with a capacity of approximately 120,000 gallons per day, constructed in the late 1940s, near the railroad crossing on Cornishville Road, Kentucky Route 1989, about 0.5 mile from the intersection with U.S. 127. Principal components included a simple bar-screen headworks, an Imhoff tank (in this instance a large cylindrical vessel with a funnel-shaped bottom that serves as a clarifier and an anaerobic digester), and sludge drying beds. The treated water was chlorinated and discharged into Town Creek before its entry into the Salt River. Some thirty-five years later, the city built a new treatment plant with headworks, clarifier basins, a series of rotating biological contactor units, several finishing ponds, chlorination, water steps for final aeration, and a

dechlorination chamber before discharge into Town Creek, within several hundred feet from the point where the creek enters the Salt River.

As with all well-managed utilities, input from good operators can improve engineering design. In the late 1990s, the collection system in Harrodsburg had twelve lift stations, costly to operate and maintain. A long-term operator's experience with the topography of the community and his genuine desire to see the system achieve the greatest efficiencies feasible prompted his persistent recommendation that engineers consider an alternative routing of system flow. Once thoroughly studied and subsequently implemented, the reroute allowed for the elimination of half of the system's lift stations, to the benefit of all. Today, with significant development in the city and its acceptance of waste from a nearby out-of-city golf resort, the system again has more than twelve lift stations, but all are truly needed to move collected waste to the treatment plant.

In sum, Kentucky's cities of every size and in every region have virtually designed the game, developed the playing fields, educated the umpires, and consistently hit the winning home runs when it comes to wastewater service in the commonwealth. Of the 335 wastewater systems in Kentucky in 2015, 213 are city owned, and many of these are key participants in regional multi-jurisdictional entities such as joint sewer agencies.

County Water and Wastewater Services

Historically, public water and wastewater services in most Kentucky counties were slower to develop than in cities for many reasons, but mostly because of the significantly higher cost of providing service in a rural, less densely populated setting. However, although many rural families can safely rely on well water for drinking water, and many have passable soils to accommodate conventional septic systems, the majority of rural Kentuckians can do neither because of the karst geology in large portions of the state, other natural impediments, and the impacts of natural resource extraction. The only drinking-water fix for rural communities directly affected by unsuitable or unavailable groundwater is water via a pipe. In the mid-1960s, a concerted effort was made to extend safe drinking water to more of Kentucky with the use of the federal financial assistance cited earlier. By the mid-1970s, Kentucky's new Coal Severance Program was tapped heavily to address the community water needs of both the eastern and western coalfield counties, but much of Kentucky was left unserved by public water.

By Kentucky statutes, counties have authority to establish any of an array of services for portions or all of their territory, including drinking-water and wastewater services in accord with KRS 67. A county judge/executive may initiate the establishment of a water district in accord with KRS 74, which can provide water as well as wastewater service in a designated portion of the county or on a county-wide basis. The county judge/executive, with approval of the fiscal court, may also establish a sanitation (wastewater) district in accord with KRS 67 to be operated in accord with KRS 220 to provide wastewater services in a portion or the entirety of the county; and via KRS 76.232, a county may participate with cities and sanitation districts in a joint sewer agency. Water service is also provided by nonprofit water associations and a few private companies. Wastewater services can be provided by water districts as well.

A Lesson of Collaboration in the Land of "United We Stand"

Beginning in the final years of the 1990s and continuing through 2016, Kentucky has been engaged in a major effort to ensure that all Kentuckians have access to safe drinking water and more have access to proper wastewater services. Informed by community-based strategic infrastructure plans developed through the Kentucky Water Resource Commission in the late 1990s, Governor Paul Patton advanced an initiative in the 2000 General Assembly, encapsulated in a Senate bill sponsored by State Senator Dan Kelly. With Senator Kelly's characteristic intense shepherding and the governor's equally intense support, the bill passed, achieving near-unanimous approval of both the House and the Senate. Amending KRS 151 and KRS 224A, the legislation created a community-based infrastructure planning

council in each of Kentucky's fifteen substate regions, the area development districts, engaging every county, every city, and all water and wastewater systems. Additionally, the Kentucky Infrastructure Authority's board was expanded to include representatives of statewide organizations of local governments and utility interests to sit alongside state agency heads. The authority's purpose was expanded, beyond being just the friendly bank with funds to loan, to include also being the community infrastructure planning nexus for Kentucky. As important, the legislation created a publicly owned and broadly shared infrastructure database, built on a geographic information system known as the Water Resource Information System (WRIS). The system maintains and supports updated mapping of nearly all water and wastewater lines and facilities in Kentucky—a national first.

Within a year of the bill's passage, a new, statewide community-based regional infrastructure planning process was inaugurated, involving all water and wastewater utilities through participation on the regional councils and engaging them to collect, vet, and prioritize proposed infrastructure projects from across the commonwealth. Each area water-management planning council continues to devise a water and wastewater infrastructure plan for its region and provide this information to the authority as a basis for statewide resource allocation focused on economies of scale achievable through regionalization and the elimination of duplication.

As a result, from 2001 through 2015, hundreds of miles of water lines were extended by local utilities throughout the state with the active support of governors and successive General Assemblies, taking the 85 percent of Kentuckians with access to public water in 1999 via 479 systems to more than 95 percent in 2015 via 373 systems. The unprecedented, multiyear, cumulative state investment of $1,909,356,450 focused primarily on water service, capitalized by bonds being retired by payments from Coal Severance Funds and Tobacco Settlement Funds. This investment, coupled with expenditures by Rural Development, Community Development Block Grants, the Appalachian Regional Commission, and the Economic Development Administration, funded many miles of lines, many new storage tanks, and construction of new and expansions of existing treatment plants, all involving nearly every local public water utility in the commonwealth.

As cited, greater emphasis was placed on drinking water, but significant gains were made in expanding wastewater service during this time as well. In 1999, 265 public community wastewater providers provided approximately 55 percent of Kentucky residents access to wastewater services. By early 2015, via an investment of $2,026,556,842, 337 systems, including 213 city systems, 1 county system, 1 metropolitan sewer district, 5 joint sewer agencies, 50 sanitation districts, 40 water districts, 25 small investor-owned for-profit corporations, and 2 nonprofit corporations, together provided approximately 73 percent of Kentuckians with wastewater services.

These successes were made possible by an unparalleled level of collaboration among state agencies and local elected and utility leadership, made possible by bipartisan support for legislation that enabled local leaders to take control of infrastructure planning and project development. Equally dramatic, the interaction between the state agencies and service organizations serving in an official capacity on the Kentucky Infrastructure Authority's board continues to grow and improve. Aided by sharing accurate community data and information via the WRIS, the fifteen area councils' planning and the authority's resource allocation for infrastructure projects are more transparent and equitable for all Kentuckians.[1]

Present and Future Infrastructure Challenges in Kentucky

The dramatic improvements in the extent of Kentucky's water and wastewater service over the previous ten years indicates that the community-based water and wastewater infrastructure-planning and project-implementation process set in motion by former Senator Dan Kelly's SB 409 is working. It also proves that Kentucky's local public utilities are open to and capable of improving utility service and have demonstrated their

capacity as responsible and efficient service providers. What is also known and easily quantified by the WRIS project-profile database is that the commonwealth still has a great need for additional investment in public water and wastewater service. The primary focus relating to drinking water infrastructure in the near term will obviously be treatment plant and appurtenance rehabilitation and system upgrades, including computer-automated management technology. As for wastewater, the road will be much longer, for even if we had access to sustained higher levels of state funding, such as those that were available for the water expansions of the past decade—which is not a reasonable expectation at this time—big-pipe solutions alone cannot solve Kentucky's present wastewater problem. Kentucky must face and deal with extensive life-cycle replacement projects for many major systems, as well as keep pace with required expansions and construct many new big-pipe systems, but it should do so with a serious first-option reliance on on-site, publicly managed, decentralized systems of every variety and combination of appropriate technologies.

Just as critical as funding, to achieve the gains necessary in wastewater service availability, Kentucky needs experienced utility mentors and new organizational tools. Kentucky needs the experiential technical operations and fiscal management skills that cities have gained over their decades of handling this vital service. Proper and timely execution of the designated tasks of each party, be it funding, engineering design, regulatory guidance, project administration, or other more specialized services, is critical to community water and wastewater infrastructure development, but nothing is as important and valuable as the guidance and counsel of the experienced mentor-partner available through a neighboring city, water district, or sanitation district in service to another city, a water district, sanitation district, or another entity confronted by problems with delivery of water or wastewater service. The role of an experienced mentor, be it manager to manager, operator to operator, or board member to board member, is unquestionably valued and has benefited many systems and their customers across Kentucky. Many have graciously risen to this challenge both formally and informally; most all are unsung and uncompensated but all the taller for it.

Additionally, rethinking and reinventing traditional public utility organizational structure at its core are critical to the challenge of how best to provide wastewater service in Kentucky. Rather than schizophrenically berating the EPA or the Kentucky Division of Water for development and enforcement of water-quality standards that seek the protection of human life and that of other life forms, time and resources are much better spent researching ways to encourage existing public utilities to take on the role of responsible management entities—namely, to own, operate, and maintain on-site wastewater systems alongside their more traditional big-pipe systems so that they can accommodate all types of best-option treatment methods, from the simplest to the more complex, all under their shingle of public utility.

Finally, while Kentucky's citizens and its public water and wastewater utilities have already benefited from the fiscal auditing and transparency initiative begun by passage of KRS 65A, greater improvement could be gained by a comprehensive legislative streamlining and recodification of current utility statutes; requiring all public utilities (city, district, and private) to participate in a uniform, automated financial accounting system; requiring all public utilities (city, district, and private) to have a periodic management audit, with copies of all audits published online and supplied to the Auditor's Office for actual, independent review and reporting; subjecting those and only those that fail to meet collaboratively established performance audit standards to a period of time under Public Service Commission jurisdiction; and legislating a clear path for a team of state agencies, aided by staff of public utilities, to take temporary operational control of utilities that jeopardize the health and well-being of customers.

Kentucky's significant accomplishments in providing citizens access to high quality drinking water and wastewater service is largely credited to its reliance on statewide community-based regional planning, continuous updating of its inclusive Water Resource Information System, built on a GIS platform, and an uncommonly strong collaboration among federal, state, and local governmental personnel across the Commonwealth.

Kentucky Water and Wastewater Management: Governmental and Nongovernmental Agencies

- Kentucky Infrastructure Authority, http://kia.ky.gov
- Kentucky Division of Water, www.ky.gov
- Kentucky Area Development Districts, www.kcadd.org
- Kentucky Public Service Commission, http://psc.ky.gov
- Kentucky Rural Water Association, www.krwa.org
- Kentucky League of Cities, www.klc.org
- Kentucky Association of Counties, http://kaco.org
- Kentucky Geological Survey, www.uky.edu/kgs
- U.S. Department of Agriculture, Rural Utilities Service, www.rd.usda.gov/rural-utilities-service
- U.S. Geological Survey, http://ky.water.usgs.gov
- U.S. Department of Agriculture, Natural Resources Conservation Service, www.ky.nrcs.usda.gov
- U.S. Environmental Protection Agency, www.epa.gov/aboutepa/region4.html
- Kentucky Division of Conservation, http://conservation.ky.gov
- University of Kentucky, Cooperative Extension Service, http://ces.ca.uky.edu/ces/
- Kentucky Attorney General, http://ap.ky.gov

Note

1. The following links to the WRIS provide graphic illustrations and tabular data of water and wastewater services in Kentucky for the years 2000 and 2015:

Water: http://wris.ky.gov/downloads/wmp/2015_WMP_Executive_Summary_Final.pdf

pages 12, 13, 14.

Wastewater: http://wris.ky.gov/downloads/wmp/2015_Wastewater_Management_Plan_Final.pdf

pages 12, 13, 14.

Further Reading

Baird, Nancy D. 1992. "Cholera." In *The Kentucky Encyclopedia,* ed. John E. Kleber, 184. Lexington: University Press of Kentucky.

Baker, M. N., and Michael J. Taras. 1981. *The Quest for Pure Water: The History of the Twentieth Century.* 2 vols. Denver: AWWA.

Bratby, J. 2006. *Coagulation and Flocculation in Water and Wastewater Treatment.* London: IWA Publishing.

Connelley, William Elsey, and E. M. Coulter. 1922. *History of Kentucky.* Edited by Charles Kerr. Vol. 1, p. 2. Chicago: American Historical Society.

Hoff, Brent H., Carter Smith, and Charles H. Calisher. 2000. *Mapping Epidemic: A Historical Atlas of Disease.* Jefferson City, MO: Scholastic Library Publishing.

Krishnamurthy, Radha. 1996. "Water in Ancient India." *Indian Journal of History of Science* 31 (4): 27–337.

Melosi, Martin V. 2000. *The Sanitary City: Urban Infrastructure in America from Colonial Times to the Present.* Baltimore: Johns Hopkins University Press.

Oldstone, Michael B. A. 1998. *Viruses, Plagues, and History.* New York: Oxford University Press.

Salzman, James. 2006. "Thirst: A Short History of Drinking Water." Duke Law School Faculty Scholarship Series, Paper 31.

CHAPTER FIVE

The American Dream and the Water Bounty in Appalachian Kentucky

John R. Burch Jr.

Since the early nineteenth century, entrepreneurs have looked to the exploitation of eastern Kentucky's natural resources as a means to enrich themselves and their business partners. In their wake, they have often left an impoverished region that suffers from chronic unemployment, economic underdevelopment, and severe environmental damage. Like other impoverished regions in the United States and around the world, the symptoms of poverty are often manifested in the region's waters. Through an examination of the Appalachian portion of the Kentucky River and its forks, one can see that the economic prosperity of the Commonwealth of Kentucky has historically been tied to the Appalachian portion of the state through its waterways. For eastern Kentucky, this connection has not resulted in long-term prosperity but rather in brief moments of development tied to the water-harvest bounty.

The stories from the Three Forks of the Kentucky River region that follow help illuminate how the residents of the Appalachian portion of the Commonwealth of Kentucky have looked to its waterways over time and envisioned the American dream. Although the specifics of each version of the dream may differ, there tends to be an underlying sense that the future will be better than present circumstances. The question that needs to be addressed is, whose future will be better? Is it that of the individual? Is it the future of the community or region at large? Is it the future of individuals positioned to amass community assets to improve their economic or political standing? Unfortunately, the stories ahead shed light on how competing visions of the American dream

have contributed to the evolution of the persistent poverty that continues to grip the watersheds, landscape, and people of eastern Kentucky.

At the dawn of the nineteenth century, Lexington was one of the largest towns in the western portion of the United States and was a key cog in the nation's economy because of its booming hemp industry.[1] Many Lexington businesses were also involved in central Kentucky's salt-pork industry. The salt pork produced was subsequently shipped to markets, such as New Orleans, Louisiana, via the Kentucky, Ohio, and Mississippi Rivers. The salt required to cure the meat originated in Clay County, which had an abundant supply of underground brine water that was easily tapped by the entrepreneurs who operated fifteen saltworks at the headwaters of Goose Creek at Burning Springs, a tributary of the South Fork of the Kentucky River.[2]

Central Kentucky elites, who included Henry Clay, had made a determination that their business needs represented the best interests of the Commonwealth of Kentucky. They argued that paying for transportation improvements that made it easier to get their goods to market was a legitimate use of the commonwealth's public funds. From their perspective, it was the commonwealth that needed to fund these projects because the individual counties had already proved that no matter how much money had been expended on transportation improvements, roads remained substandard, and waterways were often impassable. Known as the "Bluegrass System," this funding mechanism for transportation-infrastructure improvement was an early model for the "American System" that Henry Clay and other Whig politicians would enact at the national level to justify the construction of toll roads and canals around the country.[3]

The salt-pork producers ensured that Kentucky's General Assembly provided the infrastructure improvements required to facilitate the transfer of salt from Clay County to Lexington for the actual meat curing. Within Clay County, the salt producers were granted the right of eminent domain in order to obtain the wood required to boil brine. They were also empowered by the state legislature to transport brine across property lines. These rights were necessary because salt production required extensive amounts of wood to separate the salt from water. The saltworks quickly used the timber located at Burning Springs, so wood had to be obtained from distant locales. It was easier to transport the brine water to the timber source and have the salt separated there than it was to haul the timber to the respective saltworks. In 1802, state legislation was passed that created a lottery for improvements on the Kentucky River and its tributaries that were intended to facilitate the transfer of salt from Clay County to Lexington. The river improvements were required because the South Fork of the Kentucky River had inconsistent water flows. It was too shallow during much of the year to allow salt to be shipped. Most of the salt was shipped during springtime, when the South Fork was often swollen by floodwaters. Despite the limitations of its local waterways, Clay County's saltworks proved so prolific that 100,000 to 250,000 bushels of salt a year were produced between 1835 and 1845.[4] Despite the amount of salt produced, Clay County's salt industry was already in decline by the 1840s. The industry's methods of separating salt from water were inefficient because of its dependence on old technologies that relied on slave labor. Other Appalachian peoples in West Virginia's Kanawha Valley subsequently became the region's primary salt producers because of a more efficient business model and the availability of dependable waterways. The Kanawha and Ohio Rivers could be navigated year-round. The availability of cheap salt produced in the Kanawha Valley allowed meat producers in locales such as Cincinnati, Ohio, to supplant Lexington's salt-pork industry.[5]

This proved to be a brief development moment for Clay County. As long as the salt producers' product was required in central Kentucky, their economic needs were addressed. Unfortunately for the salt producers at Burning Springs, the combination of the decline of salt-pork production in central Kentucky and the rise of the Kanawha salt industry resulted in the beginning of what Dwight Billings and Kathleen Blee have termed Clay County's "road to poverty."[6]

Another development moment on the Appalachian portion of the Kentucky River came during the 1880s in Beattyville, within Lee County, near where the three forks of the Kentucky River come together to create the Kentucky River's main course. Up to that time, travel on the Kentucky River was a risky venture because of variable water flow. Portions of the river were too shallow to ship goods during low-water periods; thus flatboats had to sit on gravel bars until the flow was high enough to float goods downriver. The river was also prone to flash flooding, which ripped boats apart and caused extensive economic losses to both individuals and businesses. Because of Beattyville's proximity to the Kentucky River and its forks, it became the primary shipping center in eastern Kentucky for both timber and coal. Politicians at the state and federal levels determined that they needed to extend slack-water navigation on the Kentucky River to Lee County to ensure that they could get both timber and coal from the mountains to markets in such locales as Louisville, Kentucky, and beyond. In the case of coal, it was believed that improved transportation infrastructure would allow eastern Kentucky coal to displace coal produced in Pennsylvania within many southern markets.[7]

Since the commonwealth could not afford to pay for the improvements on the Kentucky River, U.S. Congressman John D. White secured $75,000 in the 1882 Rivers and Harbors Act for the construction of a lock and dam in Beattyville. Pending improvements on the Kentucky River were accompanied by the arrival of capitalists endeavoring to profit from the region's natural wealth. For example, the Three Forks Investment Company's officers and directors were William Cornwall Jr., who was affiliated with the Louisville manufacturing firm Cornwall and Brother; John H. Leathers of the

Louisville Banking Company; J. W. Stine of the Richmond, Nicholasville, Irvine and Beattyville Railroad; Bennett H. Young of the Louisville Southern Railroad; C. D. Chenault of the Madison National Bank in Richmond, Kentucky; and James B. McCreary, a U.S. congressman and former governor of Kentucky. Although investors such as those named provided employment for eastern Kentuckians, they were also the primary beneficiaries of the profits realized from the region's natural resources because they quickly acquired large swaths of land that included coal mines, stands of timber, sawmills, and riverfront property. The Beartrap Dam constructed by the U. S. Army Corps of Engineers opened on October 30, 1886. It quickly became obvious that the artificial wave it created to ship goods downstream was too swift to be navigated safely. The dam then collapsed within a month because the river lifted the wooden pilings that made up the dam's outer support structure. After this costly and embarrassing debacle, the Corps of Engineers determined that rebuilding the dam was not an option because the transportation infrastructure needed to bring the required supplies to Beattyville did not exist at the time. The Beartrap Dam was the last lock or dam on the Kentucky River to be constructed of wood. All subsequent structures were constructed using stone masonry.[8]

In 1902, the Louisville and Atlantic Railroad arrived in Beattyville. Logically, its arrival should have ended the need for the extension of slack-water navigation to Lee County. The reality was that the outside entrepreneurs who had acquired much of the land around Beattyville had tied their economic fates to the Kentucky River and had cultivated powerful friends in Washington, D.C., to protect those interests. The arrival of the railroad thus provided a convenient excuse to once again begin building dams in Lee County because the needed materials for dam construction could be easily shipped via railroad to construction sites.[9]

The U.S. Congress subsequently provided funding to the U.S. Army Corps of Engineers for two locks and dams in Lee County. Lock and Dam 13 on the Kentucky River was constructed between 1909 and 1914. Lock and Dam 14 construction commenced in 1911 and was completed in 1917.[10] This was definitely a population- and economic-

growth era in Beattyville because workers streamed into the community. In 1900, Lee County's population stood at 7,988. By 1920, its population had grown to 11,918.[11]

Unfortunately for Lee Countians, the completion of the final lock and dam effectively marked the end of their moment of development because the nature of shipping on the river had changed. The river improvements around Beattyville had been designed to accommodate the flatboats in use during the nineteenth century. The towboats and barges that dominated shipping by the second decade of the twentieth century were too large for use in the newly constructed locks and dams. Lee County's population began to shrink as people out-migrated in search of new economic opportunities. Although there was a brief revival in river traffic from Beattyville during World War I because of the need for coal to fuel the war effort, the community's place as the shipping center of eastern Kentucky disappeared just as the Great Depression was commencing.[12]

The Appalachian portion of the Kentucky River and its forks was largely forgotten by the commonwealth's economic and political leadership until the Great Flood of 1937. The January flood affected the Ohio River basin and its tributaries, including the Kentucky River. Floodwaters in Louisville crested at 57.1 feet, exceeding the previous record set in 1884 by approximately 10 feet. Frankfort, the state capital, also suffered severe flooding. In total, flooding on the Kentucky, Ohio, and Mississippi Rivers in Kentucky alone displaced 50,000 families and caused $250 million in damage.[13] The U.S. Congress's response to the natural catastrophe was the passage of the Flood Control Act of 1938.[14]

The Flood Control Act of 1938 proposed the construction of forty-five reservoirs in the Ohio River basin, including one on each of the three forks of the Kentucky River. It was believed that damming the respective forks would help control the floodwaters that regularly arose within the mountains. After initial surveys by the U.S. Army Corps of Engineers, it was determined that eastern Kentucky communities, such as Beattyville and Irvine, would receive significant amounts of floodwater protection from the proposed dams. However, since the proposed dams promised only to reduce the threat of future floodwaters by several inches in Frankfort, it was determined that their cost outweighed their benefit. Eastern Kentucky political and business elites continued to advocate for dam-construction projects to protect their respective communities from flooding, to no avail. It was not until 1951 that the idea of dam construction on the three forks of the Kentucky River gained serious traction. The protection of eastern Kentuckians from floods was not what brought the issue to the fore, but rather the need of central Kentucky for potable water supplies. Once again, outside interests turned to the development of eastern Kentucky's waterways to advance their economic and personal gains.[15]

The residents of Richmond and Lexington had been concerned about their water supply since 1930. In that year, eastern Kentucky was stricken by a drought so severe that water was not spilling over the existing dams on the Kentucky River. Downriver communities that depended on the Kentucky River for their water needs were in dire straits. Richmond residents grew so desperate that they required the delivery of water to their community via railroad cars.[16] Central Kentuckians turned first to the construction of the Jessamine Dam to alleviate their water issues. Under the Flood Control Act of 1944, the proposed dam in Jessamine County would have allowed the U.S. Army Corps of Engineers to abandon its responsibilities on the Kentucky River above Lock 7. That dam was ultimately derailed by the public relations campaign waged by the Pioneer National Monument Association, whose members included Tom Wallace, editor of the *Louisville Times,* and historians Thomas D. Clark and Hambleton Tapp. The organization led the campaign to protest construction of the Jessamine Dam in order to protect Fort Boonesborough from being flooded out of existence. Founded by Daniel Boone, Fort Boonesborough is commonly viewed as the birthplace of the Commonwealth of Kentucky. It was thus too historically significant to disappear underneath a man-made lake. With a local solution to their water issues out of reach, central Kentucky political and business leaders needed to find a way to ensure that the water needs of their growing communities would be met in both the present and the future. The central Kentucky leaders found allies in eastern Kentucky who saw an opportunity to address the ever-present threat of flash floods in their mountain communities.[17]

It is important to point out that flood control was not the sole reason that eastern Kentuckians wanted dams constructed in the region. It was believed that the dams promised sustainable employment opportunities through tourism. Jobs were desperately desired because eastern Kentucky had been suffering through a major population decline as more than a third of the people had left during the 1940s for cities such as Cincinnati, Ohio, and Detroit, Michigan, in search of employment. It was hoped that new jobs would not only stem the continuing out-migration of people but also actually reverse the trend and allow kinfolk to return home.

The Kentucky River Development Association (KRDA) was formed in June 1951 for the expressed purpose of developing the entire course of the Kentucky River and its major tributaries. Its plan for development of the region was modeled on the Tennessee Valley Authority (figure 5.1). The KRDA made it a priority to build dams on each of the three forks. In this effort, it found allies in Kentucky's U.S. congressional delegation, including Congressmen Carl Perkins and James Golden, as well as Senator Earle Clements, who managed to secure all the federal funding required to construct a dam on the Middle Fork of the Kentucky River by 1954. Construction on the Buckhorn Dam began in 1956 and was completed in 1960. By the time the dam was constructed, the KRDA had reached the end of its effectiveness. Central Kentuckians involved in the organization had achieved their primary goal and thus no longer needed the eastern Kentuckians. They had one reservoir already constructed for their needs, with the dam on the North Fork a near certainty for construction since it was Congressman Perkins's pet project, and he had the political power to get it funded. Perkins had dam construction under way at Carr Creek by 1966.[18]

Since the proposed dam on the South Fork of the Kentucky River was not the primary priority of the KRDA or politicians at the federal level, the responsibility for getting it constructed fell to local political and business elites from eastern Kentucky, led by Beattyville residents. With the demise of the KRDA, these eastern Kentuckians formed the Middle Kentucky River Area Development

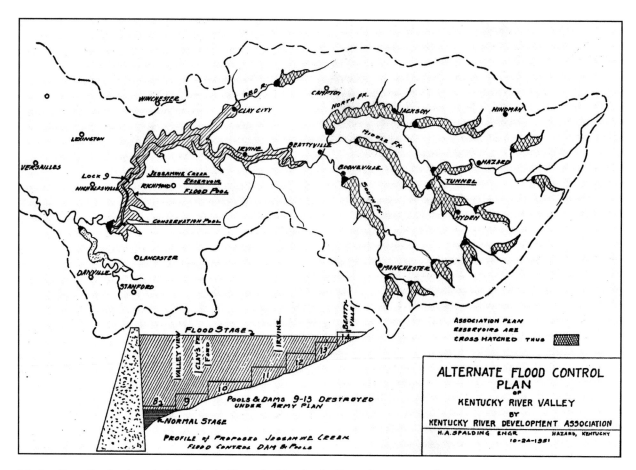

5.1. An alternative flood-control plan for the Kentucky River by the Kentucky River Development Association in 1951.

Council (MKRADC) to push their development agenda. They advocated for the construction of the Booneville Reservoir as both a solution to flood problems and the centerpiece of a tourism industry built around what would later be known as Lake Daniel Boone. They endeavored to create another development moment in Owsley and Clay Counties designed to provide economic opportunities for themselves at the expense of the people who were going to lose their homes when their property was inundated by the creation of the Booneville Dam's reservoir.[19]

In order to understand the history and legacy of the Booneville Reservoir, one has to understand the history of proposed dams in Owsley County, Kentucky. In 1925, the Louisville Hydro-electric Company proposed building a dam on the South Fork of the Kentucky River. It promoted the project as the first example of "foreign" in-

vestment in the county. Thousands of jobs would be created, which would soon be followed by the tax revenue required to improve roads and schools. Owsley Countians did not fall for the promises because they recognized that the project would inevitably result in the depopulation of the county as the land was flooded. With the aid of outside capitalists with extensive landholdings in the county, such as Henry Ford, Owsley Countians were able to defeat the designs of the Louisville Hydroelectric Company.[20]

The passage of the Flood Control Act of 1938 brought the specter of a dam on the South Fork once again to the fore because the act authorized the construction of the dam by the U.S. Army Corps of Engineers. The authorization was reaffirmed by the U.S. Congress in 1944. The presence of U.S. Army Corps of Engineers employees conducting surveys on local waterways resulted in the phenomenon known as "dam talk," informal conversations about dam construction that may or may not have had any basis in fact, but that occurred whenever engineers visited the area. These conversations may have been harmless when they began during the late 1930s but would have profound consequences for the people and the landscape in later decades.

The ability of the KRDA and its supporters in the U.S. Congress to secure funding for construction of the Buckhorn Dam during the early 1950s resulted in dam talk hitting a fever pitch. Suddenly a dam was not theoretical; it was an actual possibility. U.S. Senator John Sherman Cooper and Congressman Carl Perkins did nothing to temper expectations because they routinely gave the impression that dam construction in Owsley County was imminent. This expectation also spread through a contiguous portion of Clay County because it was determined that Lake Daniel Boone would extend from the vicinity of Booneville to the campus of Oneida Baptist Institute, near Manchester. The creation of the Carr Creek Lake in the 1960s only confirmed the inevitability of the Booneville Reservoir's construction because the South Fork of the Kentucky River was the only one not dammed.[21]

The U.S. Army Corps of Engineers also helped convince locals that dam construction was inevitable when it held a public meeting, at the urging of the MKRADC, about the Booneville Reservoir on January 4, 1963, in Irvine, Kentucky. The meeting should never have been held because it was the policy of the Corps at the time not to hold formal meetings until dam construction was about to commence. At the time of the meeting, the dam had only been authorized for construction, which simply meant that the dam's construction was up to the discretion of the Corps, which never wanted to build the dam in the first place. The Corps of Engineers recognized that there were numerous daunting engineering challenges that it would have to overcome to construct the dam. Most important, the transportation infrastructure needed to get materials to the work site simply did not exist. Constructing the roads required promised to be a time-consuming and expensive venture. In the end, even the Corps of Engineers could not justify the cost of the dam in comparison to even the most optimistic tabulation of its benefits. Although Congress had authorized construction on two different occasions,

it never appropriated any funding even for preliminary studies. The studies done by the Corps since 1938 on the South Fork had actually been conducted as part of general surveys of the Kentucky River and its tributaries.[22]

In 1966, representatives of the MKRADC, at the invitation of Senator John Sherman Cooper, testified before the Senate's Public Works Appropriations Subcommittee in the hopes of finally getting funding to begin the preliminary studies required for dam construction. They were unsuccessful in their efforts. Despite the setback, the MKRADC and its supporters continued to clamor for the federal financing required for the preliminary studies to be conducted. Their efforts were rewarded when Kentucky's U.S. congressional delegation finally secured initial funding in 1970. The project was awarded $230,000 for preconstruction planning. This was followed in 1972 by an additional $245,000. By that time, despite the funding, it was obvious that the Booneville Reservoir was never going to be constructed. The U.S. Army Corps of Engineers finally declared the Booneville Reservoir an inactive project in 1976. At no point throughout the Booneville Reservoir's theoretical existence did it ever move beyond the authorization stage.[23]

Although the Booneville Reservoir was never constructed, that does not mean that it was not real and did not have impacts to the landscape and the people. Individuals made life-altering decisions because of the proposed Booneville Reservoir. Some sold their property for whatever they could get and out-migrated because the land they loved was going to become a lake, or so they thought. Other people refused to improve their farms because, in their eyes, doing so was only a waste of money. What infrastructure existed, such as roads, was allowed to deteriorate. The resulting decades of neglect in portions of Owsley and Clay Counties that were expected to become Lake Daniel Boone (Booneville Reservoir) compounded the region's poverty when the lake never materialized. As the poorest portions of these two counties, these contiguous areas are the primary reason that U.S. census data released in 2003 rated Owsley as the second-poorest county in the United States, followed closely by Clay at number six.[24]

The repercussions extended beyond Owsley and Clay Counties. Officials in neighboring counties had been anticipating tourism-industry development and had planned for it accordingly by making investments in the infrastructure that was going to be required for that endeavor. Although making decisions in regard to a structure that never existed proved misguided, their poverty was not exacerbated to the degree that it was in Owsley and Clay Counties because economic development of any kind was an improvement over allowing economic infrastructure to deteriorate over decades.

A common thread running through all these development moments concerns how various versions of the American dream were prioritized over others. The dreams of people within the region were not favored over those of people from beyond. Needs of people outside the region fueled the development efforts that occurred on the Kentucky River and its forks. Salt was needed in Lexington for the salt-pork industry, so river improvements were made to Burning Springs in Clay County. Coal and timber needed to be shipped from eastern Kentucky, so suddenly federal funding was secured in Washington, D.C., for the construction of dams and locks in Beattyville and its environs. Central Kentucky business and political leaders wanted to ensure the availability of potable water for their communities; thus dams were built on the Middle and North Forks of the Kentucky River. During these moments, all efforts were directed toward a specific economic engine that was primarily intended to fuel interests outside Appalachian Kentucky. There was no effort to diversify the respective local economies during the brief moments of development. Thus as each economic engine disappeared, there was nothing sustainable to build on. Depopulation and impoverishment historically followed. These trends continue today.

Notes

1. Stephen Aron, *How the West Was Lost: The Transformation of Kentucky from Daniel Boone to Henry Clay* (Baltimore: Johns Hopkins University Press, 1996), 133–35.

2. Dwight B. Billings and Kathleen M. Blee, *The Road to Poverty: The Making of Wealth and Hardship in Appalachia* (New York: Cambridge University Press, 2000), 61–78.

3. Aron, *How the West Was Lost,* 133–37; Charles Sellers, *The Market Revolution: Jacksonian America, 1815–1846* (New York: Oxford University Press, 1991), 63–65; Craig Thompson Friend, *Along the Maysville Road: The Early American Republic in the Trans-Appalachian West* (Knoxville: University of Tennessee Press, 2005), 254–61.

4. *Kentucky Acts, 1801* (Lexington: William Garrard, Printer to the Commonwealth, 1802), 93; *Kentucky Acts, 1810* (Lexington: William Garrard, Printer to the Commonwealth, 1811), 134; *Kentucky Acts, 1811* (Lexington: William Garrard, Printer to the Commonwealth, 1812), 113; Billings and Blee, *Road to Poverty,* 69–73; Mary Verhoeff, *Kentucky River Navigation* (Louisville: Filson Club, 1917), 153–54; John R. Burch Jr., *Owsley County, Kentucky, and the Perpetuation of Poverty* (Jefferson, NC: McFarland, 2008), 23–25.

5. Billings and Blee, *Road to Poverty,* 69–71, 74–79.

6. See ibid.

7. Leland R. Johnson and Charles E. Parrish, *Kentucky River Development: The Commonwealth's Waterway* (Louisville: Louisville Engineer District, United States Army Corps of Engineers, 1999), 98.

8. Ibid., 98, 258–60; *Three Forks Enterprise,* March 30, 1885; *Prospectus of the Three Forks Investment Company: Its Beattyville Town Site, with Maps, Plats, Illustrations, Etc.* (Louisville: Courier-Journal Printing, 1889), 5, 10–12; Rosemary Kilduff and Mary Helen McGuire, *Peoples Exchange Bank, 1912–1987* (Beattyville, KY: Peoples Exchange Bank, 1990), 1–2, 8–11; Burch, *Owsley County, Kentucky, and the Perpetuation of Poverty,* 41–43.

9. Kilduff and McGuire, *Peoples Exchange Bank, 1912–1987,* 15.

10. Leland R. Johnson, *The Falls City Engineers: A History of the Louisville District, Corps of Engineers, United States Army, 1970–1983* (Louisville: United States Army Engineer District, 1984), 265, 270–75.

11. U.S. Census Bureau, *Kentucky Population of Counties by Decennial Census: 1900–1990,* http://www.census.gov/population/cencounts/ky190090.txt (accessed October 4, 2010).

12. Johnson and Parrish, *Kentucky River Development,* 137–38; Johnson, *Falls City Engineers,* 154.

13. An amount of $250 million in 1937 equates to approximately $3.7 billion today.

14. "January Rainfall More Than 5 Months in 1936," *Beattyville Enterprise,* January 28, 1937; William E. Ellis, *The Kentucky River* (Lexington: University Press of Kentucky, 2000), 107–12; Jerry Hill, *Kentucky Weather* (Lexington: University Press of Kentucky, 2005), 58, 74–79.

15. U.S. Army Corps of Engineers, Louisville District, *Booneville Reservoir, Ohio River Basin, South Fork Kentucky River, Kentucky, Design Memorandum No. 3: Structure, Site Selection* (Louisville: United States Army Engineer District, Louisville, 1970), 1-1; Johnson and Parrish, *Kentucky River Development,* 151, 153.

16. Ellis, *Kentucky River,* 31.

17. Burch, *Owsley County, Kentucky, and the Perpetuation of Poverty,* 66–67; Johnson and Parrish, *Kentucky River Development,* 153; U.S. Congress, House of Representatives, *Creation of Pioneer National Monument, KY,* 73rd Cong., 2nd Sess. (Washington, DC: Government Printing Office, 1934), n.p.; U.S. Congress, House of Representatives, *Kentucky River and Tributaries, Kentucky,* 87th Cong., 2nd Sess. (Washington, DC: Government Printing Office, 1962), 93.

18. Burch, *Owsley County, Kentucky, and the Perpetuation of Poverty,* 67–77, 80; Ellis, *Kentucky River,* 30–31; "River Development Association Recommends 11-Point Program," *Owsley County News,* December 11, 1953; Johnson and Parrish, *Kentucky River Development,* 160.

19. Burch, *Owsley County, Kentucky, and the Perpetuation of Poverty,* 79–90, 119–46.

20. Ibid., 50–51; "Protest Heard on Owsley Dam," *Courier-Journal,* March 19, 1925.

21. Burch, *Owsley County, Kentucky, and the Perpetuation of Poverty,* 69, 71–90.

22. Ibid., 83–90; "Public Meeting for Review of the Status of the Authorized Booneville Reservoir in the Kentucky River Basin," Thruston Ballard Morton Collection, 1933–1969, Box 4, Special Collections, King Library, University of Kentucky; *Lexington Herald,* January 8, 1963.

23. "Booneville Dam Awaits Federal Funds; Much Depends on Vietnam," *Lexington Leader,* February 18, 1966; "Carl R. Reynolds Heads Owsley Group Asking Funds for Booneville Dam; Col. Begley Represents Lee County," *Peoples Journal,* May 5, 1966; "Statement of Senator John Sherman Cooper before Senate and House Appropriations Subcommittees on Fiscal 1967 Budget Recommendations for Kentucky Corps of Engineers Projects, May 3–4, 1966," Thruston Ballard Morton Collection, 1933–1969, Box 10; Burch, *Owsley County, Kentucky, and the Perpetuation of Poverty,* 123–27, 138–46.

24. "Kentucky Counties among the Poorest—'03 Census Data Ranks Median Income, Poverty Rates," *Lexington Herald-Leader,* November 30, 2005.

Further Reading

Billings, Dwight B., and Kathleen M. Blee. 2000. *The Road to Poverty: The Making of Wealth and Hardship in Appalachia.* New York: Cambridge University Press.

Burch, John R., Jr. 2008. *Owsley County, Kentucky, and the Perpetuation of Poverty.* Jefferson, NC: McFarland.

Ellis, William. 2000. *The Kentucky River.* Lexington: University Press of Kentucky.

Johnson, Leland R., and Charles E. Parrish. 1999. *Kentucky River Development: The Commonwealth's Waterway.* Louisville: Louisville Engineer District, United States Army Corps of Engineers.

CHAPTER SIX

Appalachia

Gathered at the River

Sam Adams

If one thinks of the Appalachian Mountains as ripples on a mill pond, then the spot where the stone struck the water is the valley at the Kentucky-Virginia border that divides the two great thrust-fault mountains of Pine and Cumberland. To the east is the range's spiny core, known as the ridge and valley region. To the west are foothills that terminate in the craggy cliffs of the Pottsville Escarpment and the rolling knolls of Kentucky's Pennyrile (Pennyroyal) region.[1] In the center, Pine Mountain and Cumberland Mountain are mirror images, with rock strata that lie slanted to the southwest and northeast, respectively.[2]

Pine Mountain, 400 million years old, dominates the eastern end of Kentucky. It is the second-highest mountain in the state, three-quarters the height of neighboring Black Mountain, but at 125 miles, it is by far the longest mountain in the state. Not a single stream passes through Pine Mountain for nearly 90 miles,[3] but myriad streams spring from its limestone depths to irrigate the surrounding valleys. While Pine Mountain is the source of those streams, they are the source of a labyrinth of ridges known as the Cumberland Plateau that forms the heart and soul of Appalachian Kentucky.

Sixty-five million years ago, what is now Appalachian Kentucky was a broad, high plain stretching westward from Pine Mountain as far as the eye could see. Rivers roared off the mountain, grinding away at the soft sedimentary rock of the plateau and carrying microscopic particles to the sea.[4] Trilobites and other fossils are still visible in the layered-sandstone rockhouses that form the ridges of the foothills.

As the North American continent drifted toward South America,[5] water from the Appalachians continued to cut relentlessly into the rock, carving the beds of torrential rivers. Wind, rain, and ice eroded the canyon walls, and what was once a magnificent plateau gradually became a region of rounded, undulating hills and deep hollows, as wrinkled as quilts on a slept-in bed.

Just as hydrology has profoundly affected the geophysical fabric of Appalachia, so also has water affected its social fabric, its environmental problems, and its economy. Appalachian Kentuckians live in the places and in the ways they do because of water. They work the way they work because of water. And in the unforgiving hollows, where trickling streams may quickly become vengeful torrents, Appalachian residents sometimes have their livelihoods and lives stripped away by water.

Residents in eastern Kentucky still gauge the water levels in the creeks by the 1957 flood, which wiped out homes and businesses across the region and left $1 billion in damage, an amount worth $8.17 billion in 2012 dollars. Between January 27 and February 2, 1957, the U.S. Geological Survey estimated that 12.5 inches of rain fell along the Kentucky-Virginia border, flooding the Kentucky, Big Sandy, Cumberland, and Tennessee River valleys.[6] Although the flood was the most devastating on record in terms of property lost, loss of life was minimal. Those a generation older took that flood in stride because it did not compare to the 1927 flood,[7] in which nearly five hundred people in Kentucky and nine other states died.[8] Twenty-six people died in Letcher County, where residents who remembered that flood spoke of coffins washed from cemeteries blocking off bridges and causing water to rise even higher in the coal camp of Carbon Glow. I have a steamer trunk that my grandmother watched float from a second-story window of the house she lived in at Isom. The steel-covered wooden trunk was recovered on a hillside after the waters receded.

Although the 1927 and 1957 floods were horrible, flooding on the Big Sandy River in 1958 stands out as the worst flood ever for Floyd County, despite being only a minor footnote in terms of property damage and water flow. Floyd County residents, however, will always remember that flood as the one during which school bus number 27 plunged into the Big Sandy River, killing twenty-seven people—the driver and twenty-six students.[9] Some bodies were not found for weeks. This accident still ranks as one of the worst school-bus accidents in U.S. history. Because of it, the Floyd County Emergency and Rescue Squad was formed, which still operates on a completely volunteer basis with no tax support. It now trains rescue squads across Kentucky in moving-water and flood rescue.

Water influences every aspect of eastern Kentucky, just as it has for eons. Water in ancient oceans and coastal wetlands laid down the layers of sediment and vegetation that are now sandstone, limestone, shale, and coal. Water carved the Appalachian Plateau into low hills and twisting valleys. Water provided the rich alluvial soil in the valleys for farming, the Devonian shale for gas and oil production, and the bituminous coal in the hills for mining.

By the landscape it created, water dictated that eastern Kentuckians live in hollows, and that industrial complexes be built next door to residential neighborhoods. And it is indirectly because of water that some residents pour sewage into the same nurturing streams that once provided sustenance to their ancestors.

Hollows (pronounced "hollers") are nothing more than narrow, steep-sided valleys. Each hollow has a creek at the bottom and space for little more than a road and perhaps a single row of houses before the land begins to slope steeply upward again. Getting from one to another requires either trekking up and down impossibly steep mountainsides or traveling out of the "mouth" of one hollow and then up or down the river to the next one.

Such travel constraints led mining companies to build everything they needed for their operations in single hollows. If a mine was located in a hollow, then the miners' housing, the coal-processing facilities, and the loading facilities for the trains were there. Local governments have for the most part refused to pass zoning laws that would affect that clash of residential and industrial uses, often using the argument that America is a free country, and that telling landowners what they could put on their property would be "communism." As a result, homeowners today still have to deal with coal dust, traffic congestion, and noise from coal tipples located within a stone's throw of their homes.

As families have grown and the older generations have given over land to their children and their children's children, the land suitable for building has become crowded, even in communities that were not built as coal camps. Today, the houses in many hollows are too close together to allow for effective septic systems, and the expense of building sewers in the rugged terrain is too great for cash-strapped local governments to afford. As a result, streams often have bacteria levels tens or even hundreds of times higher than the safe limits for bodily contact.[10]

Still, in a place where some families have lived for more than two hundred years, the connection to the past is too strong for many to consider moving away. Many residents still make a living much as their fathers and grandfathers did—through the extraction of natural resources such as timber and coal.

About twenty thousand years ago, as the great ice sheets advanced, dirt and rock were pushed ahead of them, moving southward to the central Appalachians, and the mountains became a seed bank for the northern two-thirds of the continent. The central Appalachian region, of which Kentucky is a part, is far enough south that it avoided the glaciers, but the elevations are high enough that plant species that required cold temperatures thrived on the highest peaks. As a result, the mixed mesophytic forest was born, and the central Appalachians became the cradle of life for a continental landscape scraped clean and barren by rivers of ice. Botanist E. Lucy Braun first formed her theory of what ecologists have nicknamed "the Mother Forest" in 1916 and

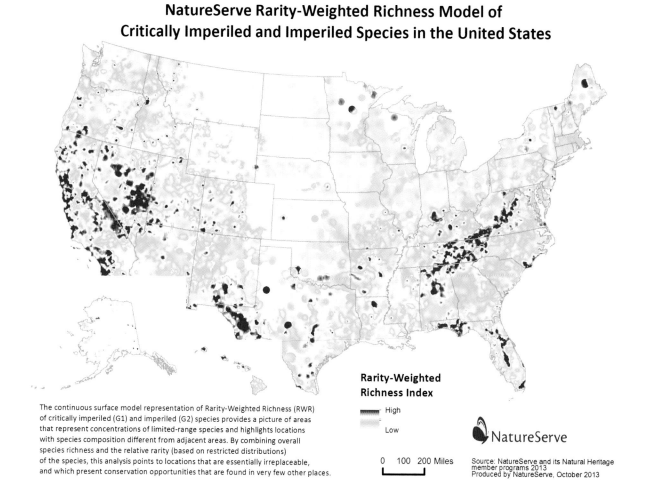

NatureServe Rarity-Weighted Richness Model of Critically Imperiled and Imperiled Species in the United States

The continuous surface model representation of Rarity-Weighted Richness (RWR) of critically imperiled (G1) and imperiled (G2) species provides a picture of areas that represent concentrations of limited-range species and highlights locations with species composition different from adjacent areas. By combining overall species richness and the relative rarity (based on restricted distributions) of the species, this analysis points to locations that are essentially irreplaceable, and which present conservation opportunities that are found in very few other places.

Rarity-Weighted Richness Index

High

Low

NatureServe

0   100   200 Miles

Source: NatureServe and its Natural Heritage member programs 2013
Produced by NatureServe, October 2013

6.1. NatureServe rarity-weighted richness model of critically imperiled and imperiled species in the United States. Map produced by NatureServe, October 2013.

solidified that theory over the next thirty years of examining the Appalachian forests. Her book *Deciduous Forests of Eastern North America* was published in 1950 and is still the bible of Appalachian forest research.

Plants that grow in glaciated areas north of the Ohio River did not provide the seeds for cold-weather plants in Kentucky and other parts of the Appalachians. The Appalachians provided the seeds for those areas where plant life was destroyed by the moving ice.

Although it is not technically a rainforest, the mixed mesophytic forest of the Cumberland Plateau historically has been every bit as important as the tropical rainforests of Central America and the temperate rain-

forests of the Pacific Northwest. Eastern Kentucky receives more than forty-eight inches of rainfall annually, on average, and has approximately 163 fair-weather days per year. This abundant rainfall, coupled with plenty of sun and the well-drained geographic area, creates ideal conditions for plant life, particularly deciduous trees. The unglaciated region now is home to the most diverse ecosystem on the North American continent and one of the most diverse temperate forests in the world. The slopes of the central Appalachians are covered in beeches, tulip trees, basswoods, sugar maples, chestnut trees, sweet buckeyes, red oaks, white oaks, hemlocks, silverbells, birches, cucumber trees, white ashes, red maples, sour gums, black walnuts, and several different species of hickory, as well as evergreens and a broad spectrum of herbaceous species. As many as thirty species of trees share dominance in the region, more than in any other forest in the United States.[11]

The rich mix of vegetation that the European settlers discovered when they moved here was like manna. The forests in Appalachia became the raw material that built much of America. Trees became houses, railroad ties, telegraph poles, and tall ships. Herbs became medicines, flax became clothing, and hemp became ropes and sails for the navy and the merchant marine. As America's hunger for building materials grew, and timber barons' hunger for money became insatiable, water once again reshaped Appalachia. Not content with waiting for nature to flood the rivers and float the logs downstream, loggers built splash dams high on the mountain slopes, filled the temporary impoundments with logs, and then dynamited the dams.[12] The water and logs rushed down the mountainsides, scouring away dirt, rocks, and trees and filling streams with silt.

In the first half of the twentieth century, much of Appalachia was barren hills, logged until there was little left to log. The great expanses of old-growth forests were mostly gone, replaced by cornfields. As water eroded the slopes, even corn stopped growing, and farmers turned to a new industry—coal mining.

Land speculators from the Bluegrass, from Appalachia itself, and from the cities of New England rode up and down the creeks, buying mineral rights and selling them to the highest bidders. John C. C. Mayo, a Pike County native and schoolteacher who moved to Johnson County in his youth and also became a lawyer, was the largest, selling more than 100,000 acres in one deal to the Elk Horn Coal Co.[13]

Much of the land purchased by Mayo, Richard M. Broas, and others was purchased through the use of "broad form deeds," also known as mineral severance deeds. These deeds purchased only the rights to the minerals, ostensibly leaving the use of the surface to the original landowner. In reality, some of these deeds gave mineral owners broad rights to disturb the surface and use it in any way necessary to remove the minerals beneath. Deeds such as these included such rights as the

rights to mine, to build roads and railroads, and to use "the rocks on the ground." Although speculators told residents that the land might be disturbed later, it is difficult to believe that landowners could imagine the extent of the disturbance. Such deeds have been used to move homes, farms, cemeteries, and entire mountains.[14]

Land speculators like Mayo knew that the railroads and coal companies did not want to deal with the small landowners, so they did. They bought up land for as little as ten cents an acre, bundled it into large contiguous tracts, and then sold it to the corporations, opening the mountains to the coal industry. Once again, water played a role. The railroads were usually built next to the creeks because doing so was easier and cheaper than building on the mountainsides. Bridges built a hundred years ago still catch floating debris, damming creeks and causing flooding upstream.

Until 1912, the rails stopped at Stonega, Virginia, to the east, and at Jackson, Kentucky, to the west. Even there, the rails were late, not arriving until 1880[15] and 1891,[16] respectively. The bankers and speculators from the North knew that there was coal in the Appalachians, and the extension of the rails from Jackson to the Virginia state line in 1912 gave them a way to get it out. With that newfound ability to transport coal, the industry boomed, and eastern Kentucky experienced growth unlike anything anyone there had seen. The coal was mined where it was found—usually on the mountainsides—but there were other uses for the level land in the river bottoms. Coal camps sprang up; some, such as Jenkins, were touted as model cities to house the miners, school their children, and provide the essentials for life. Houses sprouted like the highly valued morel mushrooms in spring, often so close together that neighbors could shake hands or steal a kiss through facing windows.

Since the railroads had already claimed much of the flat land in the narrow valleys, the houses were squeezed in around them and the coal tipples used to take coal from the mines to the train gondolas. Every day during which the houses were being built was a day in which the laborers were not mining coal, so the houses were built quickly. There was no time to install indoor plumbing in many houses, and even if there had been, sewage treatment was uncommon even in large cities of the period.

Immigrant workers from Europe and African Americans from the Deep South poured into the region to work jobs in the coal camps. The population exploded, with counties such as Letcher and Harlan—the heart of metallurgical coal production—increasing by tens of thousands of residents. In 1940, Harlan County, with a population of 75,275, was the fourth-largest county in Kentucky, behind Jefferson, Kenton, and, barely, Fayette (home of Lexington), which boasted only 78,899 residents. Harlan County had grown by 646.7 percent in thirty years, compared to a growth rate of only 65.35 percent for Fayette County over the same period. Pike County was the fifth largest with a population of 71,122. Daviess County, now home to Kentucky's third-largest city, was sixth with a population of 52,335, one of only two western Kentucky counties in the top ten. McCracken County (population 48,534) was seventh, Perry (population 47,828) was eighth, and Letcher County, now a tiny, tight-knit community of about 25,000 people, was the ninth-largest county in Kentucky with a population of 40,592, up from only 10,623 in 1910, two years before the arrival of the railroad. Pulaski County rounded out the top ten with a population of 39,863.[17]

In Letcher County alone, nearly eighty coal camps were operated by various mining companies between the turn of the twentieth century and 1958, most between 1917 and 1958. A list of coal camps shows only the headquarters town of the company, not the names of all the camps. Many of these camps, such as Elsie Coal, Faraday, and Mater, remain today only as a few foundations and rusty rails, but some still exist, although the houses are now owned by the residents or by landlords. Those surviving camp houses still stand shoulder to shoulder, many with straight pipes leading from their bathrooms to the nearest creek.[18] Too close together for septic systems and too far from what passes for urban centers for sewer systems, the houses often still pump raw waste directly into the streams. Today, the biggest single environmental problem in Appalachian streams in Kentucky is fecal coliform bacteria from human excrement, closely followed by sedimentation and siltation from mining disturbance and acid mine drainage from the interaction of water and air with sulfur and iron pyrite that are common in coal seams.[19]

Because the erosion of the Cumberland Plateau left so little flat land, coal waste is often piled on hillsides or dumped into hastily dug ponds on the sides of steep slopes. Because of water, those same refuse heaps and silt ponds sometimes break away, sending mud, sludge, and torrents of water down on the residents below. In the early years of strip mining, landslides were common and would push houses from their foundations or bury them under loose rock. Because of the frequency of these occurrences, Congress passed the Surface Mining Control and Reclamation Act of 1976 and began requiring compaction of mining sites to prevent slides.

That did not always help. Although it greatly reduced the number of slides, it made the ground so hard that trees could not grow, exacerbating the runoff problem. It also did little to prevent problems caused by the interaction between surface mines and old underground mines.

Perhaps the most notorious of these in Kentucky was the Martin County sludge spill, which occurred on October 11, 2000, when the mining company either miscalculated or ignored the amount of rock strata between a coal-sludge impoundment and a worked-out mine below. The impoundment owned by Massey Coal Co. near Inez broke into the abandoned underground mine, sending an estimated 306 million gallons of gooey black coal slurry into Wolf Creek and Coldwater Fork, tributaries of the Tug Fork River.[20]

It is because of water—specifically the filling of headwater streams with dirt and rock—that many people are so opposed to mountaintop-removal mining. Opponents of mountaintop removal say that more than 1,200 miles of Appalachian streams have been lost to the process.[21]

During mountaintop removal, coal companies use explosives and heavy equipment to scalp dirt and rock strata from mountain peaks and then push that material, known as "spoil," into adjacent valleys. These areas are called "hollow fills." Although federal law requires land to be returned to "approximate original contour," most companies obtain waivers to return the land instead to a "higher and better use." In most cases, that means pastureland, leaving what was once a mountain peak looking more like a central Kentucky savanna. Trees are rarely returned to sites, partly because the compaction used to prevent slides makes the ground too hard for seedlings to survive. Highly invasive plants used to fix nitrogen in the soil also outcompete slow-growing trees and leave any trees that do grow stunted. As early as 2005, the U.S. Office of Surface Mining, in partnership with state mining agencies, began encouraging companies to practice lower compaction and to plant native species of tree seedlings instead of the nitrogen fixers.[22] The partnership, called the Appalachian Regional Reforestation Initiative (ARRI), is also seeking out abandoned mines that are in a state of suspended natural plant succession. Those sites are being "ripped" using equipment similar to that originally used to mine the sites, and trees are being planted in an attempt to reduce runoff and jump-start natural succession.

Opponents of mining have responded cautiously to the ARRI project. Some have expressed the feeling that the project will make people believe that it is okay to continue mountaintop removal because it can be "fixed." Others—notably the Sierra Club—have volunteered extensively to help plant trees on abandoned mines in an effort to mitigate the damage already done.

In a fight that has gone on for so long, it is difficult to find any middle ground. Strip mining has been a point of contention in eastern Kentucky since the beginning, largely because of the broad form deeds used by speculators like Mayo and the methods some companies used to enforce their mineral rights. Residents were sometimes physically removed from their land so that mining could take place, and many homes, farms, and cemeteries have been moved or destroyed. The practice was so reviled that in 1988, Kentucky voters approved by a four-to-one margin a constitutional amendment requiring companies to use only the mining methods currently in use at the time the broad form deed was signed.[23]

Supporters counter that the people who live here have to work, and that mountaintop removal does not do any damage because the land is put back to approximate original contour. They also cite the proliferation of elk that were introduced by the Kentucky Department for Fish and Wildlife Resources as evidence that the mountaintop-removal mines are good for wildlife.

So-called Astroturf organizations funded by coal companies[24] and through state license plates have mounted an all-out public relations effort, with T-shirts and bumper stickers proclaiming such slogans as "Coal Mining, Our Future," "If you don't like coal, don't use electricity," "Power, Progress, Coal," and "Save a miner, shoot a tree hugger." In a region where nearly a third of the population lives in poverty and there is little opportunity for employment, the effort has resonated with some people. However, others cite the loss of aquifers and streams as more important.

Limited land, pollution problems that were unforeseen or ignored, a topography that makes construction of modern infrastructure an engineering nightmare, and an industry that left little room for any other have turned Appalachian Kentucky into a virtual third-world country. These difficult living conditions can all be traced to a substance that is essential to life itself. Water influences lives of eastern Kentuckians in ways that most never stop to think about. The eastern Kentucky mountains—and the mountaineers—truly are shaped by water.

Notes

1. Sergey Tokarev, "Appalachian Mountains," in *Encyclopedia of World Geography,* February 15, 2012, http://world-geography.org/mountain/38-appalachian-mountains.html.

2. Preston McGrain, *Scenic Geology of Pine Mountain in Kentucky* (Lexington: Kentucky Geological Survey, 1975). Ser. 10, Special Publication 24, 34 p at ftp://rock.geosociety.org/pub/Memorials/v29/mcgrain.pdf.

3. Ibid.

4. "Cumberland Plateau: Facts, Discussion Forum, and Encyclopedia Article," *AbsoluteAstronomy.com,* February 16, 2012, http://www.absoluteastronomy.com/topics/Cumberland_Plateau.

5. "Pangea Supercontinent," *Geology.com: News and Information for Geology & Earth Science,* February 16, 2012, http://geology.com/articles/supercontinent.shtml.

6. *Floods of January–February 1957 in Southeastern Kentucky and Adjacent Areas,* Rep. no. 1652A (Washington, DC: Government Printing Office, 1964); *Floods of January–February 1957 in Southeastern Kentucky and Adjacent Areas,* U.S. Geological Survey, February 21, 2012, http://pubs.usgs.gov/wsp/1652a/report.pdf.

7. "Flood of 1927 in Letcher County, KY," *RootsWeb.com Home Page, Mountain Eagle,* June 2, June 9, June 16, June 23,

1927, February 16, 2012, http://www.rootsweb.ancestry.com/~kyletch/articles/flood.htm.

8. Kayla Webley, "Top Ten Historic U.S. Floods," *Time .com, Time,* May 11, 2011, February 21, 2012, http://www.time.com/time/specials/packages/article/0,28804,2070796_2070798_2070780,00.html.

9. Sharon Young Jebavy, "Big Sandy Bus Accident 1958," *RootsWeb.com Home Page,* Ancestry.com, February 21, 2012, http://www.rootsweb.ancestry.com/~kyjohnso/Bus.htm.

10. Kentucky Division of Water, comp., *Integrated Report to Congress on the Condition of Water Resources in Kentucky, 2010,* Rep. Vol. I. 305(b) (Frankfort: Kentucky Division of Water, 2010).

11. E. Lucy Braun, "The Forest," in *Deciduous Forests of Eastern North America* (1950; repr., New York: Hafner, 1964), 6–7.

12. Fred Coy Jr., Tom Fuller, Larry G. Meadows, and Don Fig, "Splash Dams Construction in Eastern Kentucky's Red River Drainage Area," *Forest and Conservation History,* 1992, 179–84; *Duane and Eva's Old Kentucky Home Page,* Duane Bristow, June 30, 1997, February 21, 2012, http://www.kyphilom.com/www/wood/splashdm.html.

13. Jeanette Knowles, *The History of Elk Horn Coal, Elkcoal.com,* Elk Horn Coal Co. LLC, February 21, 2012, http://www.elkcoal.com/History.pdf.

14. Ronald D. Eller, "A Magnificent Field for Capitalists," in *Miners, Millhands, and Mountaineers: Industrialization of the Appalachian South, 1880–1930* (Knoxville: University of Tennessee Press, 1982), 55–63.

15. "Coal Camps of Early Southwest Virginia," *The Southwest Virginia Museum Historical State Park,* Commonwealth of Virginia, February 21, 2012, http://www.swvamuseum.org/coalcamps.html.

16. "The Railroad Comes to Breathitt," *Breathittcounty .com,* Charles Hayes, July 20, 2001, February 21, 2012, http://www.breathittcounty.com/BreathittWeb2/Railroad.html.

17. "Population 1900 through 2000, Kentucky and Counties," *Kentucky State Data Center,* University of Louisville, January 22, 2012, ksdc.louisville.edu/sdc/census1990/copop1900_2000.pdf.

18. "Raw Sewage Poses Threat to Kentucky County," *New York Times,* December 22, 1996, February 15, 2012, http://www.nytimes.com/1996/12/22/us/raw-sewage-poses-threat-to-kentucky-county.html.

19. "Acid Mine Drainage," *Water: Polluted Runoff,* U.S. Environmental Protection Agency, February 21, 2012, http://water.epa.gov/polwaste/nps/acid_mne.cfm.

20. "Martin County Sludge Spill," *SourceWatch,* Center for Media and Democracy, November 18, 2009, February 22, 2012, http://www.sourcewatch.org/index.php?title=Martin_County_sludge_spill.

21. "What Is Mountain Top Removal Mining?," *Mountain Justice,* February 22, 2012, http://mountainjustice.org/facts/steps.php.

22. "ARRI—Forestry Reclamation Approach (FRA)," *ARRI,* U.S. Office of Surface Mining Reclamation and Enforcement, February 22, 2012, http://arri.osmre.gov/FRA/FRApproach.shtm.

23. "Kentuckians for the Commonwealth," *History & Accomplishments,* Kentuckians for the Commonwealth, February 22, 2012, http://www.kftc.org/about-kftc/history.

24. "Friends of Coal," *SourceWatch,* Center for Media and Democracy, February 22, 2012, http://www.sourcewatch.org/index.php?title=Friends_of_Coal.

Chapter Seven

Life on the River

Jamey Wiglesworth

For those who choose life on the river, the river becomes an ever-present force, a metaphor that informs our way of thinking in all matters. We become as in tune with the seasons and the variation of the weather as the old farmers. We learn how to work with nature and not against it, so that the river may join its force to ours and not diminish it. The sight of the river brings us the joy of a child's face.

When the rains fall and the river rises in late winter and early spring, we watch anxiously, knowing, at least on the Kentucky River, that sleep will not come easy until the first of June. We live a combination of high tech and low tech, with cell phones and computers that help gauge the river's level, and a yardstick at the water's edge. But we still have no control over the river. It is a powerful giant lumbering by, staring up at us on our front porch, and we must let it do what it wants. (My father says that the front of a river house always faces the water, even though it may look like it faces the road to those from the city.) Like a pilot fish cleaning dead skin from a shark, we have a symbiotic relationship with the river, but we still know who the boss is.

Most people who continue to stay on the river move to higher ground or put their houses on stilts or over an unfinished basement. Although there still can be quite a mess when the river floods, you are not out of your house and away from the river as long. I spent ten years in a house that was the lowest on the pool from Lock 5 to Lock 4. It flooded three times, in 2002 and 2003 with about a foot of water in the house, and in 2010 where only a foot of the gable showed above water. As the river rose, I would measure its rate of rise on a stick sunk into the mud at water level. This would give me a good idea of how long I had before I needed to move out. If the rate

of rise lessened and it was not a measuring error, the river would soon crest. I could usually calculate my safety with a bit of leeway. After being narrowly flooded in the early part of this century, I figured, statistically, that I should be narrowly missed once in a while. In 2007, the river crested lapping at my back porch and proved my theory.

Although the river is usually quiet from the house, it takes on an ominous rumble during floods, especially at night. The occasional splash that would have been undetectable during the rest of the year seems too loud. It is tough to sleep when one does not know where the water will be in the morning. Roads have become impassable, and the power is out. The people sounds that are few in the country anyway dwindle away to near silence. The only sounds are the river rumbling and the higher pitch of water flowing around the trunk of the sycamore in the yard. Occasionally, a loud crash sets nerves on edge as somewhere a weakened bank gives way, sliding into the river with half a dozen trees. As morning breaks, the calls of blue jays and robins take over as they feast on hordes of worms, centipedes, millipedes, beetles, and other insects that have taken to the tops of posts and bannisters to escape the rising waters. The sounds hearken back to the dawn of time before humanity had tamed the planet.

River life has to be accepted whole, like life itself. The floods serve their purpose as they did in ancient Egypt, nourishing the Nile valley. There will never be condos down the block or a new shopping center. The hardiness bred by these frequent disasters builds community and keeps people on their toes.

If spring does not bring high water, the summer people begin to arrive in droves around Memorial Day weekend. Some put their boats in only for the weekend,

7.1. Sharing with neighbors on the river. Pileated woodpeckers are frequent winter visitors.

while locals tie their boats to homemade docks for the summer, knowing that they should be safe from rising water. In the weeks leading up to summer, we shovel the spring mud into the river from stairs leading to homemade docks of Styrofoam pontoons or plastic barrels and relaunch the gangplank for access. We replace broken boards and missing barrels, as well as pull a few docks out from the muddy shore. The few public docks on the river get moving around then too.

Public boat docks provide beer, the occasional cheeseburger, and gas for boats. They are also places to get news of the river. You will find out about a poker run or the Kentucky River clean sweep, as well as the hours and days on which the locks are open. These also offer the opportunity to socialize with other boaters whom you know or to whom you may have given only the mandatory passing wave.

Running a dock seems like idyllic living; you are on the river and it is beautiful, and the pace is relaxed. Nevertheless, not many docks have much longevity around here. Most can be found only in the Frankfort pool between Locks 4 and 5 and in the Boonesboro pool between Locks 9 and 10. With the supervision required during the spring and the lack of business nine months out of the year, as well as most weekdays, people tend to lose their taste for running a boat dock. It is a labor of

love that keeps you from being out on the river cruising during the summer, and if you need it to actually provide income, you are looking in the wrong place.

Above Lock 4, the locks are permanently closed these days, but supposedly, work being done on Lock 3 will once again allow access to the Ohio River from Frankfort. Until the 1980s, the river could be totally traversed from Beattyville to Carrollton, but as the last of the commercial traffic and barges disappeared, the money to maintain the locks dried up too.

Summer on the river gives most transplants their first taste of river life as the river puts on its best face of the year. Beaches that lie underwater for most of the year are revealed, and boats fill the water, which had been lucky to see one boat in a week during the off season. Now, there are pontoons cruising lazily from lock to lock, houseboats seeking a beach to camp for the weekend, and runabouts pulling skiers, tubers, skurfers, wakeboarders, and kneeboarders, as well as the occasional jet boat flying down a straight stretch of the river. Bass boats and other fishing boats are on the water too, but with the traffic on summer weekends, more successful fishing is found during the off season or during the week.

Eventually, most people on the river seek pontoon ownership. Pontoons dominate the river's motorized vehicles. They can be relatively inexpensive, although the nicer ones with enclosed cabins and upper decks may run a bit higher. Upkeep is minimal. For the most part, they do not sink (at least, you do not need to bail them), and they run on an outboard motor, which is much less trouble than an inboard. They also give you a great amount of unencumbered space for the money, so there is room for the whole family and the dogs.

Life on the river is leisurely; there is no place that needs to be gotten to that quickly. You just want to be out in the sun absorbing the beauty around you. It is pure joy. The river flows slowly and warm in the summer and is the earthy green of a green worm dug from its banks for catfish bait. As you cruise the river, kingfishers fly with you sometimes for miles, twittering loudly as they go and diving for fish and frogs. Wet and sandy dogs stand on the front deck with noses in the air like bowsprits. High in sycamores along rock bars, great blue herons make their nests and occasionally swoop down, while buzzards circle even higher on the updrafts at the top of the river valley. As the sun sets, the moist tree-lined banks teem with lightning bugs like a spectral city. It feels like the endless summer of youth.

Summers on the river make you want to stay forever. The river has given you its best sales pitch, and you are hooked.

After Labor Day, the summer people depart, and the flow of boats thins to a drip. Days in fall when the leaves turn and few people travel the river are some of the best to be on it. River folk have a tendency to leave their boats in the water a couple of months later than the summer people, and often they leave their boat docks in year-round. On a warm October day, cruising on the river among falling leaves conveys the beauty of Kentucky and causes a sense of pride to well up in you. You are right where you need to be. This is your place in the universe.

The river flows cold and quiet into the winter, and you see the majesty of the old king getting his well-deserved rest. The river can take on a dark forest green that reflects the opposite bank with beautiful clarity. The silence of the crisp winter air is serene. This is what inspired Paul Sawyier, impressionist known for his Dix and Kentucky River watercolors. The birds that dominate the summer give way to the river birds that, like us,

7.2. Satchmo dog-paddling.

make their homes here year-round: the finches, the chickadees, the titmice, and all the multicolored woodpeckers. It is a time to be alone with the river, and, as the river knows, it is a time for reflection.

As you age on the river and get accustomed to its ever-changing nature, it becomes a guide in life and also sufficient to you in itself. During childhood, the river is a place to swim, to fish, and to water-ski. It is home to rope swings and mud slides. It provides weekend camp-outs, ghost stories, marshmallows, and hot dogs. To a teenager, the river gives mobility and freedom before you obtain a driver's license. At this time, you begin bringing girlfriends around the river or maybe even find them there, and with your newfound mobility and impressive knowledge of boats, the river becomes one long lover's lane.

Early adulthood may require a brief absence from the river to deal with the cares of the world, but you still hear it calling to you, and you find your way back. Then the river becomes the home of the barbecue and coolers of beer, and of varied attempts to make a living on the river

7.3. My treehouse, Wu Wei (Taoist "effortless doing"), over-looking the river, built in the summer of 2006.

7.4. Three generations of the Wiglesworth family cruising the Kentucky River, July 2011.

so you never have to leave. For some, this means commercial fishing or taking a turn running the boat dock for the summer. For a brief time, mussel fishing was a viable alternative (spheres from mussel shells were used to induce oysters to make cultured pearls)—and let me tell you, it was the life. Odd jobs also pop up here and there, like construction of a boat dock or ramp or scuba diving for a sunken boat or car or to fix the locks (as my family has done). Scuba diving in the river is a bit challenging because even with a powerful flashlight you can see only about six inches. To some, this silent darkness can be scary, but I am the Brer Rabbit of this briar patch, and to me it is home. Eventually, you too become the teller of ghost stories and the engineer of rope swings for your own children, and you see your life on the river coming full circle.

But at this point, the river itself is enough. The boats, the food, the beer, and the camaraderie are nothing compared with the happiness you derive from simply being around an ever-changing source of natural beauty. It becomes a religion. The river is forever new and different; it is in a constant state of birth, death, and resurrection. As Heraclitus said, "You never step into the same river twice." Instead of clinging to security and wishing for an unattainable permanence that dries the soul like a raisin, the river teaches acceptance of the changes to which we are all heir, the droughts and the floods, the meanders and the rapids. It teaches us to accept these changes with tranquility, knowing that this is the way of the world. The rhythm of the flowing river attunes us to the constant flow of life. It raises us from childhood and acts as a constant teacher. It is grandfather to us all.

CHAPTER EIGHT

The Martin County Coal-Waste Spill and Beyond

Reflections and Suggestions

Shaunna L. Scott and Stephanie M. McSpirit

Sometime after midnight on October 11, 2000, the Big Branch coal-waste impoundment owned by Martin County Coal Company ruptured in Martin County, Kentucky. Over 300 million gallons of coal waste leaked through a series of abandoned underground mine tunnels and exited through Coldwater and Wolf Creeks (US DOI OSM 2002; US DOL MSHA 2001). The spill killed all aquatic life in both watersheds and disrupted public water supplies in Martin County and other neighboring communities. The black, lava-like substance flooded people's residences and properties, temporarily displacing many Coldwater and Wolf Creek residents from their homes. Although no human life was lost, unlike the 1972 Buffalo Creek, West Virginia, coal-impoundment disaster (Erikson 1976), the ecological damage in this impoverished, coal-mining Appalachian county was severe. The Martin County coal-waste spill was the largest environmental disaster in the southeastern United States, dwarfing the well-known *Exxon Valdez* disaster spill of 11 million gallons of oil.

In spite of the size and seriousness of the disaster, neither the corporation nor the government notified or evacuated the residents of the coal reservoir area. Instead, local residents woke up to the shocking discovery that the creeks were black or, worse, that their houses were surrounded by a thick, black, potentially toxic substance (Scott et al. 2005). The local water district was not drawing water from the river when the disaster occurred, so disaster-related damage to the local water plant was minimal; other communities along the rivers shut down their water intake as well to give time for the waste to dissipate as it flowed down the Tug, Ohio, and Mississippi Rivers. Nevertheless, the public was concerned that heavy metals and carcinogens (e.g., arsenic, cadmium, selenium, and mercury) had permanently contaminated the water supply because it proved impossible to remove completely the black tar-like substance from the riverbeds and banks (Scott et al. 2005).

Two months after the October 2000 coal-waste spill, emergency water lines froze, forcing the local water plant to return to water intake from the Tug Fork of the Big Sandy River. Soon thereafter, the water plant had to close its water intake after black water from heavy December rains complicated pumping from the Tug River (Turner 2001). During this short period, residents had already begun to complain that their tap water had a strange smell and left a powdery residue; in some cases, skin rashes were reported (Scott et al. 2005). The Kentucky Division of Water (DOW) responded to the complaints by inspecting the facility in late December; inspectors found serious problems at the water plant. These included irregular filter washing and jar testing, malfunctioning pH meters, and the plant being "junked up" (Ball 2001a).

Although Martin County Coal Corporation representatives held public meetings at local churches and schools to reassure residents that they would clean up the mess, some community members remained skeptical. In the coming weeks, people raised questions about the effectiveness of government regulatory agencies, including the Mine Safety and Health Administration (MSHA) and the Environmental Protection Agency (EPA), as well as the competence of the management of the local water district. When the EPA set up its command post on gated and guarded corporate property, it appeared that it was evading the public and aligning itself with Martin County Coal Corporation. That the EPA refused to sue the corporation, allowed the corporation to draft daily press releases on behalf of the agency, and collected its water-quality data from samples provided to it by the corporation supported the conclusion that the EPA had been "bought off" by Martin County Coal (Adkins 2001). Behind the scenes, the EPA agreed to the coal corporation's request to have the disaster redefined as a violation of the Clean Water Act rather than as an imminent public danger under the Comprehensive Environmental Response, Compensation, and Liability Act (CERCLA), thereby surrendering the authority to require the corporation to pay for environmental and public health assessment and public technical assistance grants. Conflicting reports were issued by different government agencies, lawyers, and public officials regarding the presence of heavy metals and other toxic substances in the water in 2001. Public confidence in government regulatory agencies and the local public water board was severely eroded as a result of these conflicts. In that context, it is no surprise that tempers flared at a March 13, 2001, public meeting when residents expressed outrage that the EPA was not fining the coal corporation. They also questioned the safety of the water supply and challenged federal EPA and state water officials to drink water from the water fountain where the meeting was held. Although the officials had assured the public that the water was safe to drink, they refused to partake of it themselves on that evening (Hickman 2001).

To their credit, MSHA officials at least set up a temporary command post at the courthouse even though they did not appear in Martin County until a week after the disaster. However, the MSHA's credibility was damaged when the public learned that the same coal-waste impoundment that caused the 2000 disaster had also leaked 100 million gallons of waste in 1994. Even though

an engineer working for the agency had warned that the reservoir was at risk of a catastrophic failure unless remedial actions were taken, the MSHA had not forced the company to repair the impoundment and, worse yet, had allowed it to add coal waste to the facility. On April 3, 2001, MSHA investigator Jack Spadaro resigned from the Martin County investigatory team and filed a complaint with the Department of Labor's inspector general. When the MSHA issued a $110,000 fine to the company for two minor violations and then released a report that failed to mention the agency's failures in enforcing safety regulations, Spadaro spoke publicly about the agency's failure. Shortly thereafter, the agency placed Spadaro on administrative leave, searched his office, and then charged him with misuse of a government credit card. Because of this harassment, Spadaro left his position at the MSHA and gained national media attention when he was featured on a 2004 episode of *60 Minutes* (Leung 2009).

Interestingly, the state of Kentucky took a stronger stand on behalf of public and environmental health than did federal agencies under the Bush administration. Kentucky invoked its right under CERCLA to sue Martin County Coal Corporation for damages caused by the disaster. In a negotiated settlement, the state of Kentucky collected $1 million in natural resource damages, $1.75 million in civil penalties, and $500,000 to compensate for state response costs (Haines 2005). This could be because the state was more responsive to local citizen complaints and activism than the federal government was, at least under the Bush administration.

During the first year after the spill, various citizen groups emerged, including a group called Health, Environment, Life, and Preservation (HELP), a local chapter of the Ohio Valley Environmental Coalition called the Big Sandy Environmental Coalition, and a citizens' advisory group, to work with the EPA and the coal company on environmental recovery from the disaster. In January 2001, local citizen groups sponsored informational presentations by nationally known environmental lawyer Jan Schlictmann, other lawyers, representatives from Kentuckians for the Commonwealth, a local community-organizing group, and the Ohio Valley Environmental Coalition; they worked with teams of Eastern Kentucky University researchers to circulate the findings of interviews and surveys conducted in the county to assess the impact of the spill on the community. HELP also tried in vain to persuade the EPA to include citizens in environmental recovery and monitoring efforts, to open an outreach office in the town of Inez, and to require the corporation to set aside funding for independent, citizen-monitored water testing. However, citizen activism may have influenced the state of Kentucky, specifically the DOW, the Public Service Commission (PSC), and the Environmental Quality Commission (EQC). Local citizens succeeded in securing $150,000 for independent water testing from Kentucky's settlement with Martin County Coal Corporation (McSpirit and McCoy 2005).

Between 2002 and 2006, public scrutiny began to focus on the management of the local water district and of local government and politics generally. Local government had not escaped criticism in the immediate aftermath of the disaster. In our spring 2001 interviews, residents told us that local officials were unwilling to protect public health because they owed money and political favor to the coal corporations. They also accused the county judge/executive of attempting to make political gain by falsely claiming credit for providing free bottled water to disaster victims. Most egregiously, some people believed that local officials were lying to them about the safety of the drinking water. Local officials attempted to restore their credibility by inviting local clergymen to monitor water testing, a maneuver that was ridiculed by the 2001 interviewees and the local press (Ball 2001f; Scott et al. 2005). By 2002, the controversy over water-plant conditions increasingly took center stage away from the EPA, the MSHA, and Martin County Coal Corporation (Ball 2001a, 2001c, 2002a, 2002b).

Local news reports continued both to facilitate and to contribute to the debate by publishing conflicting reports, editorializing, printing letters to the editor, and soliciting public opinion about the safety of the water, often under provocative headlines such as "Fiscal Court's Water Sample 'Bandaid' for Life-Threatening Problem" (Ball 2001d), "Who Do We Believe?" (Ball 2001g), "Have I Lied about the Water?" (Ball 2001e), and "Don't Trust the Division of Water, Says Alpha Branch Couple" (Ball

2001b). The *Mountain Citizen*'s investigation and publication of the water district's management problems raised the ire of the water board's chair, who tried to take advantage of the publisher's having missed a filing deadline in order to strip the newspaper of its corporate name, Mountain Citizen, Inc. This ploy failed and only caused public opinion to shift further away from the water district's management. Gary Ball, editor and reporter for the *Mountain Citizen,* was awarded a 2002 Kentucky Press Association award for excellence in journalism as a result of his investigation of the Martin County water district (Smith 2003).

The water plant's inadequacies led to citations by the PSC, a temporary takeover of water management by Kentucky/American Water Corporation, the eventual ouster of the water board's chair, and the county judge/executive's failed reelection bid. After the following administration appointed a new manager and secured funding to upgrade the water plant (Hickman 2002), confidence in the water board was somewhat restored. A 2005–6 EKU- and citizen-led study concluded that the spill had no long-term effects on the water source or the treated public drinking water. It further found that the water plant had improved in its facilities, management, and ability to produce and deliver good drinking water (LaSage and Caddell 2006; McSpirit and Wigginton 2006). The findings of this study were delivered to every household in Martin County. As a result, by 2009 the newspaper editor reflected on the coal-waste spill as "a blessing in disguise" because "it focused a lot of attention on our water plant" (Scott interview of Ball, March 30, 2009). After water testing found no long-term contamination of the water, citizen engagement and social activism in environmental and disaster recovery diminished (S. L. Scott 2012).

Surveys conducted in June 2011 also found that that public confidence in the water supply had increased in the ten years since the coal-waste disaster of October 2000 (Scott, McSpirit, and Howell 2012; Scott et al. 2012). Whether that means that confidence in the water has been restored is debatable inasmuch as half of those surveyed still reported that drinking water was a serious problem in their community. As a retired Kentucky DOW environmental scientist pointed out, public concerns are slow to dissipate. "Do they have concerns about the water? You bet they do! And it's lingering probably to this day. It'll probably be lingering for 50 years in somebody's head, even if it's fixed. And I think it is fixed," he said (Scott interview of Withrow, April 13, 2009).

That fear seems to be lingering more in the heads of Martin County residents who have higher education levels and those who distrust coal companies. Our 2011 survey found that people who are educated are less likely to express confidence in the water than those who have less education, and those who distrust coal companies are less likely to trust the local water supply (Scott et al. 2012). This could indicate that Martin County residents have still not recovered from the negative impacts of the October 2000 disaster inasmuch as they still lack trust in coal companies and see a connection between coal mining and poor water quality. In addition, it could be that educated residents are more likely than those with less education to recognize the ways in which coal mining, preparation, and burning affect the environment and threaten water quality. Even though the citizen-led water tests indicated that the water is safe and newspaper editorials have argued consistently that water problems were caused by the local water plant rather than the coal-waste disaster (S. L. Scott 2012), some sectors of the public do not accept the newspaper editor's claim that the coal-waste disaster was a "blessing in disguise." In addition, our 2011 survey indicates that, compared to neighboring counties where low trust in water quality is related to low trust in water-plant management (McSpirit and Reid 2011), Martin County residents emphasized the link between coal companies and poor water quality. In the face of a consistent effort by local government and the newspaper to convince residents that the previous administration's poor management rather than the coal-waste disaster had caused poor water quality, the public's persistent belief that coal mining corrupts water quality is worth emphasizing.

There are lessons to be learned from the experience of the Martin County coal-waste disaster, lessons of importance to other places. First, the threat of a catastrophic coal-waste disaster is real. There are approximately 313 coal-waste impoundments in the coalfields of central Appalachia; this figure includes only permitted impoundments regulated by the federal MSHA and the U.S. Department of Surface Mining (Coal Impoundment Project, Locator Information System). Of these, 113 are located in Kentucky, primarily eastern Kentucky. There are at least 350 coal-waste and ash impoundments nationwide. In 2009, the EPA listed 44 of them as "high-hazard" facilities, meaning that their failure would likely result in the loss of human life (Sourcewatch.org 2011). Six of the 44 high-hazard facilities are in Kentucky, specifically in Louisa, Harrodsburg, Ghent, and Louisville. Neither the Martin County coal-waste impoundment nor the Tennessee Valley Authority's Kingston coal-fly-ash plant, which in 2008 spilled 5.4 million cubic yards of coal ash in Harriman, Tennessee (Dewan 2008), was classified as a high-hazard facility at the time of its failure. Both caused significant property damage and polluted the water. In both cases, the authorities reassured the public that it was not placed at risk by this polluting event in spite of studies that demonstrate a link between long-term exposure to coal-waste products and various cancers and other health problems (Gottlieb 2010). According to reports released by the EPA and studies done by EarthJustice, the Environmental Integrity Project, and the Sierra Club, there have been 157 damage cases related to coal-waste and coal-ash contamination in the United States (Sourcewatch.org 2011).

Given the high number of coal-waste facilities that endanger human life (44), it seems prudent to mandate that coal-waste-threatened localities have an emergency warning and evacuation plan in place. West Virginia has required such planning since the 1972 Buffalo Creek disaster, which killed 125 people. Kentucky does not require such planning, a fact that shook Martin County residents who realized that they could have been killed in their beds just as the 125 residents of the Buffalo Creek area were in 1972. Since the 2000 Martin County disaster, we have been working with citizen and environmental groups to promote emergency-action-planning legislation for coal-waste impoundments in Kentucky. A bill has circulated in the Kentucky General Assembly since 2004 to require owners of high-hazard dams and impoundments to file emergency procedures with state and local communities. During the 2008 legislative year, the bill passed the Senate but was recommitted for a second hearing. As part of the negotiation process, it was made more restrictive to apply only to those dams with human populations directly downstream. During negotiation, the public notification and comment period was conceded to coal industry representatives in an attempt to get a plan that would at least provide Kentucky county emergency officials with contact numbers, inundation plans, and evacuation strategies. In spite of this, the bill died in the Natural Resource Committee of the House of Representatives.

On the basis of our subsequent published and unpublished research and our reading of other studies (Fitchen, Heath, and Fessenden-Raden 1987; Rich et al. 1995; Gunter, Aronoff, and Joel 1999; Murdock et al. 1999; Wulfhorst and Krannich 1999; Picou and Gill 2000; Tretten and Musham 2000; Gregory and Satterfield 2002; Zavestoski et al. 2002; Walls et al. 2004; Pandey and Okazaki 2005; Paton 2007), we now conclude that concession on the public comment and review clause in Kentucky was not wise. Emergency planning—indeed, all planning—should involve a wider variety of stakeholders and social institutions. A call for more community-based planning rests on the assumption that local communities have more intimate knowledge about the environment and a greater stake in solving and monitoring the problems caused by disasters (Kingsley 1996; J. C. Scott 1998; Kellert et al. 2000). Locally produced plans should be more effective because they reflect local knowledge and have local buy-in. Such community-based planning is also said to result in local empowerment and a more vibrant civil society, both of which reinforce and enhance democracy (Western, Wright, and Strum 1994; Hinsdale, Lewis, and Maxine 1995; Couto 1999; Gibson, McKean, and Ostrom 2000; Li 2002; Lane and McDonald 2005).

Similarly, we also recommend that the Martin County and other local water districts incorporate citizen oversight into their management and water testing. Our 2001 survey in Martin County indicated that there was broad-based public support for citizen oversight in the management of the water district (McSpirit et al. 2007). In the aftermath of the disaster, citizens organized themselves, at least temporarily, into various community groups and allied themselves with regional and state groups in order to secure public oversight of water testing and public participation in the environmental recovery process. The primary area of concern to the public after the disaster was the safety of the water. Once noncorporate experts and citizens conducted independent water testing and determined that the drinking water was safe, public concern and citizen activism decreased dramatically. However, our 2009 interviews and 2011 surveys show that a decrease in activism and public critique of the water district management does not necessarily indicate that Martin County residents fully trust the water. Half of those we surveyed said that public drinking water was a serious problem in the county. Citizen oversight of the water district management could improve the water service there and help restore confidence in the water. Such input could be incorporated by creating a citizens' board or task force, by holding citizens' workshops, or by allowing some seats on the district management board to be elected by popular vote rather than appointed by the county judge/executive. This is done in municipalities and districts in other states and should be practiced more widely, especially in Martin County and its neighboring counties, where concern about water quality is linked to a lack of trust in the water district management.

Another lesson for citizens is that it is difficult to determine whose side the government is on: the side of corporate profit or public and environmental health and safety. In the aftermath of the 2000 Martin County disaster, for example, interviewees were not surprised that the coal corporation had caused an accident and put public health at risk. They were surprised and disappointed that the MSHA had known about the problems at the impoundment for six years before the disaster and had failed to correct them. As time went on, the EPA and county governments allied themselves with the coal corporation, while the Kentucky state government and the Division of Water seemed more willing to contest the corporation. Ten years later, however, the tables had turned. Kentucky's Energy and Environment Cabinet partnered with the coal industry (Kentucky Coal Association) to sue the EPA (under a Democratic administration), calling its regulation of coal-mining permits under the Clean Water Act "arbitrary" (McCann 2010).

Finally, Kentucky communities should recognize that a strong culture of civic engagement and an independent, local press will help them in many ways—in creating democracy, securing social justice, guarding against corruption, conserving and managing natural resources, and responding to disasters. In Martin County, local activism and strategic alliances with state and regional citizens' groups played some part in attracting state attention to the water problems in 2001–2. However, it is equally obvious that the local media were crucial in drawing both the public's and the state's attention to problems at the water plant. Newspapers are an important means of communication and, when they operate well, are watchdogs for the public interest. With the decline in newspaper readership, global corporate takeover of locally based newspapers, and the difficulties in generating revenue from noncorporate sources, the future of independent, small-town newspapers like the *Mountain Citizen* is in jeopardy. It is possible that alternative business models and new communication technologies can help compensate for the decrease in local newspapers. Certainly we have seen examples of the effective use of the Internet and alternative media in the Occupy movement, which emerged in the United States during the fall of 2011, the Arab Spring movements of 2011, and the 1994 Zapatista revolution in Chiapas, Mexico. A community that supports an independent, local media and has high levels of civic engagement is a resilient community that can cope more effectively with disaster.

References and Further Reading

Adkins, Lily. 2001. Citizens Outraged When EPA Say Water "Safe" and MCC Won't Be Fined. *Martin County Sun,* March 14.

Ball, Gary. 2001a. Don't Drink the Water: Health Risk Hazards at "Junked Up" Water Plant. *Mountain Citizen,* January 17.

———. 2001b. Don't Trust the Division of Water, Says Alpha Branch Couple. *Mountain Citizen,* March 28.

———. 2001c. DOW Engineer Visits Water Plant: Supervisors Call In Consultant, Consider Bringing In Others to Address Problems. *Mountain Citizen,* February 7.

———. 2001d. Fiscal Court's Water Sample "Bandaid" for Life-Threatening Problem. *Mountain Citizen,* February 14.

———. 2001e. Have I Lied about the Water? *Mountain Citizen,* February 21.

———. 2001f. Water Is Safe! So Saith the Flock. *Mountain Citizen,* February 7.

———. 2001g. "Who Do We Believe?" *Mountain Citizen,* February 14.

———. 2002a. Editorial: Water Quality: A "Health" or "Political" Issue? *Mountain Citizen,* March 27, 6.

———. 2002b. Martin County Water District in Crisis: PSC Issues Emergency Order. *Mountain Citizen,* April 3, 1, 10.

———. 2002c. Myths and Facts of Local Water Quality. *Mountain Citizen,* June 6.

Coal Impoundment Project, Location and Information System (CIP, LIS). 2009.

Couto, Richard A., with Catherine S. Guthrie. 1999. *Making Democracy Work Better: Mediating Structures, Social Capital, and the Democratic Prospect.* Chapel Hill: University of North Carolina Press.

Dewan, Shaila. 2008. Tennessee Ash Flood Larger Than Initial Estimate. *New York Times,* December 26. http://www.nytimes.com/2008/12/27/us/27sludge.html (accessed October 4, 2016).

Erikson, Kai. 1976. *Everything in Its Path: Destruction of Community in the Buffalo Creek Flood.* New York: Touchstone Books.

Fitchen, Janet M., Jenifer S. Heath, and June Fessenden-Raden. 1987. Risk Perception in Community Context: A Case Study. In *The Social and Cultural Construction of Risk,* edited by B. B. Johnson and V. T. Covello, 31–54. Boston: D. Reidel.

Gibson, Clark C., Margaret A. McKean, and Elinor Ostrom. 2000. Explaining Deforestation: The Role of Local Institutions. In *People and Forests: Communities, Institutions, and Governance,* edited by C. C. Gibson, M. A. McKean, and E. Ostrom, 1–26. London: MIT Press.

Gottlieb, Barbara. 2010. *Coal Ash: A Toxic Threat to our Health and Environment.* A Report from the Physicians for Social Responsibility and Earthjustice. With Steven G. Gilbert and Lisa Gollin Evans. http://www.psr.org/assets/pdfs/coal-ash.pdf (accessed October 4, 2016).

Gregory, Robin S., and Teresa A. Satterfield. 2002. Beyond Perception: The Experience of Risk and Stigma in Community Contexts. *Risk Analysis* 22 (2): 347–58.

Gunter, Valerie, Marilyn Aronoff, and Susan Joel. 1999. Toxic Contamination and Communities: Using an Ecological-Symbolic Perspective to Theorize Response Contingencies. *Sociological Quarterly* 40 (4): 623–40.

Haines, Mike (Kentucky Environmental and Public Protection Cabinet). 2005. Email message to Stephanie McSpirit, January 6, 2005.

Hickman, Ray. 2001. Tempers Flare at EPA, Division of Water Meeting. *Mountain Citizen,* March 21.

———. 2002. Water Contract a Done Deal. *Mountain Citizen,* August 21.

Hinsdale, Mary Ann, Helen M. Lewis, and Waller S. Maxine. 1995. *It Comes from the People: Community Development and Local Theology*. Philadelphia: Temple University Press.

Kellert, Stephen R., Jai N. Mehta, Syma Ebbin, and Laly L. Lichtenfeld. 2000. Community Natural Resource Management: Promise, Rhetoric, and Reality. *Society and Natural Resources* 13: 705–15.

Kingsley, G. Thomas. 1996. Perspectives on Devolution. *Journal of the American Planning Association* 62 (4): 419–26.

Lane, Marcus B., and Geoff McDonald. 2005. Community-Based Environmental Planning: Operational Dilemmas, Planning Principles and Possible Remedies. *Journal of Environmental Planning and Management* 48 (5): 709–31.

LaSage, D., and M. J. Caddell. 2006. Chemistry in Bottom Sediment of Crum Reservoir, Martin County, Eastern Kentucky Compared to a Reference Reservoir in Central Kentucky.

Leach, William D., Neil W. Pelkey, and Paul A. Sabatier. 2002. Stakeholder Partnerships and Institutions as Collaborative Policymaking: Evaluation Criteria Applied to Watershed Management in California and Washington. *Journal of Policy Analysis and Management* 21 (4): 645–70.

Leung, Rebecca. 2009. A Toxic Cover-up? http://www.cbsnews.com/stories/2004/04/01/60minutes/main609889.shtml (accessed October 4, 2016).

Li, Tanya Murray. 2002. Engaging Simplifications: Community-Based Resource Management, Market Processes and State Agendas in Upland Southeast Asia. *World Development* 30 (2): 265–83.

McCann, B. 2010. Kentucky Enters Lawsuit against EPA over Coal Mining. CivSource. October 19. http://civsourceonline.com/2010/10/19/kentucky-enters-lawsuit-against-the-epa-over-coal-mining/ (accessed October 5, 2016).

McSpirit, Stephanie, and Nina McCoy. 2005. Addendum: The Commonwealth of Kentucky Releases Monies for Independent, Outside Assessment of the Martin County Watershed, May 2005. *Journal of Appalachian Studies* 11 (1 & 2): 59–63.

McSpirit, Stephanie, and Caroline Reid. 2011. Residents' Perceptions of Tap Water and Decisions to Purchase Bottled Water: An Analysis of Survey Data from the Appalachian, Big Sandy Coal Mining Region of West Virginia. *International Journal of Society and Natural Resources* 24 (5): 511–20.

McSpirit, Stephanie, Shaunna L. Scott, and Sharon Hardesty. 2012. "Risky Business: Coal Waste Emergency Planning in West Virginia and Kentucky." *Journal of Appalachian Studies* 18 (1 & 2): 149–77.

McSpirit, Stephanie, Shaunna L. Scott, Duane Gill, Sharon Hardesty, and Dewayne Sims. 2007. Public Risk Perceptions after an Appalachian Coal Waste Disaster: A Survey Assessment. *Southern Rural Sociology* 22 (2): 83–110.

McSpirit, Stephanie, Shaunna L. Scott, Sharon Hardesty, and Robert Welch. 2005. EPA Actions in Post-disaster Martin County, Kentucky: An Analysis of Bureaucratic Slippage and Agency Recreancy. *Journal of Appalachian Studies* 11 (1 & 2): 30–58.

McSpirit, Stephanie, and Andrew Wigginton. 2006. Assessment of Finished Water, Public Water System Martin County, Kentucky. http://www.martincounty.eku.edu/reports_pdf/Water_Fnl.pdf.

Murdock, Steven H., Richard S. Krannich, F. Larry Leistritz, S. Spies, Jeffrey D. Wulfhorst, K. Wrigley, R. Sell, S. White, and K. Effah. 1999. *Hazardous Wastes in Rural America: Impacts, Implications, and Options for Rural Communities*. Lanham, MD: Rowman and Littlefield.

Pandey, Bishnu, and Kenji Okazaki. 2005. Community-Based Disaster Management: Empowering Communities to Cope with Risk. *Regional Development Dialogue* 26 (2): 52–57.

Paton, Douglas. 2007. Preparing for Natural Hazards: The Role of Community Trust. *Disaster Prevention and Management* 16: 370–79.

Paton, Douglas, and Michael K. Lindell. 2000. Politics of Hazard Mitigation. *Natural Hazards Review,* May, 73–82.

Picou, J. Stephen, and Duane Gill. 2000. The *Exxon Valdez* Disaster as Localized Environmental Catastrophe: Dissimilarities to Risk Society Theory. In *Risk in the Modern Age: Social Theory, Science and Environmental Decision-Making,* edited by M. J. Cohen, 143–70. New York: St. Martin's Press.

Rich, Richard C., Michael Edelstein, William K. Hallman, and Abraham H. Wandersman. 1995. Citizen Participation and Empowerment: The Case of Local Environmental Hazards. *American Journal of Community Psychology* 23 (5): 657–76.

Scott, J. C. 1998. *Seeing like a State: How Certain Schemes to Improve the Human Condition Have Failed*. New Haven, CT: Yale University Press.

Scott, Shaunna L. 2012. Martin County Citizens Speak: Public Trust in Water and Government Ten Years after the Martin County Coal Waste Disaster. *Kentucky Journal of Anthropology and Sociology: Special Issue on the Environment* 2 (2): 114–32.

Scott, Shaunna L., Stephanie McSpirit, Patrick Breheny, and Britteny Howell. 2012. The Long-Term Effects of a Coal Waste Disaster on Social Trust in Appalachian Kentucky. *Organization and Environment* 25 (4): 401–18.

Scott, Shaunna L., Stephanie McSpirit, Sharon Hardesty, and Robert Welch. 2005. Post Disaster Interviews with Martin County Citizens: "Gray Clouds" of Blame and Distrust. *Journal of Appalachian Studies* 11 (1 & 2): 7–29.

Scott, Shaunna L., Stephanie McSpirit, and Britteny Howell. 2012. A Blessing in Disguise (?): Reform of Public Water Management in Post-disaster Martin County, Kentucky. *Journal of Appalachian Studies* 18 (1 & 2): 149–77.

Smith, Dorothy. 2003. Mountain Citizen Editor Receives Herald-Leader Lewis Owens Award: "Best Traditions of Hell-Raising Mountain Journalism." *Mountain Citizen,* January 29.

Sourcewatch.org. 2011. Coal Issues. A Project of Coalswarm and the Center for Media and Democracy. http://www.sourcewatch.org/index.php?title=Coal_waste#EPA.27s_List_of_44_High_Hazard_Potential_Units.

Trettin, Lillian, and Catherine Musham. 2000. Is Trust a Realistic Goal of Environmental Risk Communication? *Environment and Behavior* 32: 410–26.

Turner, Cletus. 2001. Water Test Study Shows Health Issues: No Reference to Concerns Made at March Meeting. *Martin County Sun,* May 9.

United States Department of the Interior, Office of Surface Mining. 2002. *Report on October 2000 Breakthrough at the Big Branch Slurry Impoundment.* Washington, DC.

United States Department of Labor, Mine Safety and Health Administration (MSHA). 2001. *Report of Investigation: Surface Impoundment Facility Underground Coal Mine; Non Injury Impoundment Failure / Mine Inundation Accident.* Washington, DC.

Walls, John, Nick Pidgeon, Andrew Weyman, and Tom Horlich-Jones. 2004. Critical Trust: Understanding Lay Perceptions of Health and Safety Risk Regulation. *Health, Risk and Society* 6: 133–50.

Western, David, Michael Wright, and Shirley C. Strum, eds. 1994. *Natural Connections: Perspectives in Community-Based Conservation.* Washington, DC: Island Press.

Wulfhorst, Jeffrey D., and Richard S. Krannich. 1999. Effects on Collective Morale from Technological Risk. *Society and Natural Resources* 12: 1–18.

Zavestoski, Stephen, Frank Mignano, Kate Agnello, Francine Darroch, and Katy Abrams. 2002. Toxicity and Complicity: Explaining Consensual Community Response to a Chronic Technical Disaster. *Sociological Quarterly* 43 (3): 385–406.

Interviews Cited

Shaunna Scott, interview of Gary Ball, March 30, 2009.
Shaunna Scott, interview of Ted Withrow, April 13, 2009.

CHAPTER NINE

Protecting Water Resources with Streamside Management Zones at Robinson Forest

Christopher D. Barton, Emma L. Witt, and Jeffrey W. Stringer

In the steep sloping hills of eastern Kentucky, water resource issues abound. As Mark Twain is credited with stating so poignantly, "Whiskey is for drinking; water is for fighting over." And fight people do. Whether the culprit is coal mining, timber harvesting, straight piping of sewage, or any number of construction or agricultural activities, discussions of the impact of land use on water quality and quantity in the region are often emotionally—and sometimes politically—fueled. Take, for instance, a study performed at Robinson Forest that created quite a stir from groups such as Kentucky Heartwood, the Kentucky Resources Council, the Sierra Club, and the Kentucky Waterways Alliance. The study's aim was to provide critical information needed to determine the effectiveness of forestry best management practices (BMPs) for eastern Kentucky, but some voiced concern that the potential degradation of water resources from the study outweighed benefits that might be gained by conducting the experiment.

Forestry Best Management Practices

The Commonwealth of Kentucky has established forestry BMPs that are designed to reduce nonpoint-source pollution (NPSP). When asked whether Kentucky's BMPs are sufficient for protecting water resources, our answer has been, "We think so." The reason for the wishy-washy response is twofold. First, few studies have been performed to examine specific BMP guidelines and test their effectiveness. Second, recommendations for many BMPs that are employed in eastern Kentucky were developed from information gathered outside the region. For example, Kentucky forestry BMPs addressing riparian streamside management zones (SMZs) were developed in part from demonstrations in New Hampshire in the 1950s. Given that the forest industry in Kentucky has experienced considerable growth over the past few decades, establishing BMPs specific to eastern Kentucky forests is essential for ensuring the protection and preservation of water resources in the region.

Forested watersheds play an important role in maintaining water quality. Nationally, forests constitute one-third of the land area but provide two-thirds of our water supply. Undisturbed forests have several characteristics that promote high surface-water quality, but forest-harvesting operations can result in negative impacts on water quality. Increases in erosion, litter disturbance, flow duration, nutrient export, temperature, and connectivity between road networks and stream channels have been associated with timber harvesting. Streamside management zones are used to provide a buffer between upland forest-harvesting operations and the stream. The importance of SMZs for filtering erosion, using nutrients, maintaining in-stream and near-stream temperatures, and providing habitat and corridors for aquatic and terrestrial fauna has been identified but not well quantified.

Most states in the Appalachian region have two specifications associated with SMZs—the first related to the distance of the nearest severe disturbance (e.g., roads or log landings) and the second related to the allowable harvest within the SMZ. Also, SMZs vary with the flow duration, or permanence, of the stream. Kentucky's forestry BMPs describe ephemeral streams as those that

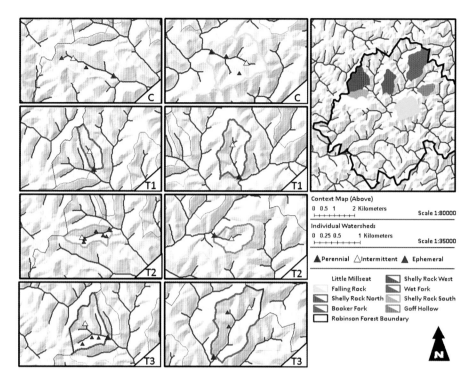

9.1. Study watersheds were located in the Clemons Fork watershed in Robinson Forest. Flow and water quality were measured at perennial (yellow circles) and intermittent (green triangles) streams within each of the treatment and control watersheds. Monitoring of ephemeral channels (red circles) occurred in selected channels that contained improved stream crossings or unimproved fords.

flow during or directly after precipitation or in response to snowmelt and conduct surface water directly or indirectly to perennial streams. Perennial streams are those that flow continuously except in extreme drought conditions, whereas intermittent streams flow primarily during the wet, or nongrowing, season. For perennial streams, the distance to severe disturbance increases as the upland slope increases because of the higher potential of surface-runoff impacts with higher upland grades. Within the SMZ, most states allow some amount of overstory removal. For example, Kentucky allows 50 percent overstory removal. Intermittent streams are not considered to have the same potential NPSP impact as perennial streams, so the distance to disturbance is shorter relative to perennial streams, and 100 percent harvest is allowed within the SMZ. Finally, no SMZ (width or canopy retention) is required for ephemeral streams in Kentucky. Other SMZ requirements vary considerably among

states. For example, North Carolina requires that 75 percent of the trees remain in the perennial and intermittent riparian zone, while West Virginia and Pennsylvania allow 100 percent harvest within the riparian zone on both perennial and intermittent streams. The differences in SMZs among states do not necessarily reflect best available knowledge but are the culmination of battles among forestry groups, environmental groups, and policy makers within each state. The region needs better information on the effectiveness of SMZs. Given these conditions, despite the protest, we moved forward with our study to provide the needed information.

SMZ Study Design

Eight headwater watersheds were included in the study. Each was located in the 3,800-acre Clemons Fork watershed at Robinson Forest (in parts of Breathitt,

9.2. Flumes monitoring discharge from two perennial streams. Storm-flow samplers were located inside the shed.

Knott, and Perry counties), and all were outfitted with a weir or flume to monitor flow continuously (figures 9.1 and 9.2). Watersheds ranged in area from 70 to 275 acres. Water-quality and quantity monitoring began in 2004. Six watersheds were harvested from June 2008 to October 2009. The remaining two watersheds were not harvested to serve as controls. Both control watersheds (Falling Rock Branch and Little Millseat Branch) are listed as exceptional waters by the Commonwealth of Kentucky. Treatment watersheds were harvested using a shelterwood-with-reserves, or two-aged-deferment, harvest method with a target postharvest basal area of approximately 15 square feet per acre (approximately five to ten mature trees were retained on-site per acre). Small–diameter trees (less than 10 inches) occurring in the watersheds were not harvested. Harvesting equipment included wheeled cable and grapple skidders, tracked dozers, and tracked feller-bunchers. Skid trails were constructed along hillslope contours, where feasible, at various intervals from the top to the bottom of slopes. The skid-trail system constituted 6 percent to 12 percent of the watershed area (figure 9.3).

The six harvested watersheds were treated with one of three SMZ combinations (table 9.1). Treatment 1 was based on the Kentucky SMZ guidelines and included a 55-foot perennial SMZ with 50 percent overstory retention and a 25-foot intermittent SMZ with no overstory retention requirement (figure 9.4). Treatment 2 maintained the 55-foot perennial SMZ but required 100 percent canopy retention and 25 percent canopy retention in the 25-foot intermittent SMZ. In addition, improved crossings were used in ephemeral stream crossings, and the nearest channel-bank tree was retained (figure 9.5). Treatment 3 increased the perennial SMZ width to 110 feet with 100 percent canopy retention and the intermittent SMZ width to 55 feet with 25 percent canopy retention and included a 25-foot SMZ around ephemeral streams. The nearest channel-bank tree also was retained, and improved stream crossings were used in the ephemeral streams (figure 9.6). For treatment 1, ephemeral streams were crossed at right angles using unimproved crossings (fords). Improved crossings (elevated over the stream) in treatments 2 and 3 included portable wooden skidder bridges, steel pipes and culverts, and PVC pipe bundles.

9.3. View of the streamside management zone (SMZ) after harvest in three watersheds at Robinson Forest. Note the retention of trees in the valley bottom. Bare soil in lines parallel to the SMZ are roads and skid trails.

	Table 9.1. Whole-Watershed Treatment Combinations Used in the Study						
Treatment	Perennial SMZ width (m)	Perennial canopy retention (%)	Intermittent SMZ width (m)	Intermittent canopy retention (%)	Ephemeral SMZ width (m)	Ephemeral canopy retention (%)	Improved crossings (yes/no)
1*	16.8	50	7.6	0	0	0	No
2	16.8	100	7.6	25	0	Stringer**	Yes
3	33.5	100	16.8	25	7.6	Stringer	Yes

Note: Each treatment was applied to two watersheds.
*Treatment 1 was based on the current Kentucky best management practice regulations.
**A stringer refers to the retention of the overstory tree nearest the channel bank on either side of the stream.

Stream Crossings and Ephemeral SMZs

Stream crossings are generally considered the primary avenue for sediment delivery to streams. Our results showed that the use of any improved crossing type significantly decreased sediment production and transport over a ford in ephemeral streams (figure 9.7). Results also indicated that limiting equipment disturbance on or directly adjacent to the stream channel could result in suspended sediment concentrations similar to those measured in unharvested ephemeral streams. Operationally this can be accomplished by increasing the amount of residual overstory trees left next to ephemeral channels or by restricting the operation of equipment next to channels. However, although limiting equipment operations and ground disturbance around channels can help in reducing total suspended solids, the importance of appropriate crossing selection, construction, maintenance, and removal cannot be overemphasized. The appropriate use of crossings is paramount in limiting

9.4. Treatment 1 was established following requirements outlined in Kentucky's Forestry Best Management Practices. Treatment 1 required a 55-foot perennial SMZ with 50 percent overstory retention and a 25-foot intermittent SMZ with no overstory retention requirement. Note the open canopy and light infiltration.

9.5. Treatment 2 maintained the 55-foot perennial SMZ but required 100 percent canopy retention and 25 percent canopy retention in a 25-foot intermittent SMZ. In addition, improved crossings were used in ephemeral stream crossings, and the nearest channel-bank tree was retained.

9.6. Treatment 3 increased the perennial SMZ width to 110 feet with 100 percent canopy retention and the intermittent SMZ width to 55 feet with 25 percent canopy retention and included a 25-foot SMZ around ephemeral streams. The nearest channel-bank tree also was retained, and improved stream crossings were used in the ephemeral streams.

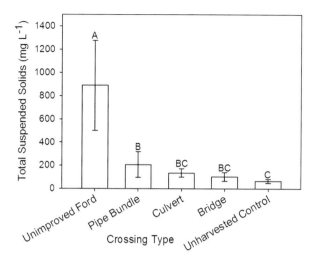

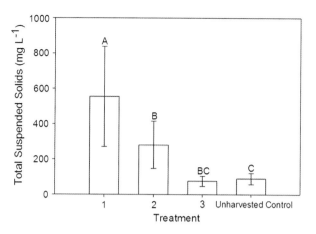

9.7. Average suspended sediment in storm flow from ephemeral streams with differing crossing treatments. Bars with similar letters are not statistically different. Unimproved fords produced significantly greater amounts of sediment than other crossing types that provide an elevated surface for equipment to travel.

9.8. Average suspended sediment in storm flow from perennial streams with differing SMZ treatments. Bars with similar letters are not statistically different. Treatment 1 yielded significantly higher sediment amounts than the other treatments, while treatment 3 exhibited suspended sediment concentrations similar to those observed in the unharvested control watershed.

sediment production, but providing canopy retention around ephemeral channels can also offer thermal protection, maintain coarse woody debris inputs, influence carbon and nitrogen dynamics, and retain some habitat characteristics.

These findings suggest that the extension of forestry BMPs to ephemeral streams is effective in reducing sediment from harvesting operations. In states that already have recommendations for ephemeral-stream protection, mandating improved crossing use for all ephemeral crossings is prudent. When further improvements in sediment reduction are warranted, as would be the case with streams containing flora or fauna particularly sensitive to sedimentation, additional canopy-retention and equipment-limiting-zone recommendations could prove valuable.

Perennial SMZ Effectiveness

Findings from the study showed that the Kentucky guidelines for SMZ width and canopy retention (treatment 1) are just as effective at maintaining nonstorm-sample suspended solid concentrations as treatments 2

or 3, which required increased canopy retention or SMZ width (figure 9.8). However, treatment 1 was found to be significantly less effective at mitigating increases in either suspended solids or turbidity from storm events. Little statistical difference in effectiveness was found between treatment 2 and treatment 3. Treatment 3 was also shown to maintain sediment levels similar to control watersheds (not harvested) in both base-flow and storm-flow conditions. Similar results between treatment 3 and the controls were observed for most parameters examined.

The differences observed between treatment 1 and treatments 2 and 3 are due to use of improved crossings at ephemeral streams and increased canopy retention in perennial, intermittent, and ephemeral segments. Although the exact contribution of improved crossings versus increased canopy retention to sediment reduction at the perennial outlet cannot be determined from these data, the combination of minimizing the hydrologic and sediment connectivity of the skid-trail system and stream network and maximizing the amount of undisturbed forest floor near streams has a definite impact on sediment transport.

Path Forward

The large watershed-scale study proved valuable for meeting our objective to assess BMP effectiveness in eastern Kentucky. Not only were we able to examine water quality and quantity responses to harvesting, but the study design also allowed us to examine many other important aspects of the forest. Ongoing studies include an examination of the influence of these treatments on biota (aquatic insects, salamanders, snakes, birds); an assessment of invasive species occupancy and pathways for colonization; sediment source tracking; and an economic and environmental examination of harvest trafficking patterns. Long-term monitoring will continue, and much more information from the study will be shared with the forestry community in Kentucky and elsewhere.

CHAPTER TEN

The Mighty Elkhorn

Our Home Creek

Zina Merkin

This chapter highlights aspects of water's call to the human spirit from a personal perspective. It raises questions about values, priorities, and policies as we look to the future. Rather than being a strictly disinterested history, this chapter describes events in which I was a participant. It offers certain details but is colored by my personal perspective and recollection.

Conflicts over water and how it should be used are common and will become more so as demands on water grow. As other chapters in this book illustrate, water has a profound influence on human activity. It affects where people travel and settle, what they grow or make, and how they transport those products. In addition to playing a fundamental role in daily life and livelihood, water also is important psychologically and culturally. From Henry David Thoreau's musings near Walden Pond to Mark Twain's stories about the Mississippi, water captures people's imaginations. But in a world of increasing water scarcity, will society be able to allocate water for habitat, for scenic beauty, and for recreation? How will these uses compare to the need for drinking water or the demand for irrigation? What will shape our priorities in the future? This story of Elkhorn Creek in Franklin County tells of people joining forces to preserve access to the creek and illustrates that love of the water, developed through recreation, drives a spirit to protect water and our access to it.

Background

Kentucky's Bluegrass region is known for its lush green landscape and rolling hills. Small waterways are everywhere, along with sinkhole ponds. The city of Lexington lies on top of a limestone dome in the center of the region, and the water drains away in several directions. Pioneer explorers followed the path of Elkhorn Creek southeast from its confluence with the Kentucky River in what is now Frankfort and founded Lexington where several freshwater springs provided an easily protected source of drinking water (figure 10.1).[1] The Middle Fork of the Elkhorn, also known as Town Branch, was the axis around which downtown Lexington was organized. Although it is now buried, Town Branch is still evident in the city's urban grid, which is offset from the cardinal directions, having been laid out with the streets parallel to the course of the creek.[2]

All three forks of the Elkhorn originate in Fayette County. The North Fork of the Elkhorn begins in northeastern Fayette County and winds north and then west. Its watershed covers approximately 276 square miles; some major tributaries are Cane Run, Lanes Run, Goose Creek, and McConnell Run. The South Fork of the Elkhorn begins near the southern border of the county and works its way northwest. This watershed of approximately 179 square miles includes the tributaries of Shannon Run, Steeles Run, and Lee Branch.[3] Town Branch rises just east of downtown Lexington. Sections were covered over throughout the late 1800s as the downtown developed, and the bed of the creek was made part of an underground storm-sewer system in the 1930s.[4] Currently, it daylights west of downtown in the parking lot of Rupp Arena and flows west-northwest. Just outside New Circle Road, another very urban stream, Wolf Run, comes in from the south. Town Branch flows into the South Fork roughly where Fayette, Scott, and Woodford Coun-

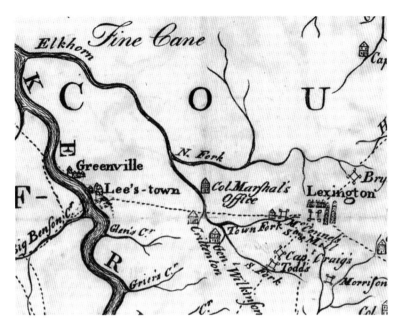

10.1. Lexington and the three forks of the Elkhorn, detail from Filson map of Kentucke, 1784.

ties meet. The South and North Forks come together at the community of Forks of the Elkhorn, in Franklin County, forming Elkhorn Creek proper.

Past and Present Impacts on Elkhorn Creek

Lexington, Georgetown, and Midway are the major cities in the Elkhorn watershed. Streams affected by effluent discharge are Town Branch, South Elkhorn, and the North Elkhorn and two of its tributaries. During heavy storms, there is a problem with combined sewer overflows sending raw sewage into the creeks.

The majority of Lexington's urban area, as well as much of its suburban land, drains to the Elkhorn. The stream is heavily affected by stormwater runoff, including sediment, pathogens, petroleum products, pet waste, and lawn chemicals. Agricultural lands in the watershed contribute sediment, pathogens, excess nutrients, and chemicals as well. The Elkhorn has occasionally been damaged by industrial accidents. One notable incident was a fire at Greenbaum Distillery in Midway in 1908, after which it was reported that "thousands of bass, newlights, perch and other game fish in the South Elkhorn creek, one of the most famous fishing grounds

in this part of the State, have been killed by the burned whiskey which poured into the stream."[5] A fire at the Pepper Distillery on Town Branch in 1934 had whiskey running six inches deep and three feet wide.[6] A paint-factory fire in May 1999 released products that washed into the storm sewer and traveled down Town Branch into the South Fork, creating a fish kill dozens of miles long.[7] Several sections of the creek have been on Kentucky's 303(d) list for nonsupport or partial support of aquatic life or swimming.

A Recreational Resource

The Elkhorn has a long history as a good fishing stream and is stocked by the Kentucky Department of Fish and Wildlife Resources. Smallmouth bass are a favorite target and have been the subject of various studies and management programs. Other species include channel catfish, sunfish, and carp.[8]

In addition to fishing, the Elkhorn is used for canoeing, kayaking, and even a little rafting. It was the first Kentucky stream to have a map published as part of Kentucky's Blue Water Trail program, on April 15, 2010.[9] At normal summer flows, most of it, including the North

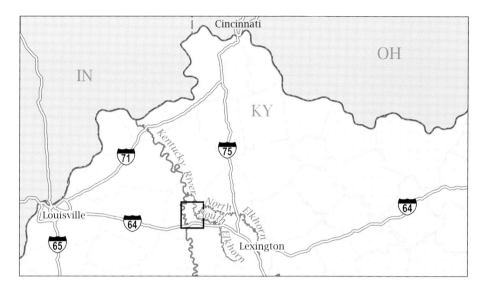

10.2. The Elkhorn is easily accessible from the metropolitan areas of Lexington and Louisville, Kentucky, and Greater Cincinnati, Ohio. Area in box is shown in detail in figure 10.3. Cartography by Jeff Levy.

and South Forks, is either "moving flatwater" or class I whitewater, with an occasional small riffle or shoal. From the Forks of the Elkhorn, the main stem flows northwest, with small riffles and one good play wave at a ledge named Church (across from Buck Run Baptist Church). A long pool leads down to the dam at the Jim Beam distillery. The dam marks the beginning of the gorge section and several miles of playful class II rapids. Although paddlers and fishermen enjoy many sections of the creek, the Elkhorn gorge is the only true whitewater section.[10]

The Elkhorn is a tremendous recreational resource because the size of its watershed yields more reliable and predictable flows than other local whitewater streams. Thus it is affectionately called the Mighty Elkhorn by local paddlers, even though its rapids are not intimidating. Almost halfway between the two large metropolitan areas of Louisville and Lexington, and less than ninety miles from Cincinnati, the whitewater section is within a two-hour drive of about four million people.[11] It can be reached in about fifteen minutes from either Frankfort exit from Interstate 64 (figure 10.2). The Elkhorn gorge draws whitewater paddlers from Ohio, Indiana, and most of Kentucky, and the easy whitewater provides a great training ground for beginners. The Bluegrass Wildwater Association (BWA), based in Lexington, and the

Viking Canoe Club in Louisville are whitewater clubs that consider the Elkhorn their "home stream," as do the Elkhorn Paddlers, who generally paddle more flatwater. All three clubs are affiliates of American Whitewater, a national organization of whitewater paddlers.

The Kentucky Department of Fish and Wildlife Resources gives pointers for fishermen on the Blue Water Trail web page for the Elkhorn and notes that "floaters may use property owned by American Whitewater as a take-out. . . . The American Whitewater property, located downstream and to the left of the bridge, has a parking lot and changing stations. This access is for boaters only—it is not for wade fishing access" (figure 10.3).[12] That access spot has been available for only a little over a decade, and the fact that it is published is indicative of changing attitudes in Franklin County toward water-based recreation. But before that story, a little more should be said about the evolution of paddle sports in the United States and locally.

Paddle Sports

Paddling, often associated with summer camp or Scouting, has never been a common form of recreation. Availability of equipment, cost of entry to the sport, and

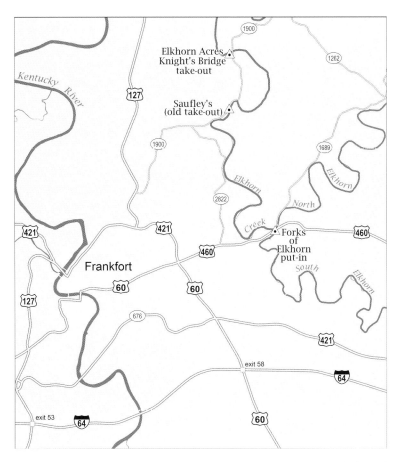

10.3. Detail of the Elkhorn Creek whitewater section put-in and take-out locations. Cartography by Jeff Levy.

access to suitable locations are all reasons that paddling is not more prevalent. Developments in technology have contributed to the growth of the paddling sports by reducing cost, making equipment more durable, and allowing advanced design features that improve the performance of boats for a variety of paddling activities. Grumman's aluminum-forming processes enabled the construction of sturdy canoes that were relatively affordable starting in the late 1940s. Kayaking for recreation was still in its infancy then, and there were few opportunities to purchase a kayak.

When fiberglass became available, more people were able to enter the sport because they could build their own boats relatively easily and cheaply. But fiberglass is brittle, and there was a constant risk of breaking a boat in the middle of a run. No one undertook a trip without

an ample supply of duct tape for short-term repairs, and many hours were spent between trips repairing equipment. There was a certain camaraderie or community created by getting together to fix one another's boats and share design ideas.[13]

The introduction of plastic rotational molding to boat building in the early 1970s revolutionized boat design. The plastic boats essentially bounced off rocks, could be shaped in a variety of ways, were relatively lightweight, and could be made in a rainbow of colors. The plastic boats could not compete in lightness with composite fiber and resin boats but were less expensive and more durable. As boats became more available commercially, the sport was opened to many more people.[14]

Roughly coinciding with the availability of more durable boats was the release of the movie *Deliverance* in

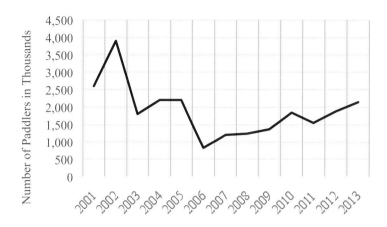

10.4. U.S. whitewater kayaking participation from 2001 to 2013, data from Outdoor Industry Association and Outdoor Foundation.

1972, starring Burt Reynolds. Although the paddling scenes are not very believable and show poor technique, the film did introduce large numbers of people to the excitement of whitewater and the beauty of the places it can take you. Local paddlers noticed that the number of boaters on the Elkhorn increased noticeably after the movie came out. Also in 1972, whitewater slalom became an Olympic sport for the first time and received national television coverage.[15]

American Whitewater and Paddling Demographics

American Whitewater (AW) began in 1954 as the American Whitewater Affiliation, a loose confederation of whitewater clubs across the country. It incorporated as a nonprofit organization in 1961, with a mission to "encourage the exploration, enjoyment, and preservation of America's recreational waterways for human-powered craft; protect the wilderness character of waterways through conservation of water, forests, parks, wildlife, and related resources; promote and celebrate safety, proficiency and responsibility in all aspects of whitewater activities . . . ; [and] promote appreciation and respect for the value of wilderness activity and whitewater sports."[16]

There are not a lot of whitewater paddlers compared with such outdoor activities as camping and hiking, although the average income of paddlers is higher than the average for participants in all outdoor activities.[17] Paddlers are well educated, too. Over 70 percent of *American Whitewater Journal* readers have a bachelor's degree or higher, and a (NSRE) survey indicated that among kayakers (all forms), 43 percent had a bachelor's degree or higher, compared with a level among all respondents of 24.7 percent.[18]

The Outdoor Industry Association began tracking types of kayaking only in 2001. Participation levels are affected by weather, especially on natural-flow rivers, and by economic factors. There is much variation, some of which is statistical error due to relatively small sample sizes. In regard to frequency of paddling, in 2001, 17 percent of whitewater paddlers went at least 11 times. In 2002, such avid participation rose to 21 percent of participants, but in 2004 only 3 percent paddled that many times. Overall participation estimates from 2001 to 2013 can be seen in figure 10.4.[19]

In 1970, it was estimated that 50,000 canoes were being sold annually, and that 4 million people went canoeing. The combined membership of paddling associations at that time, including AW, was only about 2,000. AW has more than 100 affiliates with club memberships, but its 2013 media kit listed a journal readership of "over 6,000 members," less than 0.5 percent of the estimated 1.88 million who went whitewater paddling in 2012.[20]

This apparent failure of paddlers to join an organization that represents their interests is puzzling. Paddlers rely on one another for information (known as beta) about the rivers and their hazards, and safety standards suggest that paddlers travel in groups of at least three people. But it is also a caveat that any paddler is ultimately on his or her own on the river. There are times when even the best-trained and most skilled boaters are unable to perform a rescue, especially in difficult rapids and among hazards such as undercut rocks, pinning spots, and strainers. Perhaps people who embrace whitewater are more independent and self-reliant than the average. Maybe people who like to plan ahead do not take well to a sport where you have to be ready to change destinations at a moment's notice to find the right water levels. But whitewater paddlers need one another even off the river; with water resources under growing pressure, numbers are important in influencing policy decisions. Sometimes organizing whitewater boaters seems difficult, but it can be necessary.

The Access Problem on the Elkhorn

People have been fishing in and paddling down the Elkhorn for many decades, but development of the creek as a recreational resource with a regional draw is a more recent phenomenon. In the mid-1970s, Viking Canoes and Sage School of the Outdoors in Lexington were offering classes and sponsoring trips. A core group of people who taught or took the Sage courses founded the Bluegrass Wildwater Association in Lexington in 1976, and the club affiliated with AW a few years later.[21]

In the winter of 1981, the BWA pool sessions for teaching and practicing Eskimo rolls were profiled in the *Lexington Herald-Leader,* and articles about recreational opportunities for canoeing and kayaking popped up occasionally during the 1980s.[22] Also in the 1980s, impacts of failed package-sewage plants from Lexington suburbs, several fish-kill incidents, including a toxic leak from a plating plant, and concerns about copper and lead releases from the Town Branch municipal sewage plant were in the news.[23] In 1982, fish and wildlife officials had noticed that there was a problem with the size of the smallmouth bass in the Elkhorn. In late 1983, the Kentucky Division of Water called a public hearing about upgrading the status of the North Fork and the main stem of the Elkhorn to "outstanding resource

waters," which would add some protections by further restricting what could be discharged into the stream. Both the Sierra Club and the BWA had representatives at the hearing, who commented that there was a need not only for stricter standards but also for better enforcement.[24] Increasing concern about water quality seemed to go hand in hand with increased interest in water-based recreation.

In 1987, the *Lexington Herald-Leader* business section ran a feature article about several outfitters who rented canoes on the Elkhorn, including Ed Councill, whose fleet began with six canoes purchased for the use of a Boy Scout troop he had led in 1979.[25] As Councill began to build up his business, he became concerned about increasing pressures on the creek from urban and suburban development, as well as industry. A planner by training, he helped found the Elkhorn Land and Historic Trust (ELHT), a nonprofit group made up of landowners, recreationists, environmentalists, and others with an interest in the protection of the stream. The ELHT was incorporated in 1989, and its membership covered Fayette, Woodford, Scott, and Franklin Counties. Members wrote grants and raised money for environmental education programs and development of creek access and recreational trails. Their efforts were most fruitful in Scott County.[26]

Councill's involvement in the ELHT came from the same love of water as his involvement in the canoe and kayak business, and the two efforts complemented each other. Protecting water quality and securing creek access would allow his business to grow, and paddlers could be a source of advocates for the creek. At the end of 1991, an agreement was reached among the elected officials of the four counties of Fayette, Franklin, Scott, and Woodford to form the Elkhorn Interlocal Planning Consortium.[27] This body was tasked to study the entire creek corridor and make a plan for protection of the creek and management of the considerable demands and impacts on it. In cooperation with this group, the ELHT worked with a consultant to produce the Elkhorn Creek Action Plan, comprising a resource assessment and suggested action items for improving water quality and fish and wildlife habitat; developing recreational opportunities and minimizing their impact; preserving and sharing historic resources; and mitigating agricultural impacts. A draft of this plan (the final report could not be located) stated, "Because of the lack of legitimate, managed public access points for boating, fishing, picnicking and hiking, increasing recreational use is resulting in trespassing, littering, and other conflicts with private property rights. At the same time, the growing regional emphasis on tourism and protection of our special Bluegrass landscape and way of life has given Elkhorn Creek the kind of attention it needs if these problems are to be resolved."[28]

This plan identified criteria for choosing canoe access sites and suggested several possible sites throughout the corridor. Knight's Bridge on Peaks Mill Road, about six miles upstream from Councill's canoe livery and about seven miles downstream from Forks of the Elkhorn, was identified as a priority location that could accommodate whitewater paddlers taking off the creek, as well as those launching for more sedate trips on the lower section. Councill bought the house right next to the bridge after the 1997 flood prompted the owners to put it on the market, using the property as a put-in for Canoe Kentucky customers.[29]

Both water quality and fishing slowly improved, and *Lexington Herald-Leader* outdoors writer Art Lander wrote a number of pieces praising the Elkhorn in the 1990s.[30] Paddle sports continued to grow in popularity, and whitewater in particular received some extra publicity from the 1996 Summer Olympics in Atlanta. Another notable influence was the development and spread of Internet technology. E-mail, websites, and cell phones made communication among paddlers quick and easy, and U.S. Geological Survey (USGS) river gauges became accessible online, all enabling paddlers to better predict where the water would be the most fun and to find out about runs in different areas.

A Very Real Threat

In the spring of 1997, it seemed as if more paddlers were on the creeks than ever. There was also a historic flood in early March.[31] Creekside landowners had to deal not only with problems caused by flooding and frequent high water but also with boaters clogging the narrow roads, trespassing, and perhaps being a bit noisy as they paddled past the backyards. The take-out for the Elkhorn Gorge had been across Peaks Mill Road from Zack and Yvonne Saufley's farm for many years. The road comes right up to the creek, with a wide shoulder next to the pool at the bottom of the last real rapid. There was not a lot of room, only enough to park half a dozen cars or so. At times the Saufleys had been quite friendly to various members of the paddling community, allowing creek cleanups to be staged from their place. They had even hosted a meeting or two in the early years of the BWA.[32] But paddlers were becoming more and more of a nuisance, and the Saufleys were no longer amenable to having paddlers cross their land.

By the 1998 paddling season, matters had come to a head. Not only were more paddlers trying to fit into that space, but also the character of the sport was changing. One arc of whitewater boating followed the general trend toward extreme sports, more of a rock-and-roll mentality than a back-to-nature focus. Another trend was rodeo boating, a gymnastic style of boating in which paddlers in short and highly maneuverable boats throw themselves into steep waves, turn cartwheels, and flip and spin the boat in dizzying combinations. These paddlers sometimes drive up to a good spot on a river, park, play for an hour or two, and leave again. Although there is crossover, one can speculate that this style of paddling may not develop the same networks that downriver paddling tends to, where paddlers are cooperating to shuttle boats and traveling downriver together. In the early years, everyone knew one another and understood the tacit arrangements with local landowners. They were somewhat respectful of the peace and quiet of the creekside landowners. With the sport growing so rapidly, it was not always possible to draw all the new paddlers into the social norms of the area.

A local boating-club member and representative of American Whitewater commented in a 1998 e-mail:

The explosion of ww paddlers is a recent phenomenon. There were many of us that predicted it would happen after the 1996 Olympics at the Ocoee—it's now become reality. What was for years 4–5 cars parked at the takeout on a sunny spring day with the Elkhorn at 1.5' has turned into 10–15 cars. What was once 8–10 reasonably behaved paddlers who knew the Saufley's [sic] sentiments and were sensitive to them has turned into 15–25 unruly . . . paddlers that have no regard for anything but themselves. It's a Clockwork Orange world and I'm deeply disturbed to see it arrive in paddledom. I must admit that even though I saw it coming I still can't believe the speed with which it has appeared.

In the spring of 1998, the gravel area next to the pool was often full, and people began to park on the narrow shoulder going back up toward the crest of the hill. This was dangerous because there was a large blind spot for drivers headed north over the hill. Crowding was made worse by people hanging around to talk after getting off the river, rather than freeing up the spots for other paddlers, and just jockeying to load up boats sometimes blocked traffic. Some people were less than discreet when they were changing out of their wet clothes, and the Saufleys and other neighbors found this offensive.

One day there were No Parking signs along the roadside. People started to get tickets. Some were even towed. Then the chain-link fence defining the property line was moved closer to the road. The space that had been inadequate for the growing number of paddlers was reduced even further. The BWA newsletter for July–August 1998 had an article detailing possible courses of action for protecting the existing take-out; paddlers were quite concerned about losing access to the creek. The Elkhorn Access Committee, with representatives from three local boating clubs, was formed to research possible solutions.[33]

At the local property-valuation office, paddlers looked up the ownership of any parcels near the end of the whitewater section, especially where the road came near the creek. Few suitable parcels even existed between the Saufleys and the bridge, since the creek flows away from the road in a large oxbow just after the big pool (see figure 10.3). Boaters with friends in Franklin County asked around to see whether property owners had any interest in allowing creek access. They spoke with local and state officials to explain that the Elkhorn brought visitors to Frankfort not only from Lexington and Louisville but also from Ohio, Indiana, and even farther, hoping somehow that money from Fish and Wildlife, Kentucky State Parks, or somewhere could be found to provide access. But without a willing landowner, the situation looked grim.

Then some land became available over a mile downstream from the end of the rapids, next to Knight's Bridge

on Peaks Mill Road, the location recommended in the 1992 Elkhorn Action Plan. Three long, narrow lots of about 1.5 acres each, all entirely in the hundred-year floodplain, with no utilities, were for sale at the high price of $30,000 each.[34]

A Possible Solution

Ed Councill contacted the Access Committee to let it know that the ELHT hoped to submit a grant for Recreational Trails funds to purchase and develop some of that land for river access and a trailhead for rural road biking.[35] The location was not ideal for whitewater paddlers because it necessitated more than a mile of downriver paddling in slow current, especially in low-water conditions. The 1992 Action Plan suggested the Knight's Bridge location for development of "secure off-road parking, a picnic area, toilets, changing facilities and a canoe launch . . . easing pressures on the Saufley's Bottom site," noting that the Saufley site was "heavily used and basically unsuitable."[36] Because the Action Plan had been endorsed by the Elkhorn Interlocal Consortium and adopted into the Franklin County Comprehensive Plan, Councill felt that there was a good chance to get grant money for development of a takeout.

Paddlers in the BWA had mixed reactions to the news of this grant proposal. Councill was one of the public faces of the ELHT, but as an outfitter, he was seen by local landowners as creating a nuisance and diminishing their quiet enjoyment of their creekside property. Paddlers were concerned about the possibility of getting embroiled in any preexisting conflict involving Councill. Some boaters were not sure about the intentions of the livery owner. The Access Committee contacted the access director for American Whitewater, Jason Robertson, to find out whether AW could help with matching funds for the grant, and for general advice. Although no funds were available, AW was supportive of the idea. Many of the local paddlers did support the proposal, especially those whose interests crossed over from paddling to bicycling.

The trust's proposal called for a combination trailhead for paddling and bicycling, along with an environmental education component. Provisions of the grant required that a local unit of government submit the application, and to the chagrin of local landowners who opposed it, the Franklin County Fiscal Court approved the proposal and signed on to the application. One neighbor to the property publicly voiced concern that there would be increased trespassing, litter, and traffic.[37] The grant was submitted on April 9, 1999, to the Department for Local Government (DLG), the state office tasked with reviewing funding proposals for the Recreational Trails program.

In addition to being on the board of the Elkhorn Trust, Ed Councill was on the review committee for Recreational Trail grants. This committee made recommendations, but the final decision was up to the head of the DLG and the governor. The ELHT sent an open letter to its membership, asking people to write to Governor Paul E. Patton or DLG commissioner Bob Arnold in support of the Knight's Bridge plan, to counteract the vocal opponents. This letter noted that DLG staff had ranked the proposal fifth of fifty-eight during their initial review (fax to author, July 12, 1999).

A fax from the DLG dated August 13, 1999, contained the final project list. The cover sheet had the terse statement "Note changes. Not funded: 1) Wilmore Farm Park 2) Knight's Bridge Landing." There had been more proposals than funding could support. Although the advisory board had unanimously recommended the ELHT's plan, the DLG decided to fund other efforts instead. A post on the BWA listserv (August 16, 1999) speculated that certain landowners had persuaded the commissioner to reject the project.

The grant had fallen through, but the land at Knight's Bridge was still on the market, and the ELHT floated the idea of seeking donations for the land purchase. Local boaters, who included a board member of AW, started talking about the possibility of a paddler initiative to buy the land. A few paddlers still supported the idea of a public park and even a demonstration project of best management practices for development of streamside recreation facilities. They figured that there would be more public support if the site could be used by bicyclists, picnickers, casual floaters, and fishermen than if it were only for whitewater access. They reasoned that exclusion of other user groups at this location might provoke such groups in other places to refuse to share access with whitewater paddlers. Some paddlers were

sensitive, too, that any problems on this site would reflect badly on future attempts to gain access to other sections of the Elkhorn or other creeks in the area.

Many people just wanted to ensure that whitewater paddling access was protected. They thought that having an agency in control of the property could be counter to paddlers' interests. But during the dry months of summer, the issue did not seem to be urgent.

A Crisis

People continued to talk sporadically about the need for dedicated access but were somewhat distracted by the traditional fall paddling releases on the Gauley and Russell Fork Rivers. One late fall day, a sign appeared on the lots advertising an absolute auction in thirty days. Those lots had sat unsold for almost two years, but suddenly, time was of the essence. Two groups of paddlers, both led by members of the BWA, started fundraising in parallel, at first without knowledge of each other. One group looked into buying the land and holding it as a limited liability corporation for the private use of whitewater boaters. It was not clear at first what this might mean in terms of legal exposure, how to control access, and other issues, but many boaters wanted control of the site in paddlers' hands. They were frustrated by public access points in other areas where various restrictions, including the prohibition of alcohol, made it difficult to enjoy socializing and unwinding with friends after getting off the water.

The other group approached a Lexington-based nonprofit group, Friends of the Parks, to see whether it could accept tax-deductible donations for the project, with the idea of donating the land to the Franklin County Parks system or the Kentucky Department of Fish and Wildlife Resources. Friends of the Parks was willing to be the umbrella organization, but although initial contacts with the agencies were promising, neither was prepared to take on the project.

The Strategy Develops

As the date of the auction quickly approached, the two groups of boaters realized that to bid against each other would be foolish. In earlier discussions with AW, the Elkhorn Access Committee had learned that the organization could take title to land, provide liability coverage under its blanket insurance policy, and set up an account to process tax-deductible donations (e-mail from access director of AW to author, August 19, 1999). AW did want to make sure that there were local people committed to day-to-day management, such as mowing and other maintenance work. Everyone agreed that AW should take title to the land. A meeting of donors was arranged to hammer out an agreement for a local management group.

Opinions about how to handle the property were strong and somewhat divergent, but the potential for losing access overrode those concerns. The group decided that management would be handled by a committee made up of those donating over $500, an amount selected as a proxy for the degree of commitment to the project and as an incentive to give generously. Each $500 share gave the donor a vote, with a cap of two votes for any individual. In a surprise move, the person selling the land pledged a significant amount toward the boaters' purchase, giving him a say in its management. The organizers fanned out, soliciting more donations from the three local paddling groups and their members, as well as paddlers unaffiliated with any group. In less than a week they raised over $32,000. The three local paddling clubs all contributed, there were thirty-three "voting" donors, and there were numerous smaller contributions as well. Several people also pledged to loan money, bringing available capital to approximately $40,000.

A few individuals had heard rumors that landowners who were against public access had been soliciting donations to buy the property. These opponents had no intention of letting paddlers purchase the lots. Whoever would bid on behalf of the boaters should not be recognized as a boater. A club member who seldom kayaked anymore happened to have a business distributing log-home kits. He agreed to bid on the land. The plan was that some of the better-known paddlers would also bid but would drop out after a short time. They reasoned that no one would have trouble believing that the paddling community was not able to raise much money. It was hoped that once these boaters gave up, whoever was trying to block the boaters would quit bidding and let some-

one else take it. They hoped that their agent would be that someone.

The Auction

The paddling representatives and the bidding agent arrived at the auction separately. As the auction started, the boaters did not have to feign shock and surprise at being outbid—the very first bid came in at $20,000, above their agreed maximum. Since this was an absolute auction, without reserve, they had expected the bidding to start low. They could only sit back and watch, hoping that the high start did not doom their chances. The bids went back and forth a few times, but the gavel fell for the last time at $29,000, to the man in the Jim Barna Log Homes hat, who was bidding for the paddling community.[38]

The paddlers thought it better if it took a few days for the landowners to realize that the paddling community had actually purchased the land. The BWA member who was on the board of AW quietly went over to the auctioneer and wrote out a check for 10 percent of the purchase price. Thirty days later, closing papers were signed, and AW took title to the land, adding it to its small portfolio of access points across the country. Having secured the property, the local boaters faced the new challenge of developing and managing it.

Development

The property was zoned rural-residential, a use under which public parks and boat ramps are prohibited. Although the ELHT had stated that its recommendations for a canoe access at Knight's Bridge had been adopted into the comprehensive plan, some on the management committee did not feel that that protected the group from a zoning action. They decided to move slowly in developing the property to avoid any legal challenge from local landowners. Canoe Kentucky was fighting a lawsuit for launching customers at its Knight's Bridge property, although it eventually won that suit.

During that winter's paddling season, people just drove onto the property and parked. In the spring of 2000, a simple driveway and a small parking area were graded and graveled. The following winter, the management group purchased a "wildlife" pack of one hundred bare-root native tree seedlings from the state Division of Forestry, as well as several dozen one-gallon trees to plant in two staggered rows along the long property line. As a neighborly gesture, the group hired one of the neighbors to mow the property. It also bought some shrubs from another neighbor to plant near the boundary of the Canoe Kentucky property. Species planted included sycamore, red maple, river birch, green ash, redosier dogwood, swamp white oak, and burr oak. Many of the small seedlings were planted in the back corner of the property as a sort of nursery, but the volunteer labor to transplant them did not get organized, and they were overrun with weeds or crowded one another out. The plantings along the property edges have grown nicely, however.

The design of a changing room was debated for over a year, with some concern about zoning regulations. A few people grew impatient and went out and built a structure one day in the fall of 2001. It had certain design flaws but has served its purpose, although after enduring several floods, it is a bit the worse for wear. Similarly, the driveway has needed regrading periodically. It took a few years to decide to place a portable toilet on the property and to figure out which months it was really needed. A few times there have been frantic phone calls to get the toilet removed before floodwaters got to the property, and one time a boater who lived nearby arrived in the nick of time to chain it to a tree so it would not float down the creek.

There have been some disagreements about how much to develop the site and about rules for use. Some people wanted to be able to camp there, but others were concerned that that might encourage local people to do the same, or that it would upset neighbors. There were different attitudes about whether to let people fish at the property, especially since wading fishermen had caused damage at the downstream neighbor's property. The management structure based on who had donated proved unwieldy. Many people felt that someone's commitment to access should not be measured by his or her financial capacity. Also, the finite set of volunteers that this represented was slowly being depleted as people moved or found other interests. A positive feature of this group, however, was that it had representatives from all the local paddling clubs, which facilitated communication about management and development issues. There were some heated discussions about how to transition to a new structure, and what principles would guide property management in the future.

AW executive director Mark Singleton facilitated discussions among representatives of the local clubs but made it clear that local interests had to drive property management. AW formalized guidelines for access properties in 2008, which went a long way toward clarifying a management direction for local volunteers. All parties recognized both AW's rights as the property holder and the organization's stake in maintaining a good reputation in order to protect future access opportunities.

Considerations for the Long Term

The purchase of the Elkhorn property has been a great success in meeting a need for safe, off-road parking and access. On warm spring days after a good rain, there easily can be sixty cars crowded onto the site. As predicted, it is used both by whitewater paddlers as a take-out and flatwater paddlers as a put-in. There have been some problems with fishermen wading from the site onto the neighbor's property, some littering, and local teens parking and partying, but generally the neighbors are not annoyed too often. Relations with other landowners calmed down as they realized that the site solved traffic and safety issues.

The differences in management goals among users of the property have caused some struggles. When people are passionate about a vision or cause, it can be hard to work out compromises, something for which any volunteer organization needs to be prepared. One might argue that the decisiveness and assertiveness that whitewater boaters need to negotiate difficult rapids may exacerbate this. Although the property has not yet become a model of best environmental practices, it has been consistently maintained.

Currently the local clubs elect or appoint people to an informal council that handles routine maintenance and project development in communication with both AW staff and club membership. Outreach to unaffiliated paddlers occurs informally, in conversations on the water or at the site. Although soliciting financial support from boating clubs and organizations from Ohio or Indiana who use the site has been discussed, to date there has not been a significant attempt to do so. New people keep stepping up with ideas and energy, although some people prefer a minimalist approach of just parking and a portable toilet, while others would like a more park-like site. To some degree, it comes down to whether people think that there will be volunteers to do the increased work that comes with a more complex site plan.

The paddlers accomplished their ultimate goal of providing dedicated paddling access. They were able to organize swiftly when it counted. The threat to something so important to them spurred them to action. But the very speed that was necessary at the moment led to problems later on, as issues surfaced on which the various parties did not have consensus. The most serious issue that was not addressed at the time was how to ensure a constant supply of new local leadership to manage the property. A parallel concern is making sure that a coalition among paddling interests continues to have influence over the property, rather than it becoming the province of a single club or group of paddlers and therefore having a smaller base of support.

The role of AW, setting broad property-management guidelines and requesting participation from affiliated clubs in the region, has helped local paddlers make the transition to a more sustainable process for taking care of the Elkhorn Acres takeout. One lesson to draw from this story is that having a larger umbrella organization involved can bring important experience and perspective to a local initiative. AW provides financial and other support but also is a neutral voice outside the inevitable personality conflicts that any local organization will experience. Its directive to keep regional boating clubs involved in the property is important. Good communication among regional paddlers over this site creates a foundation that positions paddlers to be able to respond to any other threat or opportunity in the area.

American Whitewater stewardship director Kevin Colburn once told me that the organization chose which access or conservation efforts to support by where there was passion "on the ground" for the project. Without strong local enthusiasm, the national organization can accomplish nothing. Water has a way of generating passion, and it is this passion for water, whether in playful waves or quiet pools, that has helped protect the Elkhorn. However, the broader vision of AW is crucial as well. A structure to channel the passion is vital to sustaining a volunteer effort.

Epilogue

It is 2016, and a few new people have stepped up to propose expanding the parking area and keeping cars a little farther from the stream edge, regrading the driveway, and planting some new trees to shade the parking. A BWA member who is a landscape architect drew up a plan which the current conservation officer is working to implement. Other people still want to keep things as simple as possible. Ed Councill has retired from Canoe Kentucky, but his daughter and son-in-law continue to operate the business. Councill remains active in environmental education efforts, although the Elkhorn Land and Historic Trust was dissolved in 2005. The neighbor who had put the land up for auction moved to Florida in 2013, and the next owner not only continued to have problems with fishermen wading or anchoring in the eddy next to the house, but also was quite unfriendly to paddlers using Elkhorn Acres. The property sold again in the summer of 2016, and the paddlers have not met the new owners yet.

A son has inherited the Saufleys' property and continues to farm. He placed the former take-out spot into the Voluntary Public Access program instituted by the State Department of Fish and Wildlife Resources in 2010. Especially in winter, with its colder, shorter days, most paddlers choose to avoid the extra half-hour paddle to the AW property when there is room at the Saufleys' site. Still, on warm days with good water levels, AW Elkhorn Acres is quite popular, and cars fill the designated gravel parking area and well beyond it. The problem of people parking along the shoulder has returned, and paddlers who experienced the site being shut down in the 1990s are concerned that this access point will be shut again. The property is also used by local paddlers for occasional club meetings, races, and staging of creek cleanups.

Kentucky is fortunate in having an abundance of water resources compared with many places. Governor Steven L. Beshear's administration has emphasized Adventure Tourism as part of the economic development strategy for this largely rural state and has included Blue Water Trails as destinations. Outdoor recreation can develop familiarity with and love of a variety of natural settings. Especially when that recreation involves close contact with water, as paddling does, it can create a demand to better protect water resources. But as water resources become more contested, will the economic contribution of water-based recreation, the goal of habitat preservation, and the simple love of water be enough to keep our streams flowing? We have to continue to actively develop an even broader network of passionate advocates to protect streams into the future.

Notes

1. Gary O'Dell, "Water Supply and the Early Development of Lexington, Kentucky," *Filson Club History Quarterly* 67, no. 4 (1993): 431–61.

2. Clay Lancaster, *Vestiges of the Venerable City: A Chronicle of Lexington, Kentucky, Its Architectural Development and Survey of Its Early Streets and Antiquities* (Lexington, KY: Lexington–Fayette County Historic Commission, 1978), 9.

3. Kentucky Water Resources Research Institute (KWRI), "Kentucky River Basin Management Plan" (Lexington: Kentucky River Authority, 2002).

4. "Historic Town Branch Now Flows in Underground, Parallel Tunnels," *Lexington Leader,* April 12, 1935, 1.

5. "Fish Killed," *Lexington Leader,* August 14, 1908, 3.

6. "15,500 Barrels of Liquor Destroyed," *Lexington Herald,* April 28, 1934, 1.

7. "Oil from Paint Spill Threatens Fish, Livestock," *Lexington Herald-Leader,* May 16, 1999, B-1.

8. Christy J. Gunderson VanArnum, Gerard L. Buynak, and Jeffrey R. Ross, "Movement of Smallmouth Bass in Elkhorn Creek, Kentucky," *North American Journal of Fisheries Management* 24 (2004): 311.

9. Kentucky Department of Fish and Wildlife Resources, "Blue Water Trail: Elkhorn Creek," http://fw.ky.gov/Education /Documents/bluewatertrail_elkhorncreek.pdf (accessed May 10, 2014).

10. Class I rapids are "fast moving water with riffles and small waves. Few obstructions, all obvious and easily missed with little training. Risk to swimmers is slight; self-rescue is easy." Class II rapids are "straightforward rapids with wide, clear channels which are evident without scouting. Occasional maneuvering may be required, but rocks and medium-sized waves are easily missed by trained paddlers. Swimmers are seldom injured and group assistance, while helpful, is seldom needed. Rapids that are at the upper end of this difficulty range are designated 'Class II+.'" American Whitewater, "Safety Code of American Whitewater," adopted 1959, revised 2005, http://www.americanwhitewater.org/content/Wiki/safety: start?#vi (accessed May 10, 2014).

11. U.S. Census Bureau, "Population Change for Counties in the United States: 2000 to 2010," 2010 Census Population and Housing Table CPH-T-1, http://www.census.gov/population /www/cen2010/cph-t/cph-t-1.html.

12. Kentucky Department of Fish and Wildlife Resources, "Elkhorn Creek," http://fw.ky.gov/Education/Pages/Elkhorn -Creek.aspx (accessed May 10, 2014).

13. Susan L. Taft, *The River Chasers: A History of American Whitewater Paddling* (Mukilteo, WA: Flowing Water Press and Alpen Books Press, 2001); Don Spangler, longtime BWA member, personal communication, February 4, 2012.

14. Taft, *River Chasers.*

15. Don Spangler, personal communication; Jim Jordan, "Outfitters Rolling on River of Success," *Lexington Herald-Leader,* August 3, 1987; Jon Nelson, "Whitewater Slalom in the Olympics," http://whitewaterslalom.us/whitewater-in-the -olympics.html.

16. American Whitewater, "About AW," http://www .americanwhitewater.org/content/Wiki/aw:about/ (accessed July 24, 2013).

17. Leisure Trends Group, *Outdoor Recreation Participation Study, Seventh Edition for Year 2004, Trend Analysis for the United States* (Boulder, CO: Outdoor Industry Foundation, 2005), 45, 250.

18. American Whitewater, "2013 Media Kit," http://www .americanwhitewater.org/resources/repository/2013 _American_Whitewater_Media_Kit.pdf (accessed July 24, 2013); American Canoe Association, "NSRE Paddlesports Participation Report," 2010, http://c.ymcdn.com/sites/www .americancanoe.org/resource/resmgr/general-documents /nsre-paddlesports-participat.pdf (accessed May 10, 2014).

19. Leisure Trends Group, *Outdoor Recreation Participation Study for the United States, 2004,* 6th ed. (Boulder, CO: Outdoor Industry Association, 2004), 212; Leisure Trends Group, *Outdoor Recreation Participation Study, Seventh Edition for Year 2004,* 247; Leisure Trends Group, *Outdoor Recreation Participation Study, Eighth Edition for Year 2005, Trend Analysis for the United States* (Boulder, CO: Outdoor Industry Foundation, 2006), 208; Outdoor Foundation, *Outdoor Participation Report, 2013* (Boulder, CO: Outdoor Foundation, 2013), 58.

20. Taft, *River Chasers,* 177; American Whitewater, "2013 Media Kit"; Outdoor Foundation, *Outdoor Participation Report, 2013,* 58.

21. Don Spangler, personal communication.

22. Sharon Reynolds, "Kayakers Getting Ready to Roll 'Em," *Lexington Herald,* February 23, 1981, B1; Beverly Fortune, "Moonlight Canoe Trip a Splendid Adventure," *Lexington Herald-Leader,* June 3, 1988, B1; "Floods of Water Fun Set for June Weekends," *Lexington Herald-Leader,* June 5, 1987, B9.

23. Mike Mayhan, "Sewage Plants Cited for Dumping Sewage in South Elkhorn Creek," *Lexington Herald-Leader,* December 1, 1989, C1; "Lexington Landfill Cited for Violation," *Lexington Herald-Leader,* July 20, 1989, B2; "Lexington Plant on List of Chief Water Polluters," *Lexington Herald-Leader,* July 7, 1988, A1; Jacqueline Duke, "Storm Drainage Improvements Are Proposed: Six-Year, $9.9 Million Project Would Upgrade Sewer System," *Lexington Herald-Leader,* April 21, 1984, B1; Merlene Davis, "More Regulations Needed for Streams, Residents Say," *Lexington Herald-Leader,* December 16, 1983, B2; "Plating Plant Is Likely Cause of Creek Poison," *Lexington Herald,* June 10, 1982, A3.

24. Davis, "More Regulations Needed for Streams, Residents Say."

25. Jim Jordan, "Outfitters Rolling on River of Success."

26. Beth Stewart, Gregory K. Johnson, and Doug Hines, "Citizens Take the Lead: Elkhorn Creek Watershed Planning and Action through Consensus," in *Proceedings: Watershed '93; A National Conference on Watershed Management*, March 21–24, 1993 Alexandria, Virginia, ed. Janet Pawlukiewicz et al. (Alexandria, VA: U.S. EPA, 1993), 249–52; Ed Councill, personal communication.

27. "Pact Signed to Preserve Elkhorn Creek," *Lexington Herald-Leader,* December 4, 1991, B2; Stewart, Johnson, and Hines, "Citizens Take the Lead."

28. Helen Powell, *Elkhorn Creek Action Plan (Draft)* (Frankfort, KY: Elkhorn Land and Historic Trust, 1992), 2. Neither I nor Mr. Councill could find a report that was not a draft. It is unclear whether a final report was ever produced, given that a small nonprofit with volunteer officers may not have had good record management.

29. Ed Councill, personal communication.

30. Art Lander Jr., "Proposal Would Help Elkhorn Reach Potential," *Lexington Herald-Leader,* June 7, 1992; "Secret Is Out about Elkhorn Creek," *Lexington Herald-Leader,* May 2, 1993; "In the Canoe It's the Only Way to See the Solitary Wonders of Elkhorn's South Fork," *Lexington Herald-Leader,* October 17, 1993; "Creek Overflows with Recreational Pleasures," *Lexington Herald-Leader,* May 29, 1994; "Canoe Business Sees a Rapids Rise," *Lexington Herald-Leader,* March 26, 1995.

31. On March 4, 1997, the USGS gauge on the Elkhorn near Frankfort showed a record stage of 17.97 feet and 35,900 cubic feet per second, http://nwis.waterdata.usgs.gov/ky/nwis/peak?site_no=03289500&agency_cd=USGS&format=html (accessed May 30, 2014); "Many two-story houses along Elkhorn Creek had water up to the second floor and at least 75 rescues were made," NOAA website, http://www.crh.noaa.gov/lmk/?n=flood97 (accessed May 30, 2014).

32. Don Spangler, personal communication; Paul Singleton, longtime BWA member, personal communication.

33. The author was a member of this committee.

34. The price was hard to evaluate since there were no comparable sales. The land was in the designated floodplain and flooded regularly; with no sewer, septic system, or city water available, it would have high costs to develop.

35. Ed Councill, letter to the author, dated March 3, 1999. The author agreed to help write the grant application.

36. Powell, *Elkhorn Creek Action Plan (Draft),* 26.

37. Greg Kocher, "Area Landowners Oppose Spot for Canoeists to Land, Launch," *Lexington Herald-Leader,* May 13, 1999.

38. Barry Grimes, "AW Acres: The Elkhorn Saga," *American Whitewater Journal,* March–April 2000, 85–88.

References

American Canoe Association. "NSRE Paddlesports Participation Report." American Canoe Association, 2010. http://c.ymcdn.com/sites/www.americancanoe.org/resource/resmgr/general-documents/nsre-paddlesports-participat.pdf (accessed September 25, 2016).

American Whitewater. "About AW." http://www.americanwhitewater.org/content/Wiki/aw:about/ (accessed July 24, 2013).

———. "Safety Code of American Whitewater." Adopted 1959, revised 2005. http://www.americanwhitewater.org/content/Wiki/safety:start?#vi (accessed May 10, 2014).

———. "2013 Media Kit." http://www.americanwhitewater.org/resources/repository/2013_American_Whitewater_Media_Kit.pdf (accessed July 24, 2013).

Filson, John. "This map of Kentucke, drawn from actual observations, is inscribed with the most perfect respect, to the Honorable the Congress of the United States of America; and to His Excellcy. George Washington, late Commander in Chief of their Army." 1784. Library of Congress Geography and Map Division, Washington, DC. Digital Id g3950 ar079200. http://hdl.loc.gov/loc.gmd/g3950.ar079200.

Grimes, Barry. "AW Acres: The Elkhorn Saga." *American Whitewater Journal,* March–April 2000, 85–88.

Kentucky Water Resources Research Institute (KWWRI). *Kentucky River Basin Management Plan.* Lexington: Kentucky River Authority, 2002.

Lancaster, Clay. *Vestiges of the Venerable City: A Chronicle of Lexington, Kentucky, Its Architectural Development and Survey of Its Early Streets and Antiquities.* Lexington, KY: Lexington–Fayette County Historic Commission, 1978.

Leisure Trends Group. *Outdoor Recreation Participation Study for the United States, 2004.* 6th ed. Boulder, CO: Outdoor Industry Association, 2004.

———. *Outdoor Recreation Participation Study, Seventh Edition for Year 2004, Trend Analysis for the United States.* Boulder, CO: Outdoor Industry Foundation, 2005.

———. *Outdoor Recreation Participation Study, Eighth Edition for Year 2005, Trend Analysis for the United States.* Boulder, CO: Outdoor Industry Foundation, 2006.

Nelson, Jon. "Whitewater Slalom in the Olympics." http://whitewaterslalom.us/whitewater-in-the-olympics.html (accessed May 10, 2014).

O'Dell, Gary. "Water Supply and the Early Development of Lexington, Kentucky." *Filson Club History Quarterly* 67, no. 4 (1993): 431–61.

Outdoor Foundation. *Outdoor Participation Report, 2013.* Boulder, CO: Outdoor Foundation, 2013.

Powell, Helen. *Elkhorn Creek Action Plan (Draft).* Frankfort, KY: Elkhorn Land and Historic Trust, 1992.

Stewart, Beth, Gregory K. Johnson, and Doug Hines. "Citizens Take the Lead: Elkhorn Creek Watershed Planning and Action through Consensus." In *Proceedings: Watershed '93; A National Conference on Watershed Management*, March 21–24, 1993, Alexandria, Virginia, edited by Janet Pawlukiewicz, Paula Monroe, Anne Robertson, and Joan Warren, 249–52. Alexandria, VA: U.S. EPA, 1993.

Taft, Susan L. *The River Chasers: A History of American Whitewater Paddling.* Mukilteo, WA: Flowing Water Press and Alpen Books Press, 2001.

U.S. Census Bureau. "Population Change for Counties in the United States: 2000 to 2010." 2010 Census Population and Housing Table CPH-T-1. http://www.census.gov/population/www/cen2010/cph-t/cph-t-1.html (accessed September 18, 2016).

VanArnum, Christy J. Gunderson, Gerard L. Buynak, and Jeffrey R. Ross. "Movement of Smallmouth Bass in Elkhorn Creek, Kentucky." *North American Journal of Fisheries Management* 24 (2004): 311–15. doi:10.1577/M02-107.

Chapter Eleven

Where Five Rivers Meet

The Far Western Waters of Kentucky

Susan P. Hendricks

The area of Kentucky west of the Cumberland and Tennessee Rivers has been shaped by natural forces over hundreds of millions of years and then by human activities over the past millennium (figure 11.1). About 50 million years ago, during the late Cretaceous (Carey 2011), there was a shallow extension of the Gulf of Mexico that inundated the landscape of far western Kentucky now known as the Jackson Purchase (Davis 1923). Today, the Tennessee and Cumberland Rivers flow into the Ohio River near Paducah and Smithland, respectively, their mouths about five miles apart; the man-made impoundments of these two rivers define the western lakes region of Kentucky. The Ohio and Mississippi confluence near Cairo, Illinois, is about fifteen miles downstream from Paducah. A fifth significant drainage, the Clarks River, originates from just south of the Kentucky-Tennessee state line in Henry County, Tennessee, and flows northward through Calloway, Graves, Marshall, and McCracken Counties and into the Tennessee River. Other significant streams in the western Purchase include Mayfield and Obion Creeks and Bayou de Chien, which drain westward to the Mississippi (Carey 2009). This chapter follows a chronological organization as closely as possible, beginning with geologic and paleontological accounts, followed by human settlement histories with respect to water resources, and ending with modern developments and future prospects.

Geologic and Paleontological History

Today's physiography has been shaped over time by geologic forces we can only imagine. The shallow sea bottom of sandstone and limestone of Mississippian origin (360–325 million years ago) created a coastal environment where sediments were deposited, becoming the precursors of the underlying geology and soils of the region (Carey 2009). As the sea receded southward, surface sediments became exposed and over subsequent geologic periods were blanketed by alluvial sands, gravels, and chert. By the late Pliocene, 5.3–1.8 million years ago, most of the stream and river valleys of the Purchase area were in place as we see them today (figure 11.1). Along the eastern banks of the Mississippi, wind-blown sediments (known as loess) were deposited upslope, creating the great bluffs of Hickman, Fulton, Ballard, and Carlisle Counties that reach as high as 150 feet above the river. One dramatic climatic event of the late Pleistocene, the Wisconsin Glaciation of 100,000–10,000 years before the Common Era (BCE), helped shape and reshape the rivers' courses, particularly during the last glacial advance of 21,000 BCE, when continental ice sheets extended all the way south to what is now the Ohio River. Although glaciers never made it into what is now Kentucky, they helped establish the present drainage patterns of the modern Ohio and Mississippi River valleys. As the glaciers melted in the north, enormous volumes of water flowed down the Mississippi, Ohio, and Illinois Rivers, further eroding the alluvium to finer-grained sediments, so that today sand, silt, and clay, in addition to ancient Mississippian deposits, are the major sediment components in western Kentucky.

Presettlement Land Cover and
Pre-1900 Water Use Sites in the Purchase Area

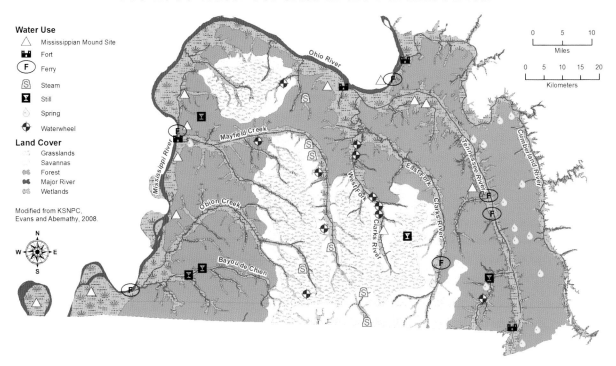

11.1. Presettlement land cover and pre-1900 water-use sites in the Purchase area. Sources used in locating important sites in the presettlement and early settlement Purchase include Perrin (1887), Hupp (2000), Fuller (1912), Mitchill (1815), Maurer (1974), MSU Special Collections (1976), Sickel (1989), Tennessee Waterway Assessment Study (2007), Kentucky GAP Analysis Program (2002), Freeman and Olds (1933), Mid-America Remote Sensing Center, Murray State University (MARC-MSU), and Bing Maps Imagery (www.bing.com). Last accessed 2/27/2012 through ESRI ArcMap, BaseMap Tool. Map courtesy of the Kentucky State Nature Preserves Commission.

The First People

The Pleistocene epoch witnessed major human migrations into the area. Each river system was there to greet and transport both paleo- and modern human travelers as the region was explored, exploited, and settled (Hickman County Historical Society 1983). What water resources did people find in the region? What drew them, and why did they stay? Paleo- and Archaic Indians (12,000–10,000 BCE) migrated from the northwest when the Ice Age created a bridge over the Bering Strait from Asia. These ancient people and those of the later Woodland culture (10,000–3,000 BCE) were boreal hunters and

gatherers who used the great rivers as highways, as did those of the later corn-based agricultural Mound Builders of the Mississippian culture (500–1500 Common Era [CE]). Archaeological excavations of prehistoric sites indicate that the Mound Builders developed villages (figure 11.1) of up to thirty thousand inhabitants along the Ohio and Mississippi Rivers (Pollack 2008). Wickliffe Mounds in Ballard County and the Goheen site in Marshall County (Fryman 1966) along the Tennessee River are among many examples (figure 11.1) of large indigenous populations tied closely to aquatic resources.

Expansive riparian wetlands provided stable sediments and soils on which to cultivate crops along with hunting, fishing (Pollack 2008, Claassen 2010), and transportation for trading and commerce. By 1350 CE, however, the larger sites were abandoned. Why did the Mississippian Mound Builder culture collapse? Did a small-scale climate shift known as the Little Ice Age from 1200 to 1700 CE cause a collapse in agriculture (Pollack 2008)? Was it because this was the "land of the shaking earth," as described by the Chickasaws (Lyell 1875)? By the time Europeans began exploring the area (e.g., DeSoto from 1539 to 1543; Marquette and Joliet in 1673; LaSalle in 1682), the Mississippian culture had all but disappeared and was replaced by the Chickasaw and Cherokee Nations in the seventeenth and eighteenth centuries.

Early American Exploration and Settlement

The period's exploration and settlement was enabled by the Mississippi River and tributaries via a number of watercraft from canoe to steamer which bounded the area that would become known as Hickman County (Hickman County Historical Society 1983). The territory was ceded by the Chickasaw Indian Nation in 1818 and described in the Jackson Purchase Treaty (Perrin 1887) almost entirely by the waterways in a counterclockwise direction. To physical geographers, this 2,300-square-mile parcel is part of the Mississippi Embayment, an extension of the Gulf Coastal Plain (Hupp 2000) of the eastern United States, bounded today on the west by the Mississippi River, on the east by the Tennessee River (Kentucky Lake), on the north by the Ohio River, and on the south by the Tennessee state line.

The convergence of major midcontinental waterways provided the highways for human migration and influenced their means of transportation, modes of self-defense, techniques for hunting, fishing, and agriculture, building styles, and construction of permanent settlements. The next 150 years saw continuous expansion by European explorers and settlers throughout the area, arriving via the Mississippi River from the Gulf of Mexico, from the north and the Great Lakes region, from the east down the Ohio River, and across land by horse and wagon. Increased contact with European explorers and settlers led to disease among the native peoples. The western waterways nonetheless remained instrumental in the continuous migration into the region.

General George Rogers Clark built Fort Jefferson in 1780 on the Mississippi River about five miles south of the convergence with the Ohio and near the mouth of Mayfield Creek (figure 11.1). However, his inability to make a land deal with the Chickasaw Indians, their hostility to the settlement, and remoteness from Virginia (of which Kentucky was part until 1792) led to the failure of the outpost. The Chickasaws used the region as a hunting ground, and the British continued to establish companies up and down the Mississippi as Euro-Americans moved westward down the Ohio and overland to the Tennessee River. Meriwether Lewis and William Clark came down the Ohio in 1803 on their exploratory trip to the Pacific Ocean. The region was eventually purchased by Andrew Jackson and then-governor of Kentucky Isaac Shelby in 1818, thereby acquiring its name. For his service in the Revolutionary War, George Rogers Clark was awarded several thousand acres in what is now McCracken County, giving names to several of the major tributaries of the area during the 1780s, including the Clarks River. The enormous wetland complexes with their fertile soils along all five major tributaries were visible to the first European explorers and settlers from their canoes and flatboats and were ideal for farming after some drainage. Lowland hardwood swamps and wetlands were referred to as "bottoms" and were major features on several of the original land survey maps of the late eighteenth and early nineteenth centuries (Kentucky Secretary of State). Flooding was a constant threat to development of the region, resulting in booms and busts of towns, farms, and businesses.

The New Madrid Earthquake

"New Madrid Earthquake" is the title of a highway marker on Kentucky 94 in Fulton County that I have encountered just before leaving the state. Myron L. Fuller (1912) in *The New Madrid Earthquake* describes the earthquake that was centered in this area as it related to the landscape and people; the tremors were felt at least several hundred miles away if not further. The occurrence of such a shock in a region like the Mississippi Valley, on the borders of a great river, is probably unprecedented in the history of earthquakes (Fuller 1912 citing Shaler 1869).

Such was one description of the major seismic activity beginning in December 1811 and continuing throughout 1812 that permanently altered the course of the Mississippi River and, therefore, the western boundary of the newly established Commonwealth of Kentucky. Archaeological and geologic evidence as far back as 6000 BCE indicates that sand blows and earthquakes, including several of very high magnitude, had occurred in the New Madrid area (Fuller 1912, Nuttli 1973, Pollack 2008). More recently, descriptions of the New Madrid earthquake of 1811-12 provided vivid accounts of the rearrangement of the Mississippi River's bed and flow patterns and the human toll in far western Kentucky (Mitchill 1815, Fuller 1912). Analyses and later interpretations of contemporary accounts of the fissuring and faulting of the New Madrid area in both scientific and colorful lay terms described the severity of the quake and its effects on water flow in the Mississippi River. According to one boatman's flowery account (Fuller 1912 citing Latrobe 1836), "A chasm opened in the Mississippi admitting great quantities of water, but immediately closed, giving rise to waves of great size. From many of the fissures . . . sand and water were forcibly extruded."

Some of the faults crossed the Mississippi, causing rapids and even waterfalls as described in Fuller (1912). "A few miles above New Madrid," a man traveling down the river reported, "he came to a most terrific fall, which, he thinks was at least 6 feet perpendicular, extending across the river. Another fall was formed about 8 miles below the town, similar to the one above, the roaring of which he could distinctly hear at New Madrid. He waited five days for the fall to wear away" before proceeding downstream (Mitchill 1815, Shaler 1869).

Eliza Bryon provided a dramatic narrative to Rev. Lorenzo Dow in 1815 which depicts a similar scene as reported in Jewell (1973):

At first the Mississippi seemed to recede from the banks and its waters gathered up like a mountain, leaving for a moment many boats . . . on bareland. . . . It then rose 15 or 20 feet perpendicularly and expanding as it were at the same moment, the banks were overflowed with the retrograde current rapid as a torrent. . . . The river falling immediately as rapidly as it had risen receded within its banks again with such violence that it took with it whole groves of young cottonwood trees.

Waterfalls six to twenty feet high on the Mississippi? Tsunami-like waves? Upheaval of sand and sediments, slumping of the riverbank, rerouting of water, and a reversal of river flow for a period of time? These scenes indeed would require extreme seismic activity in order to rearrange the Mississippi River, to create the "sunk lands" of Reelfoot Lake and surrounding wetlands, many of which reach into extreme southwest Kentucky, and to isolate the New Madrid Bend, an oxbow loop of the Mississippi, from the rest of Kentucky (figure 11.1). A U.S. Geological Survey (USGS) analysis of isoseismal maps of other recent earthquakes has estimated that the New Madrid Earthquake(s) of 1811–12 ranged in magnitude from 7.5 to 7.7 (Nuttli 1973, U.S. Geological Survey, Historic Earthquakes).

The Civil War

Beriah Magoffin, governor of Kentucky, on April 15, 1861, indicated no willingness to provide troops in order to fight the Southern states (Hickman County Historical Society 1983). Although Kentucky was technically neutral during the Civil War, the Purchase area identified culturally more with the South, and for good reason, because many settlers arrived by boat from the Southern states via the Mississippi and Tennessee Rivers. The inland waterways provided many strategic sites for construction of forts by both the Union and the Confederate armies (figure 11.1). Columbus, Kentucky, on the Mississippi River, was the site of a massive chain-and-anchor system created by the Confederates to block Union gunboats. Ulysses S. Grant moved five thousand troops down from Cairo, Illinois, to occupy Paducah's Fort Anderson and eventually took Columbus as well. Fort Smith in Smithland was another Union stronghold on the Ohio River, providing a staging area for Grant's army to eventually take Forts Donelson and Henry in Tennessee and Fort Heiman in extreme southeast Calloway County (figure 11.1). All three forts had been constructed by the Confederacy to prevent the Tennessee and Cumberland Rivers from supplying shipments to the Union. The Union army took Fort Heiman in 1862 and held it until General Nathan Bedford Forrest, one of the more notorious Confederate "generals," retook it and used it as a base

Post-settlement Land Cover and Post-1900 Water Use Sites in the Purchase Area

11.2. Postsettlement land cover and post-1900 water-use sites in the Purchase area. Shapefile created at USGS Kentucky Water Sciences Center, Louisville, KY, using data from KYGEONET and http://kygeonet.ky.gov, last accessed February 22, 2012. Other sources of information, particularly the recently accessed websites, can be found in the References and Additional Resources section. Basic map courtesy of the Kentucky State Nature Preserves Commission.

for his assault on Johnsonville, Tennessee. Forrest also attempted to recapture Paducah in 1864, but by then, the war was almost over, and clashes in the Purchase region were reduced mostly to minor raids, skirmishes, and petty horse theft.

Post–Civil War to Twentieth-Century Dam Construction

Continued settlement during the late 1800s and early 1900s depended on an abundance of clean, fresh water and determined where many villages and towns were built. All five rivers had considerable riparian wetlands associated with them before European settlement (figure 11.1). The frontier town settlers' relationship to wa-

ter was recognized very early, especially in Wadesboro, the first land office in the Purchase area and the first natural stop after crossing the Tennessee River. Pollution of the water resource by these settlers also led to the eventual downfall of the town (Jennings and Jennings 1973). Several natural springs large enough to service small communities were noted by settlers throughout the area (figure 11.1). Although excess taxation on whiskey instigated the Whiskey Rebellion of 1791 in western Pennsylvania, with subsequent migration of farmers and their stills down the Ohio River on flatboats, it was the Prohibition era, 130 years later, that made the bootlegging industry notorious in Land-Between-the-Rivers (now Land-Between-the-Lakes, LBL) (Maurer 1974). Springwater was clear, clean, and abundant in

11.3. The Ohio River flood of 1937, downtown Paducah, Kentucky. The first floor of the Irvin Cobb Hotel is flooded. Patrons are being rescued from the second floor onto a raft floating in the street below. The automobile to the left is almost completely submerged. Photo courtesy of the McCracken County Public Library.

the Purchase area and in LBL because of their underlying limestone geology, making both areas conducive to moonshine production (figures 11.1 and 11.2). Either copper or steam stills were set up near the cold-water limestone springs and perennial streams well hidden within the remote hollows and along isolated, sparsely populated riverbanks accessible only by flatboat or canoe. Golden Pond was the second-largest moonshine center in Kentucky and supplied whiskey to northern cities such as Detroit, Dayton, and Cleveland in the 1920s (Maurer 1974).

Waterborne diseases were quite common in the region because of the prevalence of wetlands and frequent flooding along the rivers (figure 11.1). Typhoid fever was a yearly scourge in Calloway and Hickman Counties (Jennings and Jennings 1978, Hickman County Historical Society 1983). The first bond issue to finance a waterworks system in Murray was passed in 1910. Other water-pollution issues came to a head around 1912 because residents needed to deal with a long cultural history of disposing of animal carcasses in the "back ditch" and with runoff from horse lots, hog pens, open-pit garbage disposal, chicken yards, and outdoor toilets,

all of which contributed to degradation of water resources. Cholera and yellow fever epidemics were common in towns on the Mississippi River (Jewell 1973). Bottomlands along all the rivers provided habitats rife with insect vectors of diseases. "Buffalo gnats" (black-flies) were documented in such abundance as to kill juvenile and occasionally adult livestock (Murray State University Special Collections 1976). These insects were also vectors of a parasitic larval nematode that caused onchocerciasis, or "river blindness," a common plague of both livestock and people.

As settlement continued, farmers drained wetlands and channelized long reaches of streams (figure 11.2) in order to plant more acreage and reduce the effects of flooding. The early twentieth century brought more efficient technology to do the job of canalization, a practice that resulted in more rapid runoff of floodwater from the land. But people were not yet aware of the positive services of wetlands in slowing floodwaters and retaining soils. Consequently, flooding was common and dramatic, and tons of topsoil were lost from the landscape.

By 1929, the Ohio River had been partially controlled by a series of locks and dams constructed by the U.S.

11.4. Construction of the Kentucky Dam was finished in 1944. Photo courtesy of the Kentucky Geological Survey, University of Kentucky.

Army Corps of Engineers (USACE). A number of devastating floods occurred in the early twentieth century; a major flood in 1937 (figure 11.3) became a turning point as people demanded congressional authorization and appropriation for the construction of a dam on the Tennessee River. The economic devastation of the Great Depression and the 1937 flood were timely events that set the stage for the creation of the Tennessee Valley Authority (TVA). The construction of a series of reservoirs would provide electricity and flood control for the South and create a reliable, navigable waterway for commercial shipping in the region.

Dam building in far western Kentucky was a testament to the political power of Senator Alben Barkley and Representative Noble Jones Gregory and the vision of Senators George W. Norris and Kenneth D. McKellar, who staunchly supported President Franklin D. Roosevelt's New Deal and the creation of the TVA in the 1930s. Dozens of books have been written about the TVA, the USACE, and the politics and construction of the dams that created Kentucky Lake and Lake Barkley,

so only brief comments will be made here. The TVA and the USACE partnered in the construction of Kentucky Dam (figure 11.4), completed in 1944, and Barkley Dam, completed in 1964.

There were unfortunate social impacts on the people who had established farms and residences over the previous 150 years along the river valleys and in the Land-Between-the-Rivers. In exchange for the electricity, the expanded economy, the better navigation, and the flood control that these dams permitted, the river-bottom people were forced to give up their land and livelihoods. These people did not necessarily view this as a fair trade. The U.S. government's exercise of eminent domain, whereby private property is ultimately under state control and can be confiscated at fair market value when a "larger public good" can be served, led to the displacement of hundreds of families and to the hatred and mistrust of government that persists to this day. In an ironic turn that contrasts with today's perception of an overly environment-friendly federal government, the river-bottom people, instead, were considered to be the earliest

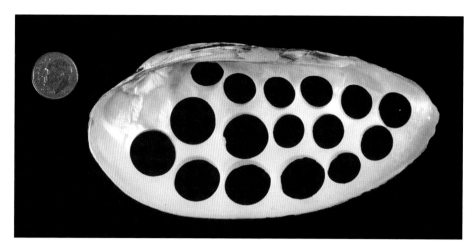

11.5. Mussel with button cutouts. Photo courtesy of the Illinois State Museum.

11.6. A brailer and his johnboat, commonly used throughout the Tennessee and Cumberland Rivers. Lines with weights and hooks (with prongs) would be lowered to the river bottom, where mussels would latch on to the prongs. Photo courtesy of the Illinois State Museum.

environmentalists and were disparaged by the politicians of that day as "radicals" who "stood in the path of progress" (Jennings and Jennings 1978). The impact of the dams on the economic development of the Purchase area, however, remains controversial to this day, depending on whom you talk to. Although the lakes have experienced enormous growth in shipping, transportation, and recre-

ation, the dreams of early twentieth-century politicians to transform the small towns of Aurora and Cadiz into entertainment and tourist meccas of the caliber of Gatlinburg and Nashville have yet to be realized.

Among the natural features of the region, the extensive freshwater mussel beds in the shallow reaches of the Tennessee, Cumberland, Ohio, and Illinois Rivers pro-

11.7. Towboat-barge pushing gravel and sand past Ginger Bay on Kentucky Lake. A towboat with barges can reach one thousand feet in length. Photo courtesy of Todd Levine.

vided a unique resource. American Indian kitchen middens (garbage pits) often contained thousands of empty shells, although how they were used is still debated (Fryman 1966, Pollack 2008, Claassen 2010). Settlers collected the shells for pearls and for the mother-of-pearl layers that later were widely used in the button industry (figure 11.5) until the mussel populations were severely depleted. By the late 1800s, this was one of the largest fisheries in the United States. In the 1950s, mussels became valuable for the Japanese cultured-pearl industry. Mussel harvests were shipped to Japan, where the shells were crushed and used to inoculate oysters for cultured-pearl production. By that time, shallow areas in the major rivers had disappeared with the construction of locks and dams, and thus brailing for mussels became the primary fishing method. Johnboat operators (figure 11.6) would lower lines with weights and hooks with prongs to the river bottom to which mussels would grapple and be dragged onto the boats. Brailing was an extremely efficient method for collecting mussels, but it was quite destructive to aquatic habitats (Sickel 1989). Cultural conflicts between brailers and fishermen in Kentucky Lake and Lake Barkley were continuous from the 1970s through the 1990s. The final collapse of the cultured-pearl industry, along with a ban on brailing, ended the industry.

Late Twentieth and Early Twenty-First Centuries

Today, the Purchase area is a region of rolling hills and flatlands where farming remains the primary human industry. The elevation of the Jackson Purchase ranges from a maximum of 590 feet above sea level in Calloway County to a minimum of 280 feet above sea level in Fulton County at the Mississippi River.

Alluvial deposits and unconsolidated sediments of the Jackson Purchase aquifer still provide excellent groundwater resources for drinking water, irrigation, livestock production, and industry. The taming of the wild Cumberland and Tennessee Rivers into a series of hydroelectric reservoirs, locks, and dams provide a new waterway by which towboats can push barges up to a thousand feet long transporting great quantities of goods (figure 11.7). Grain, coal, wood chips, minerals, aggregates, sand, gravel, and petroleum products are the most common commodities moved from the Ohio River to Chattanooga and Nashville or down to the Ohio and Mississippi Rivers to the Gulf of Mexico. Transportation by barge is more energy efficient than by rail or truck (Hanson Professional Services 2007). As of 2004, 52 and 25 million tons per year of goods were shipped up and down the Tennessee and Cumberland Rivers, respectively, worth over $1

11.8. Murphy's Pond, Hickman County, Kentucky. Bald cypress at the northern limits of their range. Photo courtesy of Ed Zimmerer.

billion (Hanson Professional Services 2007). The Mississippi River in this region transported over 310 million tons of goods. Terminals and ports were developed along the way to receive goods for transport by rail or highway, thereby increasing commerce and the tax base for local economies. Recreational use (fishing, boating) in the rivers and reservoirs brings additional dollars into the eight counties of far western Kentucky, supporting marinas, campgrounds, bait shops, sport shops, restaurants, resorts, and other family-owned businesses (figure 11.2).

Purchase-area waterways are known for their biological diversity. The area is a crossroads not only for climate but also for unique flora and fauna more similar to those found along the southern coastal plains and midwestern prairies and quite different from those found in the central and eastern parts of Kentucky. The rivers are a natural flyway for migrating whooping cranes, pelicans, loons, and many other birds. Murphy's Pond (figure 11.8) in Hickman County is a remnant of the bald cypress swamps more commonly found farther south, hence its nickname, Kentucky's Little Okefenokee (Hickman County Histori-

cal Society 1983). Several modern ecological classification schemes describe the Purchase area today: the Mississippian Embayment (Ecoregion 25) (Abell et al. 2000), Mississippi Valley Loess Plains Bioregion (Omernik 1987, U. S. Environmental Protection Agency), and the Mississippi Valley–Interior River Ichthyoregion for fish species (Compton et al. 2003). The National Ecological Observatory Network (NEON 2011) has indicated that the area represents a climate crossroads representing three climate domains: the Appalachian and Cumberland Plateau, the Ozark Complex, and the Prairie Peninsula, all of which possess characteristic precipitation and vegetation patterns.

With only 20 percent of the original wetlands left since settlement, the Clarks River National Wildlife Refuge (U.S. Fish and Wildlife Service 2010) and other wildlife management areas preserve what remains (figure 11.2). Sixteen priority watersheds (Cicerello and Abernathy 2004) within the Purchase area have been designated as hot spots for imperiled/endangered mussels and fishes. Although the five rivers of far western Kentucky are home to many

11.9. Water-monitoring buoy on Kentucky Lake. Water-quality sensors suspended underwater from the center of the buoy collect and relay water-quality data every fifteen minutes to Hancock Biological Station. Measurements include water temperature, dissolved oxygen, pH, conductivity, and turbidity. Weather data are transmitted from instruments mounted on the top of the buoy. Photo courtesy of the author, Susan Hendricks.

nonnative species, they remain sites of high biodiversity (White et al. 2005).

The Jackson Purchase is well positioned for expanded economic growth in the twenty-first century because of the development of ports, terminals, barge transportation, recreational facilities, and other waterway infrastructure. Citizens today are concerned about pollution and aware of conservation measures and water-management practices that conserve soils, reduce runoff and pollution, and mitigate human degradation of water resources. Cultivation is around 70 percent no-till practices in the Purchase. At least one stream-restoration project is under way on Obion Creek to correct a century of channelization (Parola et al. 2005). During the past fifteen years, volunteer organizations such as Four-Rivers Watershed Watch and Purchase Area Water Watch have provided outlets for citizen scientists to get in-

volved with water monitoring and local conservation issues. These organizations are supported or coordinated by the Kentucky Division of Water and the Jackson Purchase Foundation. Citizen scientists have partnered with water scientists at Murray State University's Hancock Biological Station (HBS) to monitor water quality in local streams and rivers. The HBS has monitored Kentucky Lake water quality for over twenty-five years and has created a long-term database to track trends in the aquatic environment over time and space (Hancock Biological Station, Kentucky Lake Monitoring Program). Recent acquisition of high-frequency monitoring equipment (figure 11.9) and the necessary cyberinfrastructure for storage and dissemination of data streaming in near real-time (every fifteen minutes) from experimental buoys has added a new, high-tech dimension to water-quality monitoring in the region (Virtual Observatory and Ecological Informatics System). Advanced twenty-first-century technology allows for almost instantaneous monitoring and evaluation of rapid changes in water quality after storms and spills that could lead to noxious algal and bacterial blooms (White and Hendricks 2010). Public access to data from HBS websites allows citizens to track changes in fishing conditions, such as water temperature, dissolved oxygen, and weather, on Kentucky Lake. Access to real-time data also serves to increase citizen engagement in and awareness of water-quality issues and resources.

Citizens today can live, work, and play because of the investment in infrastructure and management practices needed for flood control, hydroelectric energy, maintenance of navigable waters for commerce, and management of water quality to ensure healthy aquatic life and safe recreation. But this has not come without sacrifices sometimes inflicted on those who had homesteaded since the late 1700s. The waterways of far western Kentucky have provided services to humans for thousands of years and will continue to do so because of equally respected private sacrifices and public investments.

Acknowledgments

I thank Jane Benson of the Mid-America Remote Sensing Center at Murray State University for her expertise in mapping and web-based geographic research. I am grateful for access to important historical collections, county histories, microfiche, and other fragile documents granted by the Pogue Library, Murray State University. Many informative conversations with and suggestions from the following people provided insight and additional resources: Greg Abernathy (Kentucky State Nature Preserves), Dan Carey (Kentucky Geological Survey, University of Kentucky), Paul Rister (Kentucky Department of Fish and Wildlife), Ken Carstens (Archaeological Consulting Services and professor emeritus at Murray State University), Art Parola (Stream Institute, University of Louisville), and David White (Commonwealth Chair of Environmental Studies and director, Hancock Biological Station, Murray State University).

References and Additional Resources

Abell, R. A., et al. 2000. *Freshwater Ecoregions of North America: A Conservation Assessment.* Washington, DC: Island Press.

Carey, D. I. 2009. *Four Rivers Basin: Cumberland, Tennessee, Ohio, Mississippi.* http://kgs.uky.edu/kgsweb/olops/pub/kgs/mc194_12.pdf (accessed October 9, 2016).

———. 2011. *Kentucky Landscapes through Geologic Time.* Kentucky Geological Survey, series 12, Map and Chart 200. http://kgs.uky.edu/kgsweb/olops/pub/kgs/mc200_12.pdf (accessed October 9, 2016).

Cicerello, R. R., and G. Abernathy. 2004. *An Assessment of "Hot Spots" and Priority Watersheds for Conservation of Imperiled Freshwater Mussels and Fishes in Kentucky.* Frankfort: Kentucky State Nature Preserves Commission.

Claassen, C. 2010. *Feasting with Shellfish in the Southern Ohio Valley: Archaic Sacred Sites and Rituals.* Knoxville: University of Tennessee Press.

Compton, M. C., G. J. F. Pond, and J. F. Brimley. 2003. *Development and Application of the Kentucky Index of Biotic Integrity.* Frankfort: Kentucky Division of Water. http://water.ky.gov/Documents/QA/MBI/KIBI_paper.pdf.

Davis, D. H. 1923. *The Geography of the Jackson Purchase of Kentucky.* Kentucky Geological Survey, series 6, vol. 9. Lexington: Kentucky Geological Survey.

Freeman, L. L., and E. C. Olds. 1933. *The History of Marshall County.* Benton, KY: Tribune-Democrat.

Fryman, F. 1966. *The Goheen Site: A Late Mississippian Site in Marshall County, KY.* Lexington: University of Kentucky, Museum of Anthropology.

Fuller, M. L. 1912. *The New Madrid Earthquake.* Department of the Interior, U.S. Geological Survey, Bulletin 494. https://pubs.usgs.gov/bul/0494/report.pdf (accessed October 9, 2016).

Griffing, B. N. 1880. *An Atlas of Graves County, Kentucky.* Philadelphia: D. J. Lake and Co.

Hancock Biological Station, Kentucky Lake Monitoring Program. Databases available at http://www.murraystate.edu/qacd/cos/hbs/WQ.cfm; http://www.murraystate.edu/wsi/wsi_database.html.

Hanson Professional Services. 2007. *Tennessee Waterway Assessment Study.* Nashville, TN: Hanson Professional Services. http://tenntom.org/TNWaterwayAssessment.pdf (accessed October 9, 2016).

Hickman County Historical Society. 1983. *Hickman County History.* Vol. 1. Clinton, KY. Dallas: Taylor Publishing Co.

Hupp, C. R. 2000. Hydrology, Geomorphology and Vegetation of Coastal Plain Rivers in the Southeastern USA. *Hydrological Processes* 14: 2991–3010.

Jennings, D., and K. Jennings. 1978. The Story of Calloway County, 1822–1976. Murray, KY 584. Murray State University Special Collections, Pogue Library, F 457 C17 J46x.

Jewell, O. 1973. *Backward Glance.* Vol. 1. A History of Fulton County, KY, and Surrounding Area. Fulton, KY: Fulton Publ. Co.

Kentucky Association of Riverports. http://kentuckyriverports.com/kentucky_riverports/port_locations/ (accessed February 22, 2012).

Kentucky Department of Fish and Wildlife Resources, Division of Fisheries and Kentucky Fish and Wildlife Information Systems. Shapefile created in 2006. KYGEONET. http://kygeonet.ky.gov (accessed February 24, 2012).

Kentucky Division of Geographic Information. 2004. *Kentucky 2001 Anderson Level III Land Cover.* Frankfort: Kentucky Division of Geographic Information. http://fw.ky.gov/kfwis/kygapweb/kyreport.pdf. KYGEONET. http://kygeonet.ky.gov.

Kentucky GAP Analysis Program. 2002. *Kentucky Gap Analysis Program Land Cover Map.* Murray, KY: Mid-America Remote Sensing Center, Murray State University.

Kentucky Heritage Commission. 1978. Survey of Historic Sites in Kentucky, Ballard County.

Kentucky Secretary of State. 2012. Geographic Materials. http://www.sos.ky.gov/admin/land/resources/Pages/Geographic-Materials.aspx (accessed October 9, 2016).

Kentucky Wildlife Management Areas. KYDFWR. http://kygeonet.ky.gov.

Latrobe, C. J. 1835. *The Rambler in North America.* 2nd ed. Vol. 1. New York: Harper & Brothers.

LBL boundary: USDA Forest Service, LBL-NRA, 2006. http://lbl.org/VCMaps.html (accessed February 23, 2012).

Lyell, C. 1875. *Principles of Geology.* 12th ed. Vol. 2. London: Spottiswoode and Co.

Maurer, D. W. 1974. *Kentucky Moonshine.* Lexington: University of Kentucky Press.

Mid-America Remote Sensing Center, Murray State University. Bing Maps Imagery. www.bing.com (accessed February 27, 2012, through ESRI ArcMap, BaseMap Tool).

Mitchill, S. L. 1815. A Detailed Narrative of the Earthquakes Which Occurred on the 16th Day of December, 1811. *Trans. Lit. Philos. Soc.* 1: 281–307.

Murray State University Special Collections. 1976. A History of Carlisle County, Kentucky, for the Years 1820–1900. Pogue Library, F 457 C25 H57 v. 1.

National Ecological Observatory Network (NEON). http://www.neonscience.org/ (accessed October 9, 2016).

Nuttli, O. W. 1973. Seismic Wave Attenuation and Magnitude Relations for Eastern North America. *Journal of Geophysical Research* 78: 876–85.

Omernik, J. M. 1987. Ecoregions of the Conterminous United States. *Annals of the Association of American Geographers* 77: 118–25.

Parola, A. C., W. S. Vesely, A. L. Wood-Curini, D. J. Hagerty, M. N. French, D. K. Thaemert, and M. S. Jones. 2005. *Geomorphic Characteristics of Streams in the Mississippi Embayment Physiographic Region of Kentucky.* Final Report for the Kentucky Division of Water. Louisville, KY: Stream Institute, University of Louisville. http://water.ky.gov/permitting/Lists/Working%20in%20Streams%20and%20Wetlands/Attachments/7/MississippiEmbayment.pdf (accessed October 9, 2016).

Perrin, W. H. 1887. *History of Kentucky.* Vol. 1. Louisville: F. A. Battey and Co.

Pollack, D., ed. 2008. *The Archaeology of Kentucky: An Update.* Vol. 2. State Historic Preservation Comprehensive Plan Report No. 3. Frankfort: Kentucky Heritage Council.

Shaler, N. S. 1869. Earthquakes of the Western United States. *Atlantic Monthly,* November, 549–59. https://www.unz.org/Pub/AtlanticMonthly-1869nov-00549 (accessed October 9, 2016).

Sickel, J. 1989. Impacts of Brailing on Mussel Communities and Habitat in Kentucky Lake. Hancock Biological Station, Murray State University.

United States Geological Survey. http://waterdata.usgs.gov/ky/nwis/inventory/?site_no=03609500.

U.S. Environmental Protection Agency (USEPA). Ecoregion Publications. https://www.epa.gov/eco-research/ecoregions-publications (accessed October 9, 2016).

U.S. Fish and Wildlife Service. 2010. Clarks River NWR Boundary: USFWS-CRNWR. CD.

U.S. Geological Survey. Historic Earthquakes. http://earthquake.usgs.gov/earthquakes/states/events/1811-1812.php (accessed January 12, 2012).

U.S. Geological Survey. National Hydrography Dataset—High Resolution. GET NHD Data. http://nhd.usgs.gov/data.html (accessed February 22, 2012).

Virtual Observatory and Ecological Informatics System (VOEIS). Databases now accessible through Hancock Biological Station at http://www.murraystate.edu/qacd/cos/hbs/WQ.cfm with the VOEIS page at http://www.murraystate.edu/qacd/cos/hbs/VOEIS/ (accessed October 9, 2016).

Wethington, K., T. Derting, T. Kind, H. Whiteman, M. Cole, M. Drew, D. Frederick, G. Ghitter, A. Smith, and M. Soto, The Kentucky Gap Analysis Project Final Report. 2003. Frankfort, KY., United States Geological Survey, Biological Resources Division.

White, D., K. Johnston, and M. Miller. 2005. Ohio River Basin. In *Rivers of North America,* edited by A. C. Benke and C. E. Cushing, 375–424. Amsterdam: Elsevier Academic Press.

White, D. S., and S. P. Hendricks. 2010. Lakes and Reservoirs as Environmental Sensors: The Value of Long-Term and High Frequency Monitoring. In *Proceedings of the 13th Conference on the Natural History of Land-Between-the-Lakes,* edited by S. W. Hamilton, A. F. Scott, and L. D. Estes, 31–41. The Center of Excellence for Field Biology, Austin Peay State University, Clarksville, Tennessee.

CHAPTER TWELVE

Karst

Shaped by Water from the Inside Out

James C. Currens

Visualize a rivulet draining a declivity. The stream drops in elevation steeply as it seeks to disperse the potential energy stored in the water when it fell as rain on the upper elevations. Visualize the rivulet joining with another rivulet, and another, and eventually a creek, and finally the flow reaches a larger creek or river. Now visualize this dendritic confluence of rivulets and creeks and rivers being underground, roofed over, if you will, with massive thicknesses of limestone and other types of rock. If you can visualize this, you are halfway to understanding karst. Karst lands have sinkholes, some as large as a city, caves that are barely big enough for an adult to enter, and some caves that are hundreds of miles long. In Kentucky, springs commonly form the headwaters of rivers, and there is a multitude of physical forms etched into the limestone both above and below the ground. Belowground, the karst is shaping Kentucky from the inside out.

Karst, the landscape that covers half of Kentucky, has had more influence on human activity in the state than any other landform. The extent of the karst areas in Kentucky are very apparent in Figure 12.1. Karst aquifers, karst topography, and karst geology collectively have had a profound effect on both the geomorphic and the cultural shape of Kentucky. Early settlers and bourbon distillers chose sites for their operations because of the springs. The mineral apatite, a residue of the dissolution of Lexington limestone, supplied the phosphates that enriched the grass that made the Inner Bluegrass the home of the fastest horses in the world. The caves have also shaped the history of Kentucky and indeed the nation, in that the saltpeter mined from Mammoth Cave and Great Saltpeter Cave supplied the gunpowder mills that armed the Kentuckians who fought in the War of 1812.

How Karst Forms

The first requirement for karst formation is a large region underlain by water-soluble rock, such as limestone, dolostone, gypsum, or salt. Most karst in Kentucky is developed in and on broad bands of Mississippian and Ordovician limestone and dolostone.

The second requirement for karst development is that the limestone be structurally and topographically positioned in such a way that precipitation will eventually pass through the rock to reach a permanent surface stream (local base level). Water seeping downward fills the joints and bedding planes in the limestone. These narrow fissures are very slowly enlarged by the seeping water as it dissolves the limestone at a barely measurable rate. Ultimately the slow-moving water will discharge into the local-base-level stream.

The third requirement for karst development is adequate precipitation. If there is no rain, caves cannot form. Around Mammoth Cave and in Pulaski County, the state has its greatest average annual rainfall (127 to 132 centimeters a year). Rainwater picks up carbon dioxide as it falls through the air. Kentucky's humid climate also promotes dense plant growth that eventually dies and decays. Decaying vegetation produces acids and adds to the carbon dioxide in the soil, both of which contribute to the acidity of the groundwater. Other factors that

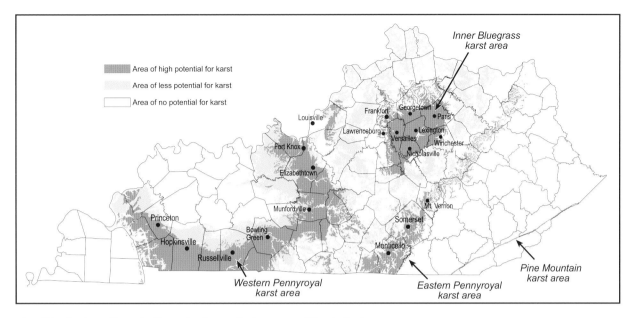

12.1. Distribution of karst in Kentucky. James C. Currens and Randy Paylor.

Generalized Block Diagram of the
Western Pennyroyal Karst

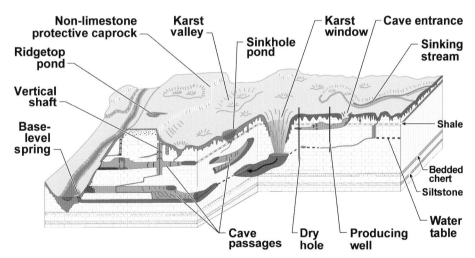

12.2. Generalized block diagram of the Western Pennyroyal. The sandstone-capped ridge to the left represents the Dripping Springs Escarpment. James C. Currens, Kentucky Geological Survey.

influence karst development are the local geologic structure, the spacing of faults, joints, and bedding planes, and the purity of the limestone.

How Karst Processes Shape the Underground

At the beginning of this chapter, we visualized a dendritic network of confluent tributary karst conduits in the subsurface. The coalescing conduits are developed by seepage along connecting sets of bedding-plane partings, joints, and faults that (if the bedrock were insoluble) would otherwise form a fracture-flow aquifer. Along the flow route and entirely by chance, one fracture path will be wider than the others. The presence of the wider fractures creates a flow route that is slightly more favorable to flowing water (figure 12.3). Decades pass into centuries, and the water relentlessly flows through the fractures in a seeping, nonturbulent, laminar flow regime. The carbon dioxide adsorbed from the air and soil by rain dissolves in the water to form carbonic acid.

$$(CO_2 + H_2O \leftrightarrow H_2CO_3)$$
$$(H_2CO_3 \leftrightarrow H^+ + HCO_3^-)$$

Carbonic acid will dissolve the calcite mineral (calcium carbonate, $CaCO_3$) that makes up most of the mass of limestone and, when compared with water with no dissolved carbon dioxide, will do so relatively quickly.

Flows of short distances (tens to hundreds of meters) over limestone are all it takes to bring the water near the point of saturation in respect to calcite. But as the water approaches saturation, the rate at which the calcite dissolves slows dramatically, and the carbonic acid never quite reaches the point where it will not dissolve any more calcite. The water remains ever so slightly aggressive over very long times and long groundwater flow paths. This phenomenon is the reason caves and the flow paths they record extend over many tens of kilometers.

The capillary-sized conduits established along the fracture networks will gradually develop the more hydraulically efficient cross section of a circular conduit when they are only a few millimeters in diameter. The headward extension of the most efficient conduit results in a minuscule depression in the water table, which creates an increased gradient from the fractures to the growing conduit. The trough in the water table creates a

self-perpetuating loop in which the extension of the conduit creates conditions that attract more flow, and the more flow that is attracted, the faster the conduit grows.

The early conduits grow slowly at first (figure 12.4), but as the chemical/hydrologic feedback loop continues to attract an ever-increasing percentage of the fracture-flow water, the conduit reaches a critical diameter of one centimeter. At that diameter, the groundwater flow rate can speed up enough that the flow becomes turbulent.

Simultaneously, the rate of calcite dissolution shifts from being controlled by the saturation of the water—how much the water can dissolve—to how long the water is in contact with the limestone and how fast the calcite mineral will dissolve. The turbulent water is moving much faster than the laminar flow through the narrow fractures. It may not stay in contact with the limestone at any one point long enough to reach saturation. Down the gradient, the water may ultimately approach saturation, but the total discharge is much greater, and a larger total mass of limestone continues to be dissolved from the conduit. The water may stay aggressive, however, all the way to the spring. At this level of development, the aquifer is no longer just a fracture-flow aquifer but a karst aquifer.

The conduit becomes a cave at an anthropomorphically defined diameter of one-half meter. During development, these passages remained filled with water most of the time, and the cave will continue to grow in diameter as long as water continues to flow through it. A sudden downcutting of the local base level or development of a more efficient conduit discharging at a lower elevation can cause the flow to seek a steeper route and the passage to be abandoned. Figure 12.5 represents the principal types of aquifers. Karst aquifers are modified from the fracture flow by dissolution of the bedrock.

There are a variety of cross-section geometries the developing cave passages can acquire as water shapes the Kentucky underground. Cave passages that remain totally filled with water during their development (phreatic passages) have a rounded, if not a circular, cross section. If conduit development was along a prominent bedding plane, the cave passage will have a nearly elliptical cross section with the long axis on the bedding plane. If the bedrock is tilted, the long axis of the ellipse will also be tilted. If joining or faulting was the domi-

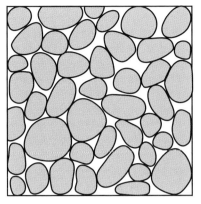
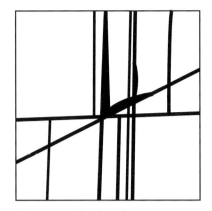
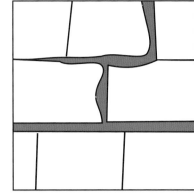

Intergranular   Fractured bedrock   Solution enhanced

12.3. Schematic drawing of the common types of aquifer porosity: unconsolidated intergranular (or primary), fractured bedrock (secondary), and solution enhanced (tertiary). James C. Currens.

nant secondary porosity, the ellipse can be along the fracture as opposed to the bedding plane. If the water table
falls, and the flow is pirated to a lower elevation, another
type of passage is formed by a free-flowing stream, a tall,
narrow, meandering canyon created by the resulting
vadose flow. Canyon passages often begin in the floors of
the elliptical phreatic conduits when the gradient is steep-
ened but frequently diverge from the course of the older
passages. Canyon cross-section passages are more common in the upper reaches of the cave system.

Among the secondary underground features, the
most important is the vertical shaft (figure 12.6). Vertical shafts are the functional equivalent of first- and
second-order streams and rapidly move water from the
surface to the water table. Vertical shafts can be as small
as one-half meter in diameter and two or three meters
high and in Kentucky as large as sixty-four meters high
and tens of meters in diameter.

Many vertical shafts are found along the edge of hilltops, where the surface outcrop of resistant lithological
units exposes the underlying limestone. The base of the
outcropping-resistant bed delineates the margin of an
area of concentrated recharge. Vertical shafts develop
where a resistant bed (commonly chert) prevents a
conduit-scale passage from downcutting. The seal of the
resistant bed is not perfect, however, and the water will
find a joint or some other leaky feature. Once the joint is
widened enough that the flow rate cannot keep the open-

12.4. A karst conduit in an early stage of development, inter-
cepted and exposed by a road cut. Photo by James C. Currens.

ing full of water, the water begins to fall. Falling water
reaches maximum velocity in a few feet, a speed at which
it dissolves or erodes any protrusions from the walls of
the shaft. Water flowing down the surface of the shaft
walls has the same effect. If the shaft drills all the way to
the water table, it cannot go deeper as a shaft. If downcutting is temporarily stalled at any time in its development, a larger-than-average passage draining the shaft
will be developed, only to be abandoned when the

12.5. Cascade Cave at Carter Caves State Park. Photo by James C. Currens.

shaft continues to deepen. The vertical shaft can also drill through higher, older cave passages without deflection. This characteristic and the stack of abandoned shaft drains allow explorers to use vertical shafts as elevators by climbing or rappelling between levels.

Other features of cave passages include come-arounds (short sections of passage that loop away from and then rejoin the main passage), natural bridges, and waterfalls. Breakout domes are chambers, commonly at the junction of two or more passages, where the ceiling has enlarged by rock fall because of the mechanical stress on the limestone beds. The room ceiling stabilizes when the broken edges of the layers of limestone step inward as the ceiling gets higher above the floor to form a natural arch. Breakout domes can be extremely large (one hundred meters in diameter). Although some rooms are nearly filled by their own fallen rock (breakdown), others have had most of it removed if a stream is flowing below the fallen rock in the joined passages.

Caves occur throughout the karst areas of Kentucky. In the Inner Bluegrass, many Middle Ordovician limestones are relatively impure, containing 20 to 60 percent insoluble material, and are also thinly bedded, thus limiting the rate of cave development. Along the deep Kentucky River gorge, thick sequences of purer, thicker-bedded Middle Ordovician limestones lower in the stratigraphic column are exposed. Close to the river, the removal by erosion of shale near the top of these formations and the large topographic relief along the Palisades have allowed the formation of some vertical shafts as much as 160 feet deep, and caves with total relief of more than 200 feet are known. On Pine Mountain, Cole Hole Cave in Letcher County has two vertical shafts in a stair-step configuration with a combined total depth of 55 meters (180 feet). The record vertical drop in the state is an open-air pit in Wayne County at 64 meters (210 feet). The Kentucky cave with the greatest overall vertical extent, more than 107 meters (350 feet), is probably Mammoth Cave (figure 12.7). Its natural upper entrances are at least as high as 235 meters (771 feet) elevation (the top of the man-made Snowball Dining Room elevator shaft is 244 meters [800 feet]), and lower levels of the cave have been explored to the level of the Green River at 128 meters (421 feet).

In the Mammoth Cave area, the geology and weather are ideal for cave formation. An area of more than eighty

12.6. A vertical shaft looking up from the bottom. Visible height is estimated at 40 feet. Photo by James C. Currens.

square miles is underlain by limestone 100 to 140 meters (330 to 460 feet) thick. The limestone is very pure, over 95 percent calcium carbonate. The individual beds of limestone are thick enough to support large openings without the bedding planes being so widely spaced that groundwater flow is restricted. The rocks dip northwestward toward the Green River at a rate of five meters per kilometer (thirty feet per mile), which promotes flow toward the river. Near the river, the limestone is protected by a sandstone caprock. Away from the river, the caprock has been eroded, and dissolution of the limestone has created a gently rolling surface riddled with sinkholes, known as the Sinkhole Plain.

All precipitation that falls there and seeps into the ground must flow through the limestone to reach a per-

manent, base-level stream. The dip of the limestone toward Green River, combined with the tendency for flow rates to be greatest near the top of the water table, has produced very long passages parallel to the bedding. This fortuitous set of geologic conditions has resulted in the longest cave in the world.

Most speleothems, or cave decorations, in Kentucky are the common flowstone varieties composed of calcium carbonate, or calcite. Flowstone is formed from drips and films of water flowing down walls or falling from ceilings. When groundwater seeps into a well-ventilated cave, the carbon dioxide dissolved in the water escapes into the air. This reduces the acidity of the solution, and calcite is precipitated on the cave walls. The shapes that flowstones take range from the well-known stalactites, stalagmites, and massive draperies to elegant cave pearls and delicate helictites. Cave pearls are formed by the tumbling action of constantly dripping water. The thin, translucent sheet of calcite slowly built out from the wall by a trickle of water running down a narrow path is called bacon. The shapes of formations such as helictites, anthodites, and heligmites are controlled by the structure of the slowly growing calcite crystals. The color of flowstone is caused by impurities. Pure calcite is white to translucent. Various shades of red and yellow are caused by iron oxides, and black is caused by manganese oxides. Sulfate minerals form another type of speleothem found in Kentucky, common in the Mammoth Cave area. Intricate snowballs, gypsum flowers (figure 12.8), angel hair, cave cotton, cave grass, and translucent stalactites are just a few of the forms. These sulfate deposits are most commonly gypsum but may also be other minerals, such as epsomite and mirabilite.

All speleothems form over the course of thousands of years. The removal or destruction of speleothems is permanent and disfigures the cave, thus denying future explorers the visual enjoyment. It is considered unethical by cave explorers, and removal of speleothems is illegal in Kentucky.

How the Karst Aquifer Shapes the Surface

The most noticeable karst landform to a casual observer is probably a sinkhole (figure 12.9). A sinkhole is a

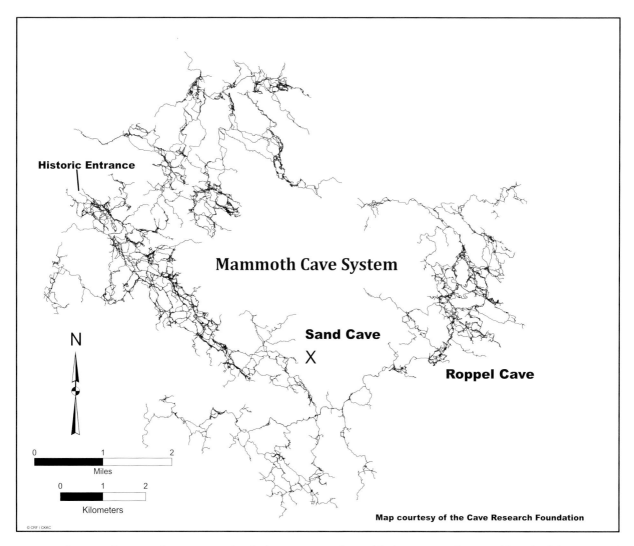

Historic Entrance

Mammoth Cave System

Sand Cave
X

Roppel Cave

N

0   1   2
Miles

0   1   2
Kilometers

© CRF / CKKC

Map courtesy of the Cave Research Foundation

12.7. Line drawing of Mammoth Cave, Kentucky, the longest cave in the world. The aggregated surveyed length of the connected and traversable passages now exceeds 400 miles. Used by permission of the Cave Research Foundation.

naturally occurring, topographically closed, internally drained depression underlain by soluble bedrock. This definition excludes collapsed storm sewers, mines, and any collapse of other man-made underground openings. "Doline" is often used as a synonym for sinkhole, but it more specifically has a cross section in the shape of a shallow bowl. The feature is caused exclusively by the dissolution of the underlying limestone. Keep in mind that all the material removed from the depression—the calcite in the limestone bedrock, insoluble

chert and silicified fossils, clay interbeds, and any unconsolidated mantle over the bedrock, such as soil, loess, alluvium, and residuum—has been eroded by water through the karst conduits in the bedrock.

Before much insoluble material can be transported through the karst conduits, they must become large enough that they have turbulent flow and can suspend and move the load of insoluble sediment. The critical diameter for the onset of turbulent flow and the transport of sediment is ten millimeters. Of course, unnatu-

12.8. A gypsum flower in Mammoth Cave will remain in place for millennia because it is well above the modern water table. One flood would destroy the delicate, water-soluble mineral. Photo by James C. Currens.

ral sediment loads from construction sites and tilled fields can overload and block even larger conduits. Because the doline has an established outlet throat, whether a single large opening or a converging number of smaller openings, soil and sediment erosion tends to be on the surface until the throat is reached. Thus, cover-collapse sinkholes seldom occur inside an existing sinkhole unless an attempt has been made to fill the sinkhole.

Sinkholes or dolines are classified by their geometry or by the process that developed them. There are three principal types of sinkholes, for example: solution, subsidence, and buried. The solution sinkhole subsides gently as the underlying limestone is dissolved away. A thin layer of residuum subsides at the same rate as the bedrock surface and is partly eroded into the subsurface. Subsidence sinkholes have a thick unconsolidated mantling layer (residuum, drift, alluvium, outwash) that has been eroded into the opening in the bedrock. The bedrock, however, has only a minor depression in its surface. Subsidence sinkholes are caused by very loose fill that crumbles into the underlying conduit without sudden

collapse. Episodic collapses of soil arches developed in the overlying material are a special case of subsidence sinkholes. The third type, buried sinkholes, is rare in Kentucky because they are commonly the result of glaciations filling in the sinkhole with drift, as in the examples for the subsidence sinkhole. The difference is that the depression is in the bedrock surface, and the mantling material has already subsided into the depression in the bedrock.

When a doline or sinkhole exceeds a loosely defined size, indicated by the presence of smaller sinkholes in its interior, it is termed a karst valley. Karst valleys are a relatively common topographic feature and may encompass a number of smaller karst features: cave entrances, estevelles (swallow hole–like opening that acts as a spring under some flow conditions), open-air pits, perched springs (wet weather), and others. Very large karst valleys (several square kilometers), with valley floors near base level that flood frequently, are termed poljes (pronounced "pol-ye"). The shape and size of true poljes are controlled by faults that cut the limestone into huge blocks that have moved downward to place the limestone

12.9. The disposal of domestic or any other kind of waste in sinkholes can threaten valuable water resources. This sinkhole could be the source of water for a well or spring. Photo by James C. Currens, Kentucky Geological Survey.

against insoluble rock. The insoluble rock limits the enlargement of the polje at the margins of the floodplain, which is on limestone. There are two karst valleys in Kentucky that approach being poljes: the Mahurin Spring karst valley near Leitchfield in Grayson County and Sinking Valley in Jessamine County. Both have broad floodplains and flood frequently, but only Mahurin Spring is bounded by faults.

Other features often called sinkholes are karst windows, estevelles, sinking springs, and unroofed vertical shafts. Karst windows (or cave sinkholes) are sections of cave that have collapsed to the surface. Although bedrock collapse is often thought to be the cause of most sinkholes, it is actually rare. If the passage has a flowing stream, most if not all of the fallen rock and other material will be eroded away. An estevelle is similar to a karst window in that it communicates proximally to the water-carrying cave, but it is generally too small to enter. When precipitation is unevenly distributed, the head or water pressure in the conduit may allow runoff from a storm close to the estevelle to sink into the feature, whereas rain occurring in the headwaters of the karst basin may result in the estevelle discharging water as a spring. Sinking springs are a near cousin of an estevelle and a karst window, but the spring is discharging from a perched aquifer. The water runs across the surface a short distance and sinks back into the ground. Unroofed vertical shafts are vertical shafts developed when the roof was intact that have fallen in to create an open-air pit.

Cover-collapse sinkholes, or dropouts, are perhaps another karst feature many people recognize as a sinkhole. These are natural phenomena and can develop with or without human influence on the concentration or route of runoff. A cover-collapse sinkhole occurs when a cavity in the soil or other mantling material is created because of internal erosion into an underlying karst conduit. The soil forms an arch over the cavity. The underside of the arch peels off in layers as the soil is wetted and dried. Both hydrostatic pressure of water in the soil pores surrounding the arch and the lubrication from the wetting in the same layer play a role in the cyclic enlargement of the arch. The fallen soil or other unconsolidated material is either stored in a large cavity in the limestone or falls into a moving stream and is carried away. The arch continues to enlarge until it becomes too thin to distribute the stress from its own weight and shears near the perimeter of the arch, creating the cookie-cutter-like opening often seen in the surface (figures 12.10 and 12.11). Cover collapse is one of the more destructive natural hazards in karst areas. It has damaged buildings, roads, and vehicles, killed livestock, and injured people in Kentucky.

Springs are perhaps the most important karst feature to people in terms of utility. The sinkholes and sinking streams that drain to a large karst spring can be many

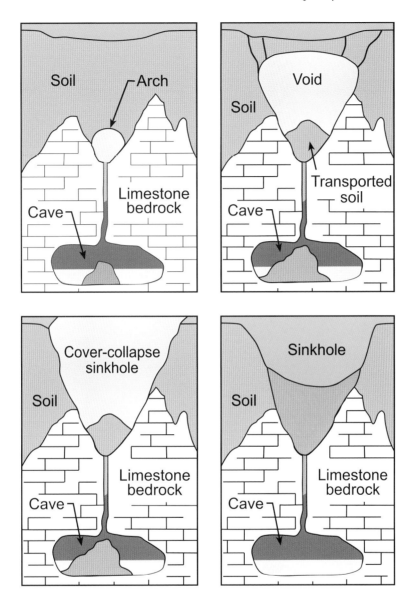

12.10. Stages in the development of a cover-collapse sinkhole. James C. Currens, Kentucky Geological Survey.

miles away from the spring. The catchment area of a spring, or springshed, can extend beneath surface watersheds, and the groundwater-basin boundary can have little relationship to surface drainage boundaries. Sinkholes and valleys shown on maps may seem to be in one watershed but drain to a faraway spring. Unlike a stream flowing on the surface, when a cave stream erodes head-

ward, it can extend beneath a ridge to capture flow from the adjacent groundwater basin. As a result, the actual springshed of a karst spring may be much larger or smaller than is apparent from topographic maps.

The major types of springs are blue-hole or underflow, overflow, intrabasin or wet-weather, and cave springs. Springs that discharge from a cave or a conduit

12.11. A cover-collapse sinkhole in Western Pennyroyal, Christian County, near Hopkinsville, Kentucky. Photo by Glenn Beck, Kentucky Geological Survey.

12.12. A blue-hole spring in the Western Pennyroyal, Logan County, Kentucky. Photo by James C. Currens.

that is only partly full, and thus the water surface is in contact with the air, are termed gravity, free-surface-flow, or cave springs. Springs that are not part of a distributary system and discharge from depth via a water-filled conduit or cave are called rise-pool or blue-hole springs (figure 12.12). The water is in effect discharging from a semiconfined aquifer because of the comparatively low resistance to flow in the main conduit.

Distributary spring systems are a branching series of conduits that discharge water to multiple springs (figure 12.13). As the water flowing in the conduits nears a permanent surface-flowing stream into which it discharges, the water seeks the lowest available exit and is constantly creating new spring openings down the gradient. Frequently, flow from one or two springs in a distributary system rises from a completely water-filled conduit as a blue-hole or underflow spring. During higher flows, the intermittently abandoned openings, or overflow springs, also discharge water. Several openings may develop almost simultaneously, resulting in many springs along a stream channel that drain a single groundwater basin. Distributaries are quite common in Kentucky.

Swallow holes are points along streams and in sinkholes where surface flow is lost to underground conduits.

Swallow holes range in diameter from a few centimeters to tens of meters, and some are also cave entrances (figure 12.14). Swallow holes are often large enough to allow large objects such as tree limbs and cobble-size stones to be transported underground. This also means that waste dumped into sinking streams can easily reach underground streams. It is not uncommon for discarded automobile tires and home appliances to be found deep within caves with flowing streams and large swallow-hole entrances. Likewise, sewage, paint, motor oil, pesticides, and other pollutants are not filtered from water entering a karst aquifer.

Many intermittent streams on karst in Kentucky are losing streams. They sink underground to reemerge further down the surface course. They may lose their flow all at once into a large swallow hole, but more commonly their flow gradually diminishes downstream as they seep through the gravel bed of the channel.

Karst natural bridges are rare phenomena compared with dolines and many other karst features. The best example of a karst natural bridge in Kentucky is Smokey Bridge in Carter Caves State Park (figure 12.15). Another example in Russell County is Rock House Natural Bridge on Kentucky 379, about two miles west of Creelsboro. The stream flowing through the karst natural bridge has pirated flow from the surface stream, Jim's Creek, directly to the Cumberland River. Others examples exist, including those in Cave Hollow, Elliott County, but are not easily accessible.

12.13. A large spring near Radcliff discharges from multiple openings on the side of a hill. The groundwater basin (aquifer) is perched (held above the true water table) on a thick bed of shale. Photo by James C. Currens.

12.14. A swallow entrance to a cave. The blue-white substance in the water is optical brightener tracer. Photo by James C. Currens.

Mapping Karst Aquifers

The springshed or groundwater basin of a karst aquifer is often poorly known. Information about it is vital, however, for people who use springs as water supplies. Water recharge to karst aquifers occurs either directly, through swallow holes and sinkholes, or indirectly through the pores in the soil overlying the limestone bedrock. Although the soil overlying a karst aquifer provides some filtration of contaminants transported by infiltrating water, almost none takes place between swallow holes and springs. For example, the number of fecal bacteria in water collected from springs draining urban and agricultural areas can reach counts of thousands of bacteria per hundred milliliters of water, making the water unsafe for people to drink and more expensive to treat so that it is safe. To protect the water quality at the spring, it is essential to know what area contaminants may come from. Groundwater tracing is the traditional tool for this task. Groundwater dye-tracing experiments are used to determine the routes groundwater takes in karst aquifers and the location, size, and shape of watersheds draining to specific springs (figure 12.16).

12.15. Smokey Natural Bridge in Carter Caves State Resort Park, a karst natural bridge. Photo by James C. Currens.

Although the concept of pouring dye into a sinking stream and waiting for the dye to reappear at a spring is very simple, completing these experiments with consistent success requires considerable skill. It would be inefficient to camp out at a spring and watch for the dye to emerge because it could take many days to resurface, it could flow from the spring at night, it could be too di- luted to see, or it could discharge from a different, unexpected spring. To monitor many springs simultaneously and around the clock, packets of material that adsorb the dye are attached to anchors and are placed in the water at the spring mouth. A commonly used material for dye detectors, also known as "bugs," is activated carbon charcoal. Bugs are placed in every spring the water could

12.16. A small swallow hole is a good location for injecting tracer dyes. Photo by David Lutz. Used with permission.

conceivably flow to, and then dye is poured into the swallow hole. The bugs are changed once a week until after the dye has come through. The bugs are taken to a laboratory where they can be carefully examined for very low concentrations of dye. A wide variety of tracers are now used, but the most common groundwater tracers are dyes that fluoresce (glow) when they are exposed to ultraviolet (black) light. The U.S. Environmental Protection Agency has approved the dyes commonly used for groundwater tracing.

Well over 2,800 groundwater dye traces have been made in the karst of Kentucky by cave explorers, students, karst hydrogeologists, consultants, and regulatory personnel since R. B. Anderson made the first reported trace in 1925. That trace was used to determine whether the site of a proposed dam on the Green River would leak. The trace showed that it would leak, and the dam was not built. Most groundwater-tracing data either are not published or are recorded only in agency and client reports, term papers, theses, and dissertations. These documents are difficult to find at best. An effort is under way to compile these data into a comprehensive and revisable series of regional-scale maps. The maps are pub-

lished in the 1:100,000-scale topographic quadrangle series and cover an area 30 × 60 minutes, the equivalent of thirty-two 7½-minute topographic quadrangles. Very few karst groundwater traces have been done specifically for the production of these maps, but one of them is shown in figure 12.17. The central groundwater basin in the figure drains to Crawford Spring. Braxton Bragg's Confederate soldiers filled their canteens here en route to fight the Battle of Perryville in October 1862. Did any of those soldiers realize that the cool, refreshing water had been diverted from the headwaters of the Salt River miles to the east? Today, many of Kentucky's karst groundwater basins have been mapped, but the karst hydrogeology of over half the karst areas of the state is still unknown.

The Shape of Kentucky Karst in the Future

It took over ten million years for the landscape of Kentucky to approach its current form, and Mammoth Cave is well over five million years old. In contrast, the activities of humans are causing rapid changes in our environment and the shape of Kentucky karst in particular. One of the most important issues is the pressure suburban and urban development, agriculture, silviculture, and transportation is putting on karst as a natural resource, both aesthetically and for practical uses such as water supply. As long as the population of Kentucky continues to grow, and more land is needed for new housing, commercial buildings, industry, transportation, and public buildings, the more frequently will negative results occur from siting these facilities on karst. Of course, the best and only certain way to avoid karst geohazards is to avoid building on the more sensitive limestone areas. Kentucky, however, is half karst, and to prohibit construction on these areas would severely damage the economy of the state and the economic well-being of many of its citizens. A better way to minimize damage to and from karst is to understand the process and learn how to safely build over and around the karst landforms that we must learn to appreciate. It is to be hoped that this will be the shape of things to come.

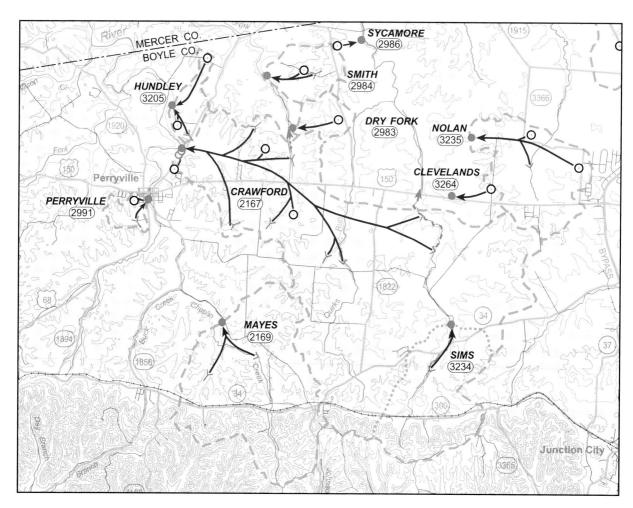

12.17. A segment of the Harrodsburg 30×60 minute topographic quadrangle showing the results of groundwater traces conducted by Kentucky Geological Survey staff. James C. Currens and Randy Paylor.

Additional Resources

Borden, James D., and Roger W. Brucker. 2000. *Beyond Mammoth Cave: A Tale of Obsession in the World's Longest Cave*. Carbondale: Southern Illinois University Press.

Brucker, R. W., and R. A. Watson. 1976. *The Longest Cave*. New York: Alfred A. Knopf.

Crain, Angela S. 2002. *Pesticides and Nutrients in Karst Springs in the Green River Basin, Kentucky, May–September 2001*. U.S. Geological Survey, FS_133-01_USGS, 4 p. http://pubs.usgs.gov/fs/2001/0133/report.pdf (accessed October 20, 2015).

———. 2010. *Nutrients, Selected Pesticides, and Suspended Sediment in the Karst Terrane of the Sinking Creek Basin,* Kentucky, 2004–06. U.S. Geological Survey, Scientific Investigations Report 2010-5167, 48 p. http://pubs.usgs.gov/sir/2010/5167/pdf/sir20105167.pdf (accessed October 20, 2015).

Currens, J. C. 1993. Caves. In *The Kentucky Encyclopedia*, ed. J. E. Kleber, 174–76. Lexington: University Press of Kentucky.

———. 1999. *Mass Flux of Agricultural Nonpoint-Source Pollutants in a Conduit-Flow-Dominated Karst Aquifer, Logan County, Kentucky*. Kentucky Geological Survey, RI_001_12, 162 p. http://kgs.uky.edu/kgsweb/olops/pub/kgs/ri01_12.pdf (accessed October 20, 2015).

———. 2001a. *Generalized Block Diagram of the Eastern Pennyroyal Karst*. Kentucky Geological Survey, MCS_017_12,

1 sheet. http://kgs.uky.edu/kgsweb/olops/pub/kgs/mc17_12 .pdf (accessed October 20, 2015).

———. 2001b. *Generalized Block Diagram of the Inner Bluegrass Karst.* Kentucky Geological Survey, MCS_015_12, 1 sheet. http://kgs.uky.edu/kgsweb/olops/pub/kgs/mc15_12 .pdf (accessed October 20, 2015).

———. 2001c. *Generalized Block Diagram of the Pine Mountain Karst.* Kentucky Geological Survey, MCS_018_12, 1 sheet. http://kgs.uky.edu/kgsweb/olops/pub/kgs/mc18_12 .pdf (accessed October 20, 2015).

———. 2001d. *Generalized Block Diagram of the Western Pennyroyal Karst.* Kentucky Geological Survey, MCS_016_12, 1 sheet. http://kgs.uky.edu/kgsweb/olops/pub/kgs/mc16 _12.pdf (accessed October 20, 2015).

———. 2001e. *Protect Kentucky's Karst Aquifers from Nonpoint-Source Pollution.* Kentucky Geological Survey, MCS_027_12, 1 sheet. http://kgs.uky.edu/kgsweb/olops/pub/kgs/mc27_12 .pdf (accessed October 20, 2015).

———. 2002. *Kentucky Is Karst Country: What You Should Know about Sinkholes and Springs.* Kentucky Geological Survey, IC_004_12, 35 p. http://kgs.uky.edu/kgsweb/olops /pub/kgs/ic04_12.pdf (accessed October 20, 2015).

———. 2005. *Changes in Groundwater Quality in a Conduit-Flow-Dominated Karst Aquifer as a Result of Best Management Practices.* Kentucky Geological Survey, RI_011_12, 72 p. http://kgs.uky.edu/kgsweb/olops/pub/kgs/ri11_12.pdf (accessed October 20, 2015).

Currens, J. C., R. L. Paylor, and J. A. Ray. 2002. *Mapped Karst Groundwater Basins in the Lexington 30 × 60 Minute Quadrangle.* Kentucky Geological Survey, MCS_035_12, 1 sheet. http://kgs.uky.edu/kgsweb/olops/pub/kgs/mc35_12 .pdf (accessed October 20, 2015).

———. 2003. *Mapped Karst Groundwater Basins in the Harrodsburg 30 × 60 Minute Quadrangle.* Kentucky Geological Survey, MCS_058_12, 1 sheet. http://kgs.uky.edu/kgsweb /olops/pub/kgs/mc58_12.pdf (accessed October 5, 2016).

Currens, J. C., and J. A. Ray. 1998a. *Mapped Karst Ground-Water Basins in the Beaver Dam 30 × 60 Minute Quadrangle.* Kentucky Geological Survey, MCS_019_11, 1 sheet. http:// kgs.uky.edu/kgsweb/olops/pub/kgs/mc19_11.pdf (accessed October 20, 2015).

———. 1998b. *Mapped Karst Ground-Water Basins in the Harrodsburg 30 × 60 Minute Quadrangle.* Kentucky Geological Survey, MCS_016_11, 1 sheet. http://www.uky.edu/KGS/pdf /mc11_16.pdf (accessed October 20, 2015).

———. 1998c. *Mapped Karst Ground-Water Basins in the Somerset 30 × 60 Minute Quadrangle.* Kentucky Geological Survey, MCS_018_11, 1 sheet. http://kgs.uky.edu /kgsweb/olops/pub/kgs/mc18_11.pdf (accessed October 20, 2015).

Dougherty, P. H., ed. 1985. *Caves and Karst of Kentucky.* Kentucky Geological Survey, ser. 11, Special Publication 12, 196 p.

Ford, D. C., and P. W. Williams. 1989. *Karst Geomorphology and Hydrology.* London: Unwin Hyman.

May, Michael T., Kenneth W. Kuehn, Christopher G. Groves, and Joe Meiman. 2005. *Karst Geomorphology and Environmental Concerns of the Mammoth Cave Region, Kentucky.* American Institute of Professional Geologists, GBS_AIPG, 49 p. http://www.uky.edu/KGS/geoky/fieldtrip/2005%20 AIPG%20Guidebooks/MammothCave.pdf (accessed October 20, 2015).

May, Michael T., Kenneth W. Kuehn, and S. Schoefernacker. 2007. *Geology of the Mammoth Cave and Nolin River Gorge Region with Emphasis on Hydrocarbon and Karst Resources, Part II: Rock Asphalt Redux and Paleovalleys Anew.* Kentucky Society of Professional Geologists, GBS_KSPG, 24 p. http://www.kspg.org/pdf/KSPGAAPG07FTday2.pdf (accessed October 20, 2015).

May, Michael T., Kenneth W. Kuehn, and Frederick D. Siewers. 2007. *Geology of the Mammoth Cave and Nolin River Gorge Region with Emphasis on Hydrocarbon and Karst Resources, Part 1: Geomorphology, Stratigraphy, and Industrial Materials.* Kentucky Society of Professional Geologists, GBS_KSPG, 24 p. http://www.kspg .org/pdf/KSPGAAPG07FTday1.pdf (accessed 20, 2015).

Palmer, A. N. 1981. *A Geological Guide to Mammoth Cave National Park.* Teaneck, NJ: Zephyrus Press.

———. 2007. *Cave Geology.* Dayton, OH: Cave Research Foundation Cave Books.

Paylor, R. J., and J. C. Currens. 2001. *Karst Occurrence in Kentucky.* Kentucky Geological Survey, MCS_033_12, 1 sheet. http://kgs.uky.edu/kgsweb/olops/pub/kgs/mc33_12.pdf (accessed October 12, 2015).

Paylor, R. L., L. Florea, M. Caudill, and J. C. Currens. 2004. *A GIS Coverage of Karst Sinkholes in Kentucky.* Kentucky Geological Survey, DPS_005_012, 1 CD-ROM.

Ray, J. A., and J. C. Currens. 1998. *Mapped Karst Ground-Water Basins in the Campbellsville 30 × 60 Minute Quadrangle.* Kentucky Geological Survey, MCS_017_11, 1 sheet. http://kgs.uky.edu/kgsweb/olops/pub/kgs/mc17_11.pdf (accessed October 20, 2015).

———. 2000. *Mapped Karst Ground-Water Basin in the Bowling Green 30 × 60 Minute Quadrangle.* Kentucky Geological Survey, MCS_022_12, 1 sheet. http://kgs.uky.edu/kgsweb/olops /pub/kgs/mc22_12.pdf (accessed October 20, 2015).

Ray, J. A., J. R. Moody, R. J. Blair, J. C. Currens, and R. L. Paylor. 2009. *Mapped Karst Groundwater Basins in the Tell City and Part of the Jasper 30 × 60 Minute Quadrangles.* Kentucky Geological Survey, MCS_196_12, 1 sheet. http://kgs .uky.edu/kgsweb/olops/pub/kgs/MCS196_12.pdf (accessed October 20, 2015).

Thrailkill, J. V., L. E. Spangler, W. M. Hopper Jr., M. R. McCann, J. W. Troester, and D. R. Gouzie. 1982. *Groundwater in the Inner Bluegrass Karst region, Kentucky.* University of

Kentucky Water Resources Research Institute, Research Report 136, 108 p.

Veni, G., H. DuChene, N. C. Crawford, C. G. Groves, G. N. Hupport, E. H. Kastning, Rick Olson, and B. J. Wheeler. 2001. *Living with Karst: A Fragile Foundation.* American Geological Institute, GIH_001_12, 64 p. http://www.americangeosciences.org/sites/default/files/karst.pdf (accessed October 20, 2015).

White, W. B. 1988. *Geomorphology and Hydrology of Karst Terrains.* New York: Oxford University Press.

CHAPTER THIRTEEN

Water Quality and Natural Resources in the Green River Basin

Brad D. Lee, Tanja N. Williamson, and Angela S. Crain

The Green River basin evolved from diverse parent materials, with the Upper Green River in karst terrain and the Lower Green River in stream deposits that overlay rocks rich in fossil fuels. The result is a basin in which people have benefited from their natural resources while, at the same time, dealing with degradation of the landscape and water quality. This chapter will begin by describing the physiographic interrelation among geology, soils, climate, hydrology, and vegetation and how they all come together as important to aquatic biological resources, with particular respect to mussels, fish, and crustaceans. We will explain how the resulting terrestrial part of the landscape was influenced by the hydrologic conditions.

Forests, agriculture, mining, and urbanization all change how water, people, and the landscape interact. Currently, approximately half of the Green River basin is involved in agricultural land uses. In a goal to restore the water quality and ecological habitat of the basin, agricultural producers, local watershed groups, and government agencies are cooperating in basin-wide programs. A successful example of this cooperation is the Green River Conservation Reserve Enhancement Program, which focused on working with willing landowners who received federal and state financial incentives for retiring land cleared for agricultural production to improve wildlife habitat and water quality in the basin. A current endeavor is the Mississippi River Basin Healthy Watersheds Initiative (MRBI), which encourages agricultural producers to implement conservation practices that avoid, control, and reduce nutrient runoff while maintaining agricultural productivity and im-proving wildlife habitat. Much of the MRBI approach is designed on the basis of a national evaluation of water quality, specifically nitrogen and phosphorus concentrations in stream water, illustrating how science can be used to understand how local water-quality issues compare with those in other regions of the country.

Physiography

The Green River basin encompasses approximately 9,230 square miles (mi^2) in Kentucky and Tennessee; 96 percent of it is in Kentucky (figure 13.1). The headwaters of the basin include the Barren and Upper Green Rivers, which flow through the karst-plain portion of the Pennyroyal and Mississippian Plateaus physiographic province. This portion of the basin covers 5,400 mi^2, with stream valleys at elevations of 390 feet above mean sea level (amsl) and uplands to 1,780 feet. The remaining 3,830 mi^2 include the Rough, Pond, Middle Green, and Lower Green basins, which flow through the Western Kentucky Coalfields physiographic province of Kentucky (figure 13.1), which is bisected by the river and has elevations ranging from 317 feet in the bottomlands to 1,182 feet amsl in the uplands.

The Pennyroyal and Mississippian Plateaus physiographic province has exposed limestone rocks at the surface that formed approximately 350 million years ago from compaction and cementation of ocean sediments rich in calcium carbonate. Because these calcium carbonate materials originally crystallized from water, water can also dissolve them, although the process is slow. The result is the Karst Plain of Kentucky (figure 13.1),

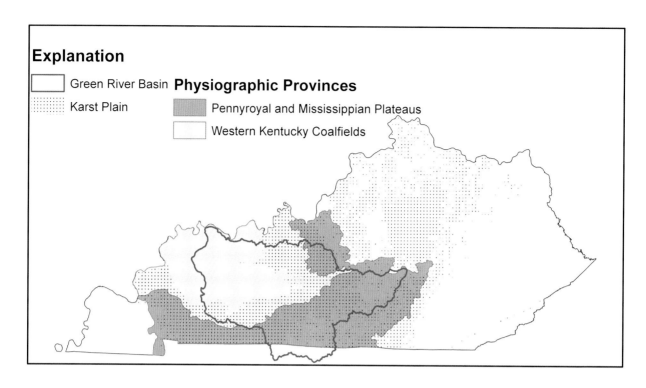

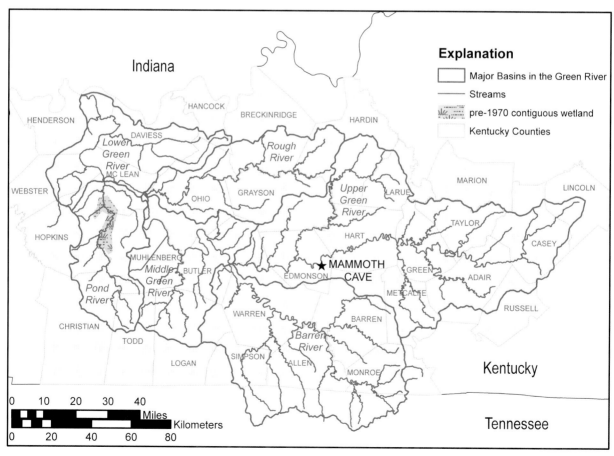

13.1. The Green River basin and Kentucky physiographic provinces (McDowell 1986).

where numerous depressions throughout the landscape are visible evidence of how water drains to underground conduits through fractures in the rocks. Some of these fractures enlarge to become caves; however, even where caves are not evident, a network of underground conduits is present that enables relatively rapid movement of water belowground. These underground conduits provide a rapid path for water to reach larger streams, resulting in two more common karst features: sinking, or "disappearing," streams and springs.

The underground conduits started as thin fractures, or cracks, in the rocks that have widened slowly over time as the edges of the cracks dissolved. Before the underground conduits were large enough to rapidly remove water from the surface, rainfall moved toward streams, as it does in other environments: a tortuous path through the soil of the hillslopes toward the nearest low point at a stream. Paths of this type of water movement are visible in any photograph or topographic map of the area, as indicated by visible stream channels that no longer have water in them (Taylor, Kaiser, and Nelson 2008). One example is the Graham Springs karst basin in south-central Kentucky (figure 13.2).

Essentially, these underground networks "pirate," or drain away, water flow from the surface, causing the surface streams to disappear. When the water in these underground networks reaches a larger stream, like the main tributary of the Green River, the water reappears as a spring. There are several implications of this subsurface water movement: (1) a shorter path and travel time to larger streams, (2) a relatively constant source of water from the underground conduits, and (3) a slowly growing network of underground conduits and sinkholes—depressions that connect the conduits to the surface. These sinkholes develop properties that are distinct from the landscape around them; for example, soil is frequently thicker and drains more slowly than that on nearby hillslopes that drain toward streams (Williamson, Taylor, and Newson 2013).

Natural Vegetation

The natural vegetation of the Green River basin is recorded in the soils, hillslopes, and river valleys of the landscape as a function of how plants grow, die, and protect the land surface from erosion. White oak was the most dominant tree species (Baskin, Chester, and Baskin 1997), but other native trees in the basin included cedar, elm, ash, hackberry, gum, poplar, dogwood, beech, ironwood, persimmon, and sassafras (Baskin, Baskin, and Chester 1994). These forest communities were supported by a frost-free growing season from mid-April to mid-October without extreme summer temperatures, average yearly temperature of 55° to 60°F, average yearly rainfall of 44 to 50 inches (in.), and a concentration of this rainfall during the winter and spring (Kerr et al. 1924). A variety of annual and perennial leafy plants flourished, including pennyroyal, the annual, minty plant that was abundant as late as the 1930s and from which the southern portion of the basin gets its name (Simon 1996).

A large portion of the Green River basin is known as the Barrens because of the abundant grassland described by explorers and settlers as early as 1784 (Baskin, Baskin, and Chester 1994; Bryant 2004). This region is marked by sinkholes of the karst plain (figure 13.1) that are usually less than 700 feet wide and 33 feet deep and, in some cases, have several feet of accumulated sediment at the bottom. However, several factors confirm that these areas were originally forests, including deep soils, a climate that is too rich in water to limit forest growth, the lack of endemic prairie plants and animals that are native to only this region, and the lack of certain animals that are found in extensive midwestern prairies, including the ornate box turtle and the Franklin ground squirrel (Baskin, Chester, and Baskin 1997). These grasslands were the result of extensive burning by Native Americans, probably to improve hunting and agriculture, and a change in vegetation is documented by an increase in both grass pollen and seeds of annual weeds around 4,000 years ago (Baskin, Chester, and Baskin 1997). Before this, there is evidence of a transition from a spruce-jack pine forest to a deciduous forest (with a seasonal leaf fall).

Distributed in the sinkhole ponds and river valleys were wetland environments that supported vegetation distinct from the uplands and slopes. Extensive cypress forests covered the bottomlands of the tributary stream valleys (figure 13.3), together with other trees like sycamore and

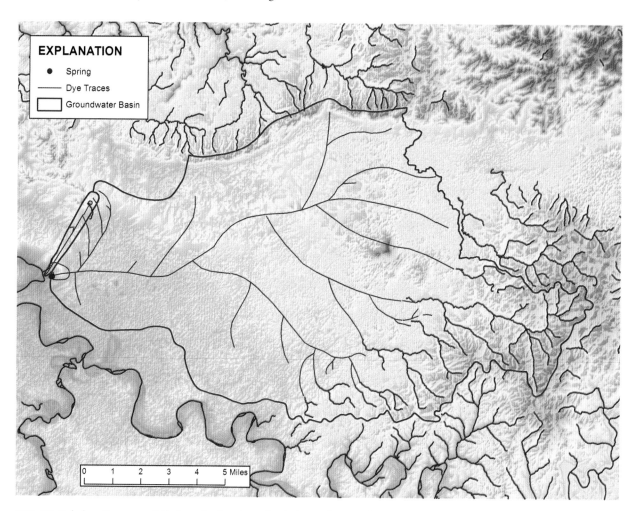

13.2. Digital elevation map of Graham Springs karst basin in south-central Kentucky. Basin boundary lines (red) are based on topography; blue lines trace the path of surface streams; purple dashed lines trace the paths of stream channels that are visible in the Digital Elevation Model (DEM) but no longer have water in them. These purple lines are underlain by a network of underground flow paths inferred by dye-tracing studies (Ray and Currens 1998, 2000). Water drained by the basin that travels in subsurface conduits is discharged from multiple karst springs along the Green River in the area of the blue dot. Illustration from http://ky.water.usgs.gov/projects/cjt_karst/index.htm, verified June 15, 2012.

elm (Burroughs 1924). Currently, wetlands are being rebuilt as mitigation for removal of bottomland hardwood forests. It is estimated that these wetlands will store about four times more carbon than the same area of cropland once they have been allowed to develop over a period of forty years. Within twenty-five years, these wetlands are expected to have developed enough to help control flooding, provide reliable habitat for aquatic life, and improve water quality (D'Angelo et al. 2005).

Soils

Much of the region is blanketed by loess from glacial outwash (material deposited by streams flowing away from a melting glacier). Loess is wind-blown, silt-sized sediment 0.002 to 0.05 millimeters (mm) in diameter (1/12,700 to 1/500 in.). In the Ohio River valley, this glacial outwash overlays limestone interbedded with sandstone and shale (compacted and cemented sediments

13.3. Kentucky bald cypress swamp. Photo by Jerry McIntosh, courtesy of USDA Natural Resources Conservation Service.

2 mm [1/12.7 in.] to 0.05 mm and less than 0.002 mm in diameter, respectively). The loess is several feet thick near the Ohio River and thins out in a southeasterly direction across the Green River basin. In some areas, lacustrine sediments from lakes that developed in glacial outwash are mixed with the loess deposits.

The Crider, Kentucky's state soil (figure 13.4), is prevalent throughout the Pennyroyal and Mississippian Plateaus physiographic province (Natural Resources Conservation Service Soil Survey Staff 2011) and covers approximately 4 percent of the Green River basin. The Zanesville soil covers approximately 5 percent of the basin and is the most abundant series in the basin, mostly in the Western Kentucky Coalfields physiographic province (Natural Resources Conservation Service Soil Survey Staff 2014). Both have a brown surface horizon that is 5 to 11 inches thick and overlies a darker brown subsoil derived from loess and weathered material. The Crider soils are developed on limestone, and this bedrock is typically 100 inches below the surface. In contrast, the Zanesville soils are developed on interbedded sandstones and shales, and the bedrock is closer to the surface, sometimes as shallow as 40 inches. Both soils develop on hillslopes, but the Zanesville soils are not as

well drained because of the presence of a fragipan (dense soil horizon that reduces vertical water movement) approximately 20 inches below the surface that acts as an impediment to both water and roots. Both soils are currently used for crops and pasture, but the thick soil profile of the Crider is indicative of the native vegetation of hardwood forest of oak, maple, and hickory.

All the soils in the Green River watershed will respond to application of lime and fertilizer. If they have never been limed, applications of ground limestone are required to raise the pH level sufficiently for good growth of alfalfa and other crops that grow only on nearly pH-neutral soils. Plant-available phosphorus and potassium levels are naturally low in all soils; however, repeated land application of animal manure over several years has increased the level of phosphorus to an excessive level in some areas of the basin. Additions of lime and fertilizer should be based on the result of a soil test, crop growth requirements, and yield expectations.

The soils used for crops in the Green River basin in general have a silt-loam texture at the surface that is light in color and contains modest organic-matter content. A loam is a soil texture with a mix of sand, silt, and clay that holds enough water for plant growth while also

13.4. Crider series soil profile. Photo by Jerry McIntosh, courtesy of USDA Natural Resources Conservation Service.

draining well enough that adequate air circulation is maintained for plant respiration. Generally, the structure of silt-loam surface horizons is weak, and intense rainfall causes the formation of a crust on the surface. The crust is hard when dry and tends to be impervious to water. Once the crust forms, it reduces infiltration and increases runoff. This crust frequently forms when land use involves removing vegetation and leaving the soil surface exposed.

Land Use

Most population centers throughout the Lower Green basin developed by relying on streams as water sources instead of groundwater. In some areas, open or drilled wells to a depth of approximately 100 feet provided a reliable source of water for individual homes (Burroughs 1924). However, farms frequently relied on cisterns in order to have a second source during the drier times of year when groundwater levels in the interbedded sandstones, shales, and limestones might be lower than the depth of the wells. Larger cities and towns relied on deep wells, reservoirs, and river water, only some of which was treated.

In the Western Coalfields physiographic province and the bottomlands of the Lower Green basin, the soils are underlain by a series of geologically recent sediments deposited by tributaries of the Green River and lake deposits near the confluence with the Ohio River. These sediments are the only reliable sources of groundwater along the larger tributaries and the Green River. In smaller valleys, the deposits are too thin and fine grained to provide a domestic water supply (approximately 500 gallons per day). Underlying these sediments is a series of sandstones, shales, and limestones that provide a reliable source of groundwater where wells intersect water-filled fractures; the more abundant the shale (compacted and cemented layers of clay), the less available the water. Within these interbedded sandstones, shales, and limestones are the coal seams of the Western Coalfields. Consequently, in some areas where coal has been mined, the water may contain iron sulfates (Sams and Beer 1999).

The entire region was originally described as a swamp. Typhoid and malaria were common in the early 1900s (Burroughs 1924) because of the abundance of mosquito-breeding habitat and contaminated water in the slow-moving streams and stagnant water of both natural and man-made ponds (for mining and agriculture). The NRCS classifies 12 percent of the basin soils as poorly drained (USDA-NRCS 2010). Large areas of land in the Lower Green were "improved" by draining the swampy bottomlands by deepening and straightening streams, creating a network of ditches to move water to the larger tributaries, and laying tile drains in land to be used as agricultural fields. The design of surface and subsurface drainage systems varies with the kind of soil, but in general, a combination of surface drainage and tile drainage is needed in most wet soils for intensive row cropping. Finding adequate outlets for tile drainage systems is difficult in some of the lower landscape positions because of the flatness of these landscapes. For example, the Pond River valley bottom is three miles wide in some places, with elevations as much as twenty feet below that of the confluence of the Green and Ohio Rivers.

Although drainage projects were popular in the past to increase agricultural land and production, some efforts are being made to reestablish natural drainage patterns and bring back wetlands. The 1967 *National Atlas* (Marschner and Anderson 1967) identifies a preserved, contiguous wetland of approximately 200 km² (48,000 acres) in the Pond River valley (figure 13.1), but the 2001 National Land Cover Database (Homer, Fry, and Barnes 2012) shows it as mostly agricultural land with a patchy 10 mi² classified as wetlands. To the north, one project along Panther Creek drained 55,500 acres in the early 1900s at a cost of $575,000 (Burroughs 1924), approximately $6.6 million in 2010 dollars.

Currently, the NRCS classifies 57 percent of the Green River basin as prime farmland (USDA-NRCS 2010). The amount of "improved" land in a county, meaning either drained or cleared, was directly correlated to the average income per farm in the early twentieth century (Burroughs 1924). Smaller farms were able to thrive by growing tobacco, the cash crop that had been brought by settlers to Kentucky from Virginia and had been commercialized as early as 1825. By the 1930s, a demonstration farm of 149 acres was established on the Ogden campus in Bowling Green, now Western Kentucky University, which was the foremost strawberry market in the commonwealth (Simon 1996). Larger farms were necessary for growing less prosperous crops like wheat, vegetables, fruits, corn, or livestock. In general, individual wealth was higher in areas of gently rolling land where agriculture could succeed than in areas that were dominated by mining. However, by 1920, problems of erosion and land degradation were already an issue because of overcropping, aggressive logging, and strip mining, and the steady increase in "improved" land acreage came to a halt. In 2010, 77,990 acres (1.4 percent of the basin) were classified as some type of gullied or degraded land (USDA-NRCS 2010). A 1983 soil-erosion and sedimentation survey showed that forty-three of the top one hundred sediment-producing streams in Kentucky were in the Green River basin. This area accounted for both the

greatest number of streams and the largest area undergoing extreme erosion and was concentrated in the Barren, Pond, and Rough basins in a combination of mining and agricultural land covers.

Soil erosion is a major concern on sloping land and pastureland in the Green River basin. If soil slope is more than 2 percent, erosion is a hazard. In general, the greater the slope, the greater the hazard, and the more difficult it is to control soil erosion. Loss of the surface-soil horizon through erosion is damaging for two reasons. First, productivity is reduced as the carbon-rich surface horizon is lost and part of the subsoil is incorporated into the plow layer. Loss of the surface horizon is especially damaging on soils that have a clay-type subsoil and on soils that have a layer in the subsoil that limits the depth of the root zone. Such layers include a fragipan or bedrock, both of which are commonly found in the Green River basin. Second, soil erosion results in sedimentation of streams, decreasing stream clarity and carrying nutrients that can lead to excessive algal growth.

Erosion-control practices provide a protective surface cover, reduce runoff, and increase infiltration. A cropping system that keeps a vegetative cover on the soil for extended periods can hold soil-erosion losses to amounts that will not reduce the productive capacity of the soils. On livestock farms, which require pasture and hay, the legume and grass forage crops in the cropping system reduce erosion on sloping land, provide nitrogen, and improve soil quality for the following crop.

Minimum tillage and a cropping system that produces substantial vegetative cover are both needed on sloping soils to slow soil-erosion processes. Minimizing tillage and leaving crop residue on the soil surface help increase infiltration and reduce the hazards of runoff and erosion. Corn and soybean production under a minimal or no-tillage management strategy (figure 13.5) is effective in reducing erosion on sloping land. The acreage planted by this method is increasing across Kentucky (Horowitz, Ebel, and Ueda 2010).

In addition to pasture and row-crop agriculture, the Green River basin provided numerous other natural resources. Abundant timber provided for industrial growth, but there was little conservation management, and this industry was in decline by the 1920s (Burroughs

13.5. No-till conservation practice of planting soybeans in corn stubble. Courtesy of the USDA Natural Resources Conservation Service.

1924). Saltpeter in the form of calcium nitrate was mined from the Mammoth Cave region in Edmonson County (figure 13.1) as early as 1811 and was in high demand during the War of 1812 (Simon 1996). The stream deposits of sand and gravel, together with the sandstone and limestone still provide raw material for different grades of cement, concrete, and building stones. Clay was mined and used locally for construction of sewer pipes, bricks, and the tile drains that were used to drain bottomlands. The sandstone provided a silica source that was 99 percent pure in some places. Some layers of the sandstone were naturally impregnated with bitumen, providing excellent rock asphalt that was used both across and outside Kentucky when paved roads were developed in the early 1900s. Iron-ore extraction began in the 1850s, primarily from siderite (iron carbonate, $FeCO_3$) associated with the interbedded limestone and coal in the Western Kentucky Coalfields physiographic province (Burroughs 1924).

Coal is not the only fossil fuel obtained from the Green River basin; oil, shale oil, and natural gas were also trapped by the structure of the rocks and were ex-tracted from the interbedded limestone and shale sequences in both the lower and upper parts of the basin as early as 1917 (Simon 1996). However, each of these resources pales in comparison with the coal extraction from the Western Coalfields from both underground and surface mining. The first commercial coal operation in this basin opened in 1820 along the Green River in Paradise, Muhlenberg County. The adjacent county, Hopkins County, was the leading coal producer in the commonwealth in the last decade of the nineteenth century and the first decade of the twentieth century (U.S. Energy Information Administration 2011). The Western Coalfields continue to be a significant source of coal; a 2009 review by the U.S. Energy Information Administration showed that a Hopkins County mine was the largest in Kentucky and twenty-sixth in the nation (U.S. Energy Information Administration 2008).

Water Resources and Water Quality

The Green River is one of Kentucky's largest, longest, and most navigable rivers. This 370-mile-long river

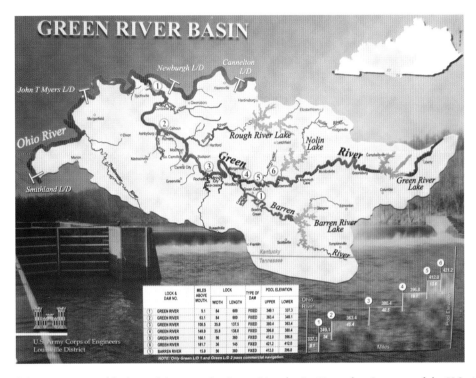

13.6. Location of the reservoirs and locks and dams in the Green River basin, Kentucky. Courtesy of the U.S. Army Corps of Engineers–Louisville District.

claims the title of Kentucky's longest river because 96 percent of it flows within the boundaries of Kentucky. Several small streams join together on ridges in Lincoln County to form the Green River. The river flows westward through Mammoth Cave (figure 13.1), where it is fed by many springs, sinkholes, caves, and the Nolin River. Downstream from the Mammoth Cave area, the Barren River flows into the Green River from the south. The confluence of the Barren and Green Rivers marks the beginning of the Green River's lower reaches. The Green River flows northward at this point and absorbs three major tributaries—Mud River, Pond River, and Rough River—before entering the Ohio River east of Henderson (Kleber 1992). According to the Nature Conservancy, the Green River gets its name from the river's depth, which gives it a green hue (http://www .nature.org/ourinitiatives/regions/northamerica/unit-edstates/kentucky/placesweprotect/green-river.xml). The Green River drains about 9,230 square miles of Kentucky, which is about 23 percent of the drainage

area of the entire state. A system of seven locks and dams enables navigation on the Lower Green River (figure 13.6).

Lakes and Reservoirs

In 2008, the Kentucky Division of Water reported that the chemical and biological health of twenty-nine lakes and publicly owned reservoirs was monitored in the Green River basin. A total of 69 percent of those reservoirs fully supported their assessed uses; this means that there were no significant limitations in terms of water quality. The remaining 31 percent only partially supported their assessed uses. Methylmercury (CH_3Hg^+) accumulation in fish tissue was by far the greatest cause of impairment. Atmospheric deposition is the dominant source of mercury in most rivers and streams. However, naturally occurring watershed features, like the wetlands and forests found in the Green River basin, can enhance the conversion of mercury to the more toxic

Table 13.1. Trophic State of U.S. Army Corps of Engineers' Reservoirs in the Green River Basin, 2008			
Reservoir	**Status of assessed uses**	**Trophic state**	**Eutrophication trend**
Barren River	Fully supporting	Eutrophic	Degrading
Green River	Partially supporting	Eutrophic	Stable
Nolin River	Fully supporting	Eutrophic	Stable
Rough River	Partially supporting	Mesotrophic	Degrading
Kentucky Department of Environmental Protection 2008.			

Trophic class	**Trophic index**	**Concentrations of chlorophyll *a* in milligrams per liter**
Oligotrophic	< 30–40	0–2.6
Mesotrophic	40–50	2.6–20
Eutrophic	50–70	20–56
Hypereutrophic	70–100+	56–155+

Oligotrophic—larger, deeper lakes with clear water, low phosphorus enrichment, low algal growth, and adequate dissolved oxygen throughout the water column.
Mesotrophic—an intermediate category with characteristics between the oligotrophic and eutrophic categories.
Eutrophic—smaller, shallower lakes with mucky bottoms, abundant accumulation of nutrients that support dense growth of algae, and low or depleted dissolved oxygen concentrations in bottom waters.
Hypereutrophic—smaller, shallower lakes with extremely nutrient-rich waters with frequent and severe algal growth.
Trophic Index: A quantitative scale based on the measurement of chlorophyll pigments and/or turbidity, and/or total phosphorus to estimate algal abundance (Carson 1977).

form of mercury—methylmercury. Other sources of impairment included (1) atmospheric deposition of toxics, (2) agriculture, (3) industrial point-source discharge, (4) other upstream sources, and (5) unknown sources (Kentucky Department of Environmental Protection 2008).

The U.S. Army Corps of Engineers plays a major role in providing for flood control and recreation in the Green River basin. In the 1960s, the U.S. Army Corps of Engineers impounded four major tributaries of the Green River to create Barren, Green, Nolin, and Rough River Lakes (figure 13.6). These reservoirs are monitored over the growing season (April–October) to determine the reservoir's trophic state (biological production) using the Carlson Trophic State Index for chlorophyll *a* (table 13.1). This index allows reservoirs to be ranked numerically according to increasing trophic state (oligotrophic, mesotrophic, eutrophic, and hypereutrophic; see table 13.1).

Rivers and Streams

According to the *2008 Integrated Report to Congress on Water Quality in Kentucky,* which includes physical, chemical, and biological analyses, fecal-indicator bacteria and sedimentation/siltation are the two most prevalent contaminants in the Green River basin (Kentucky Department of Environmental Protection 2008). Fecal-indicator bacteria are contaminants assessed only when direct contact of people with streams and water bodies is expected (i.e., swimming and fishing). Sedimentation/siltation is more associated with aquatic-life habitat. Causes of fecal-indicator bacteria affected 33 percent of assessed stream miles (555) in the Green River basin in 2008. Sedimentation/siltation affected 29 percent of assessed stream miles (486) in 2008 (table 13.2). The leading known sources of these contaminants include loss of riparian habitat, agri-

Table 13.2. Top Causes of Impairments and Probable Sources for Assessed Stream Miles in the Green River Basin, 2008			
Causes of impairments	**Assessed miles**	**Sources of impairments**	**Assessed miles**
Fecal-indicator bacteria	555	Source unknown	719
Sedimentation/siltation	486	Loss of riparian habitat	556
Habitat alteration	283	Agriculture	416
Nutrient/eutrophication	197	Channelization	392
Cause unknown	177	Nonirrigated crop production	334
Top five total	1,698		2,417

Source: Kentucky Department of Environmental Protection 2008.

culture, and channelization. Loss of riparian habitat is extremely detrimental to a stream's biological integrity, is often the direct result of land-use practices, and is a major contributing factor to sedimentation/siltation.

Water quality and water supply, stormwater management, flash flooding, and changes associated with surrounding development are all prominent local and regional issues that affect the state's surface waters. The 2000–2001 *State of Kentucky's Environment* report cited agricultural activities as a leading source of water pollution in Kentucky's waterways (Cole, Segal, and Lyle 2001). Some of these agriculture-related impacts are from

- excessive sedimentation,
- excessive nutrients,
- decreased biological oxygen availability,
- pesticides, and
- pathogens (e.g., fecal coliform, *Escherichia coli*).

In the Lower Green River basin, downstream from the confluence with the Barren River, confined animal-feeding operations that raise poultry (figure 13.7) and hogs produce large quantities of waste that is stored and applied to land as fertilizer because of its high nutrient value. If this manure is not stored and applied properly, runoff can pollute streams and water supplies, or the manure can be inundated by floodwaters. Dairy animal-feeding operations, in which animals seasonally roam in pastures, are predominant in the upstream portion of the Green River basin. These operations have significant impacts where cattle are not kept out of the streams by fences (figure 13.8) in terms of both pollution (animal waste) and physical damage to stream banks (Kentucky Department of Environmental Protection 2001).

Wetlands

Impairment of rivers and streams is an ongoing issue; however, the loss of Kentucky's wetlands also is a problem. Although only a portion of the wetlands that originally existed in Kentucky (estimated to have been 1.5 million acres) remains, loss of wetland acreage has slowed with federal and state regulations and disincentives in place for altering wetlands (Kentucky Environmental Quality Commission 1995). National Wetlands Inventory maps indicate that the Green River basin has the largest portion of remaining wetland acres in Kentucky with approximately 54,000 acres. This represents about 1 percent of the total land area in the watershed. The Pond Creek basin, located in the western portion of the Green River basin, contains the largest percentage of wetlands (table 13.3; U.S. Army Corp of Engineers 2011), but this is still less than half of the 1960s acreage (Marschner and Anderson 1967) and likely a smaller proportion of the original extent.

Groundwater

Impacts on groundwater quality occur more frequently in the karst areas of the basin and are most often caused by agricultural runoff, trash and hazardous

13.7. Poultry confined-animal feeding operation. Courtesy of the University of Kentucky.

13.8. Dairy animal-feeding operation allowing direct access to the stream. Photo by Kylie Schmidt, courtesy of the University of Kentucky.

Table 13.3. Number of Wetland Acres by Basin Name

Basin name	Wetland acres
Barren River	1,019
Lower Green	11,322
Middle Green	14,653
Upper Green	2,307
Pond Creek	19,717
Rough River	4,892
TOTAL	53,910

Source: U.S. Army Corps of Engineers 2011.

Contaminants and Their Effects

Sediments—the result of soil erosion causes reservoirs to lose capacity as silt settles out, reduces suitable habitat in streams, and transports attached contaminants (nutrients and metals).

Nutrients—nitrogen and phosphorus can cause an increase in algal growth; when algae die, their decomposition removes oxygen from the water, resulting in low dissolved-oxygen concentrations. The breakdown of some nitrogen compounds by bacteria also lowers dissolved oxygen.

Pesticides—pesticide runoff into streams can result in harm or death to beneficial plants and animals that live in or use the water. Some pesticides can remain in the environment for many years and pass from one organism to another (e.g., DDT).

Pathogens—a high concentration of pathogens can cause illness in humans and other animals and, in some instances, can cause death.

Source: Kentucky Department of Environmental Protection 2001.

materials in sinkholes, and runoff of nitrates from fertilizers. A groundwater study by the U.S. Geological Survey (USGS) (Crain 2001) found the following:

- Nine different pesticides were detected in eight karst springs sampled in the Green River basin.
- The five most frequently detected pesticides at all springs were atrazine (100 percent), simazine (93 percent), metolachlor (80 percent), tebuthiuron (66 percent), and prometon (58 percent).
- The pesticides detected were not necessarily the pesticides most heavily applied in the Green River basin.
- Nitrite plus nitrate-nitrogen concentrations did not exceed the U.S. Environmental Protection Agency's drinking-water standard (10 milligrams per liter) at any of the eight springs.

Conservation Programs

Although many water bodies are impaired, others with good water quality are afforded various levels of protection. The Nationwide Rivers Inventory (NRI) is a listing of more than 3,400 free-flowing river segments in the United States that are believed to possess one or more "outstandingly remarkable" natural or cultural values judged to be of more than local or regional significance. Under a 1979 presidential directive and related Council on Environmental Quality procedures, all federal agencies must seek to avoid or mitigate actions that would adversely affect one or more NRI segments (http://www.nps.gov/ncrc/programs/rtca/nri/hist.html). The segment of the Green River listed in the NRI extends through Edmonson, Hart, and Green Counties from river mile 189 at Mammoth Cave National Park and Lock 6 to river mile 290 at Greensburg (figures 13.1 and 13.6). The Green River was listed in the NRI in 1982 and is also listed with the commonwealth as a Kentucky Wild River and a State Outstanding Resource Water. The Kentucky Wild River Program is aimed at protecting sections of rivers identified as having exceptional quality and aesthetic character. Some activities that are strictly prohibited within a Wild River corridor include surface mining, clear-cutting timber, construction of dams, and disturbances to the land adjacent to the wild river

(Kentucky Department of Natural Resources and Environmental Protection, 1980).

The Green River, as celebrated in John Prine's song "Paradise," is an ecological wonder, being one of the most biologically diverse and rich branches of the Ohio River system. Its characteristic landscape features, such as karst topography and habitat diversity, make it unique. In 1996, the Green River Bioreserve project (Upper Green River only) was created to reduce nonpoint-source pollution and eliminate other stressors on aquatic organisms and habitats (Butler, Kessler, and Harrel 2003). The Green River Bioreserve covers 1,350 square miles and consists of the Upper Green River, its tributaries, and sections of Mammoth Cave National Park (Nature Conservancy 2010). The Green River Bioreserve extends 110 river miles from the tailwater of Green River Lake to downstream from Mammoth Cave National Park at Lock and Dam 6 on the Green River at Brownsville, Kentucky (figure 13.6).

In the United States, the most imperiled organisms are freshwater clams or mussels. The Green River is home to 71 of Kentucky's 103 mussel species (out of 300 species in North America). Of the 71 state mussel species, about 60 have been collected and identified in the Green River Bioreserve (Nature Conservancy 2010). Several of the 60 freshwater mussel species in the Green River Bioreserve are listed as federally endangered, such as the ring pink, clubshell (figure 13.9), fanshell, and rough pigtoe. In addition, the Kentucky creekshell is one of the 60 freshwater mussel species found only in Kentucky and only in the Upper Green River basin. In addition to freshwater mussels, the Green River Bioreserve is home to some 151 fish species (figure 13.10), or 19 percent of the U.S. freshwater fish fauna (Barnes 2002). Many of these mussel and fish species have been negatively affected by human influences in the basin; key threats are nonpoint-source pollution from agricultural runoff (nutrients and sediment), incompatible land development, competition with exotic species, and altered hydrologic (streamflow) and thermal (water-temperature) regimes (Green River Lake Dam) (table 13.1).

Improvements in conservation practices and decreases in nutrient applications have been supported by many cost-share programs for Kentucky producers as a financial incentive to encourage protection and restoration of wildlife and aquatic habitats. One such cost-share program was the Green River Conservation Reserve Enhancement Program (CREP) (U.S. Farm Service Agency 2010). In 2001, the United States Department of Agriculture (USDA) and the Commonwealth of Kentucky launched this CREP along a hundred-mile section of the Green River that flows from the Green River Reservoir Dam in Taylor County through Mammoth Cave National Park in Edmonson County (figure 13.1). In 2006, the CREP region was expanded thirty miles downriver to include environmentally significant areas downstream from the original project area, including Warren County and adjacent parts of the surrounding counties. The CREP program focused primarily on working with willing landowners who received federal and state financial incentives for reforesting cleared lands to improve wildlife habitat in the basin. The program had reached its cap of 100,000 acres in the selected sections of the Green River by 2009, meaning that the USDA funds used for the financial incentive were exhausted. Permanent conservation easements on many of these tracts were secured by the Nature Conservancy. This type of conservation effort improves water quality in the Green River and the Mammoth Cave system, benefiting water supply for people and habitat for aquatic organisms and wildlife.

Another cost-share program that looks to improve water quality by addressing nutrient loading through conservation practices is the Mississippi River Basin Healthy Watersheds Initiative (MRBI; http://conservation.ky.gov/Pages/Mississippi-River-Basin-Initiative.aspx). This initiative builds on the cooperative work of the NRCS and its conservation partners in the basin and offers agricultural producers in focus-area watersheds the opportunity for technical and financial assistance. Several focus-area watersheds in Kentucky were selected for this initiative, including the Lower Green River, where the NRCS focused conservation efforts in 2010 (figure 13.11). Participants associated with the MRBI in Kentucky are expected to actively participate in the development of nutrient management and Kentucky Agri-

13.9. Clubshell mussels. Courtesy of the U.S. Fish and Wildlife Service.

13.10. Blind cavefish. Courtesy of the National Park Service.

culture Water-Quality Plans and participate in the installation of on-the-ground best management practices.

The focused MRBI watersheds in the Mississippi River basin were selected, in part, using the USGS Spatially Referenced Regressions on Watershed Attributes (SPARROW) model. This model identified Kentucky as one of nine states in the Mississippi River basin that contribute a significant proportion of nutrients to the northern Gulf of Mexico. In addition, results from the SPARROW model were used to estimate and rank total nitrogen and total phosphorus incremental yields from approximately 25,000 subbasins of the Mississippi River. The model's findings show the Lower Green River basin being reliably placed in the top 150 contributing subbasins for total nitrogen and total phosphorus delivered to the northern Gulf of Mexico. The Lower Green River basin was ranked number 42 for total nitrogen and number 19 for total phosphorus. Although the watershed rankings are a useful tool in identifying watersheds with high nutrient yields, this large-scale approach

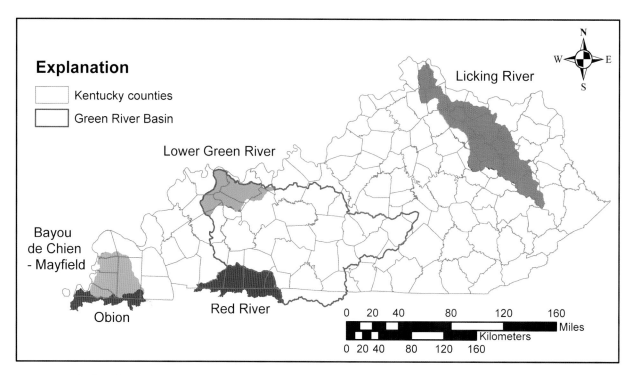

13.11. Watersheds in the Mississippi River Basin Initiative. Illustration from the USDA Natural Resources Conservation Service.

may not address nutrient-management needs to protect streams and reservoirs at a local scale (Robertson et al. 2009).

Opportunities

The challenges identified in the Green River basin present opportunities for collaboration on comprehensive strategies in addressing needs of this area. Multiple federal and state agencies and nongovernmental organizations, as well as other stakeholders, are actively working together in the basin to protect water quality. The knowledge and expertise of the agencies and stakeholders in the basin will be of value in identifying specific opportunities and areas for the installation of best management practices. Identifying these opportunities in the basin will help protect the water quality and natural resources of the Green River basin.

References and Additional Resources

Barnes, Thomas G. 2002. *Kentucky's Last Great Places*. Lexington: University Press of Kentucky.

Baskin, Jerry M., Carol C. Baskin, and Edward W. Chester. 1994. The Big Barrens Region of Kentucky and Tennessee: Further Observations and Considerations. *Castanea* 59, no. 3: 226–54.

Baskin, Jerry M., Edward W. Chester, and Carol C. Baskin, 1997. Special Paper: Forest Vegetation of the Kentucky Karst Plain (Kentucky and Tennessee): Review and Synthesis. *Journal of the Torrey Botanical Society* 124, no. 4: 322–35.

Bryant, William S. 2004. Botanical Explorations of André Michaux in Kentucky: Observations of Vegetation in the 1790s. *Castanea* 69 (sp2): 211–16.

Burroughs, W. G. 1924. *The Geography of the Western Kentucky Coal Field: A Study of the Influence of Geology and Physiography upon the Industry, Commerce, and Life of the People*. Kentucky Geological Survey, ser. 6, Geologic Reports, vol. 24.

Butler, R. S., R. Kessler, and J. B. Harrel. 2003. Down by the Green River. *U.S. Fish and Wildlife Service Endangered Species Bulletin* 28: 20–21.

Carlson, R. E. 1977. A Trophic State Index for Lakes. Limnology and Oceanography 22 (2):361–69.

Cole, L., E. Siegel, and L. W. Lyle. 2001. *State of Kentucky's Environment, 2000–2001: A Report on Environmental Trends and Conditions*. Frankfort: Kentucky Environmental Quality Commission. http://eqc.ky.gov/State%20of%20the%20Environment%200001/state_of_ky_envir_2001.pdf (accessed October 20, 2015).

Converse, H. T., and F. R. Cox. 1967. *Soil Survey of Henderson County, Kentucky*. U.S. Department of Agriculture, 108 p.

Crain, A. S. 2001. *Pesticides and Nutrients in Karst Springs in the Green River Basin, Kentucky*. U.S. Geological Survey Fact-Sheet 133-01, 4 p.

D'Angelo, Elisa, Anastasios Karathanasis, Earl Sparks, Sloane Ritchey, and Stephanie Wehr-McChesney. 2005. Soil Carbon and Microbial Communities at Mitigated and Late Successional Bottomland Forest Wetlands. *Wetlands* 25, no. 1: 162–75.

Froedge, R. D. 1980. *Soil Survey of Christian County, Kentucky*. U.S. Department of Agriculture, 147 p.

Homer, C. H., J. A. Fry, and C. A. Barnes. 2012. *The National Land Cover Database*. U.S. Geological Survey Fact Sheet 2012-3020, 4 p. http://pubs.usgs.gov/fs/2012/3020/ (accessed October 20, 2015).

Horowitz, J., R. Ebel, and K. Ueda. 2010. *No-Till Farming Is Growing Practice*. U.S. Department of Agriculture Economic Information Bulletin no. 70, 22 p.

Kentucky Department of Environmental Protection. 2001. *Green and Tradewater Basins: Status Report*. 23 p. http://water.ky.gov/watershed/Documents/Green%20and%20Tradewater%20Rivers/GTBasinStatusComplete.pdf (accessed October 20, 2015).

———. 2008. *2008 Integrated Report to Congress on Water Quality in Kentucky*. http://water.ky.gov/waterquality/pages/integratedreport.aspx (accessed October 20, 2015).

Kentucky Department of Natural Resources and Environmental Protection. 1980. Kentucky Revised Statutes §146.270. Frankfort, KY.

Kentucky Environmental Quality Commission. 1995. *The State of Kentucky's Environment: 1994 Status Report*. Frankfort, KY. http://www.e-archives.ky.gov/pubs/Natural/environment/state_of_ky_envir94.pdf (accessed October 20, 2015).

Kerr, J. A., Grove B. Jones, S. W. Phillips, and P. E. Karraker. 1924. *Soil Survey of Muhlenberg County, Kentucky*, 939–64. USDA Bureau of Soils, http://www.nrcs.usda.gov/Internet/FSE_MANUSCRIPTS/kentucky/muhlenbergKY1924/muhlenbergKY1924.pdf (accessed September 20, 2016).

Kleber, J. E., ed. 1992. *The Kentucky Encyclopedia*. 3rd ed. Lexington: University Press of Kentucky.

Marschner, F. J., and J. R. Anderson. 1967. Major Land Uses in the United States. In *The National Atlas of the United States of America*. U.S. Geological Survey, Reston, VA. http://water.usgs.gov/GIS/metadata/usgswrd/XML/na70_landuse.xml (accessed September 18, 2016).

McDowell, R. C. 1986. *The Geology of Kentucky: A Text to Accompany the Geologic Map of Kentucky*. U.S. Geological Survey Professional Paper 1151–H, 76 p.

Natural Resources Conservation Service Soil Survey Staff. 2011. Crider. Official Soil Series Description. Natural Resources Conservation Service, United States Department of Agriculture. https://soilseries.sc.egov.usda.gov/osdname.aspx (accessed September 20, 2016).

Natural Resources Conservation Service Soil Survey Staff. 2014. Zanesville. Official Soil Series Description. Natural Resources Conservation Service, United States Department of Agriculture. https://soilseries.sc.egov.usda.gov/osdname.aspx (accessed September 20, 2016).

Nature Conservancy. 2010. *Kentucky: The Green River*. http://www.nature.org/ourinitiatives/regions/northamerica/unitedstates/kentucky/placesweprotect/green-river.xml (accessed October 20, 2015).

Ray, J. A., and J. C. Currens. 1998. *Mapped Karst Ground-Water Basins in the Beaver Dam 30×60 Minute Quadrangle*. Kentucky Geological Survey, Map and Chart Series 19_11, Lexington, KY. http://kgs.uky.edu/kgsweb/olops/pub/kgs/mc19_11.pdf (accessed October 20, 2015).

———. 2000. *Mapped Karst Ground-Water Basins in the Bowling Green 30×60 Minute Quadrangle*. Kentucky Geological Survey, Map and Chart Series 22_12, Lexington, KY. http://kgs.uky.edu/kgsweb/olops/pub/kgs/mc22_12.pdf (accessed October 22, 2015).

Robertson, Dale M., Gregory E. Schwarz, David A. Saad, and Richard B. Alexander. 2009. Incorporating Uncertainty into the Ranking of SPARROW Model Nutrient Yields from Mississippi/Atchafalaya River Basin Watersheds. *Journal of American Water Resources Association (JAWRA)* 45 (2): 534–49.

Sams, J. I., III, and K. M. Beer. 1999. *Effects of Coal-Mine Drainage on Stream Water Quality in the Allegheny and Monongahela River Basins—Sulfate Transport and Trends*. Water-Resources Investigations Report 99-4208, 23 p.

Simon, F. K., ed. 1996. *The WPA Guide to Kentucky*. Lexington: University Press of Kentucky. Original publication: Federal Writers' Project for the Work Projects Administration for the State of Kentucky. 1939. *Kentucky: A Guide to the Bluegrass State*. New York: Harcourt, Brace and Company.

Taylor, C. J., W. P. Kaiser, and H. L. Nelson Jr. 2008. Application of Geographic Information System (GIS) Hydrologic Data Models to Karst Terrain. In *Sinkholes and the Engineering and Environmental Impacts of Karst* (American Society of Professional Engineers), 146–55.

U.S. Army Corps of Engineers. 2011. *Green River Watershed Section 729 Initial Watershed Assessment*. http://www.lrl

.usace.army.mil/Portals/64/docs/CWProjects/GreenRiver Study.pdf (accessed October 20, 2015).

U.S. Department of Agriculture, Natural Resources Conservation Service (USDA, NRCS). 2010. Soil Survey Geographic (SSURGO) Database by State - Kentucky: Soil Survey Staff, Natural Resources Conservation Service, United States Department of Agriculture. http://websoilsurvey.nrcs.usda .gov/app/ (accessed September 20, 2016).

U.S. Energy Information Administration. 2008. *Annual Coal Report, 2009.* 73 p. http://205.254.135.7/coal/annual/ (accessed October 20, 2015).

———. 2011. *Annual Coal Report, 2010.* 63 p. http://205.254 .135.7/coal/annual/ (accessed October 20, 2015).

U.S. Farm Service Agency. 2010. *Kentucky Green River Conservation Reserve Enhancement Program: Annual Program Accomplishment Report.* CEP-68R, 68 p. http://www.fsa .usda.gov/Internet/FSA_File/crep2010annualreport.pdf (accessed October 20, 2015).

Williamson, T. N., C. J. Taylor, and J. K. Newson. 2013. Significance of Exchanging SSURGO and STATSGO Data When Modeling Hydrology in Diverse Physiographic Terranes. *Soil Science Society of America Journal* 77: 877–89.

CHAPTER FOURTEEN

Wetlands of Kentucky

Connecting Landscapes and Waterways

Stephen C. Richter, Michelle Guidugli-Cook, and David R. Brown

Wetlands are one of the most important ecosystems on earth because of the functions they perform and the unique habitat they provide to natural systems. Additionally, they are among the most economically valuable ecosystems in the world because of the services they provide specifically to humans (Costanza et al. 1997). Wetlands are crucial for the cycling of water from the atmosphere to land and streams, rivers, lakes, and oceans and back to the atmosphere. Wetlands reduce flood severity by impeding flow and absorbing surge, help maintain streams during drought periods by gradually releasing water into them, filter sediments from surface runoff before it reaches streams, and decompose or sequester pollutants (Mitsch and Gosselink 2007). These functions improve water quality of our streams, rivers, lakes, and oceans, which in turn provide much of the drinking water and food that humans depend on daily. Beyond the hydrologic cycle, wetlands provide unique habitat for highly diverse biological communities in addition to their importance for nutrient cycling, including carbon sinks and climate stabilizers (Mitsch and Gosselink 2007). It should be obvious that wetlands serve critical local and global functions, but somehow people tend to overlook the critical services provided to human communities at watershed, landscape, and global scales.

We begin this chapter by describing characteristics that define wetlands. We then describe wetlands of Kentucky, including their status and history. Because this is primarily a story of wetland loss, considerable attention is devoted to describing the indirect and direct impacts

on wetlands and the consequences for humans and natural systems. Following this, we review legislation meant to protect wetlands and water quality and its implications for preservation of wetland functions. History shows that humans have treated wetland habitats poorly, but recent conservation efforts provide reasons to be positive. We conclude the chapter with recommendations on how to move forward to preserve what remains and to replace what was lost more effectively in the midst of ever-increasing human demands on land use.

What Are Wetlands?

Despite the acknowledged importance of these habitats, a definition of the term "wetland" that is mutually agreed on among scientists, the public, and resource managers has been elusive (Luenn 1994; Mitch and Gosselink 2007). Mitsch and Gosselink (2007) devoted an entire chapter to wetland definitions in which they described physical and biological characteristics and provided legal wetland definitions. In the simplest terms, wetlands are distinguished from other habitat types by the presence of surface or subsurface water for extended periods of time that results in unique soils and vegetation suited for wet conditions. Wetlands develop along the continuum between terrestrial and aquatic systems and often occur as transitional areas between water and land. Some wetlands have permanent to semipermanent water, whereas others are ephemeral and hold water only for part of the year. Fluctuating water level is generally considered a key process in most wetland types. In examining one wetland across time or multiple wetlands across space, it is not uncommon to find that they sometimes appear and function as aquatic ecosystems, sometimes as terrestrial ecosystems, and most often as an intermediate between the two (Kusler, Mitsch, and Larsen 1994; Mitsch et al. 2009). Hydrology is the primary driver of wetland functions and the resulting species of microbes, plants, and animals inhabiting wetlands (Wellborn, Skelly, and Werner 1996; Mitsch et al. 2009).

Wetland diversity is driven by factors at both regional (physiographic, geologic, and soil characteristics) and local (landscape position, soils, and water source and duration) scales to which microbes, plants, and animals respond. For example, wetlands occur on slopes where groundwater surfaces to form seeps, such as on Pine Mountain; along the edges and adjacent to streams, rivers, and lakes; in bottomlands; in flat areas that are not in floodplains but where runoff is slow or not possible; and in depressions that have no obvious hydrologic link to groundwater or surface water (figure 14.1). These depressional wetlands are sometimes referred to as isolated wetlands; however, the name is misleading because they have watersheds and their hydrology is sufficient to support wetland plant species, and they might be connected to other waters of the United States (Mitsch and Gosselink 2007; Mushet et al. 2015). These wetlands are important habitats for wildlife such as mole salamanders, wood frogs, multiple species of bats, and deer and should not be undervalued, despite judicial rulings that removed this wetland type from federal regulatory jurisdiction (Brown and Richter 2012).

History and Status of Kentucky Wetlands

Kentucky is characterized by profound changes in ecophysiological regions as one moves from the Mississippi River alluvial valley in the west to the Appalachian Mountains in the east (Woods et al. 2002). The largest and most concentrated wetlands occur in the Mississippi Embayment and Western Kentucky Coalfields regions. This is mostly because of their low elevation and location within major river floodplains of the Mississippi, Ohio, and Tennessee Rivers. Wetland types in these regions include forested bottomland swamps (often dominated by forest communities of bald cypress and water tupelo),

wet bottomland hardwood forests, sloughs, and riparian or floodplain forests (KSNPC 2009). Not surprisingly, smaller and fewer wetlands occur in the central and eastern portion of the state because of higher elevation, smaller river systems, and more rugged landscapes that restrict the formation of large wetland areas (Jones 2005). The more common wetland types in central and eastern Kentucky include forested flats and depressions, seeps on slopes, and headwater wetlands.

Wetlands are relatively scarce across Kentucky's current landscape, and remaining wetlands tend to be geographically isolated or are often restricted to public lands or areas owned by nongovernmental organizations (e.g., the Nature Conservancy). We do not know the actual acreage of remaining wetlands in Kentucky because an accurate state-level survey has not been conducted; however, current estimates range from 300,000 to 500,000 acres (Dahl 1990). Several wetland types have become rare or extinct because of development, although these types were probably relatively uncommon even before settlement (figure 14.2). Some examples of these rare or extinct wetland communities are cypress-tupelo swamps, wet meadows and prairies, and seeps and bogs (KSNPC 2009). Several areas containing rare wetland types and rare wetland plant species have been purchased and protected by the Kentucky State Nature Preserves Commission and other public and private organizations to ensure that they remain for wildlife, as well as for the enjoyment of future generations of people. However, the amount of wetland area protected is small, and most wetland areas in Kentucky are not protected from further degradation or destruction.

Drivers of Historic and Contemporary Wetland Loss

Wetland loss is devastating from ecological, environmental, and emotional points of view. For example, when Aldo Leopold returned to his childhood home in Wisconsin after spending several formative years learning forestry at Yale and then in the field in Arizona and New Mexico, he found that his childhood hunting and exploring grounds, which were bottomland hardwood wetlands along the banks of the Mississippi River, had been ditched and drained into a massive canal. "I

14.1a, 14.1b, 14.1c, and 14.1d. Examples of wetlands from across the state of Kentucky: 1a. Appalachian seep bog from the Cumberland Plateau, Daniel Boone National Forest, Whitley County, Kentucky. Photo taken by Brian Yahn. 1b. Fringe wetland on Lake Barkley at Crooked Creek in Trigg County, Kentucky. Photo taken by Jesse Godbold. 1c. Murphy's Pond wetland in Hickman County, Kentucky. Photo taken by Michelle Guidugli-Cook. 1d. Depressional wetland in Morgan County, Kentucky. Photo taken by Stephen Richter.

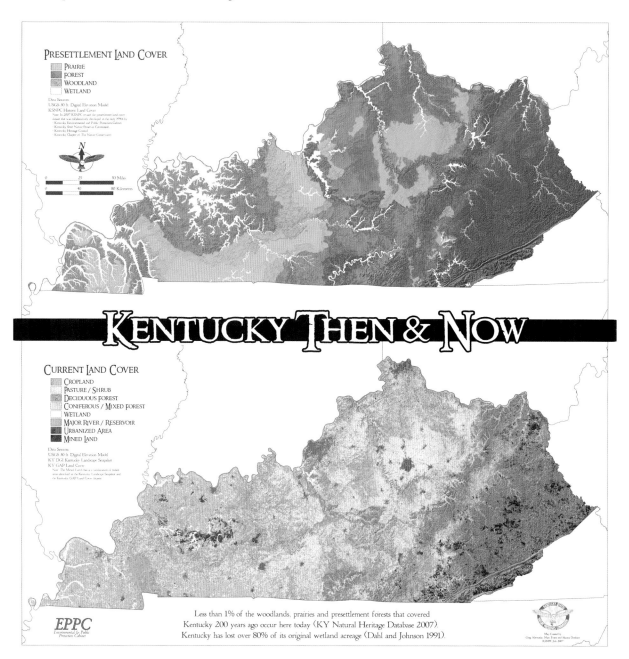

14.2. Map produced by the Kentucky State Nature Preserves Commission depicting presettlement habitat and postsettlement land cover in Kentucky, available at http://naturepreserves.ky.gov/data/Documents/ky_thenandnow_web.pdf.

came home one Christmas to find that land promoters, with the help of the Corps of Engineers had dyked and drained my boyhood hunting grounds on the Mississippi river bottoms. . . . My hometown thought the community enriched by this change. I thought it im-

poverished" (Aldo Leopold, from 1947 draft foreword of *A Sand County Almanac*, published in Callicott 1987, 282). The canal was part of a large network of water drainage, conveyance, and storage structures that had been built with the intention of creating new land for development and reducing floods lower in the Mississippi River basin.

During the first half of the twentieth century, much of the Mississippi River basin was diked, dammed, drained, and channelized (Barry 1997). The process was similar in Florida, California, and elsewhere; Kentucky was no exception. Kentucky was historically rich in wetlands; at the time of settlement, Kentucky is thought to have had 1.5 to 2.3 million acres of wetlands (Dahl 1990; Evans and Abernathy 2008). Alarmingly, it is generally understood that over 80 percent of these wetlands have been destroyed, largely because of human causes, including urban and suburban development, forestry, agriculture, and mining (Dahl 1990). This places Kentucky among the states with the greatest proportion of original wetland loss in the Mississippi River basin and in the United States broadly.

The massive water-control projects that drained much of the Mississippi River basin were often tied to other perceived benefits, such as mosquito control, navigation, and agriculture (figure 14.3). In fact, considerable wetland area has been lost at smaller scales, but almost invariably at the hands of human development. Although wetlands are not as integral to the social fabric of Kentucky as they are in some other states, such as Louisiana, they still form an important landscape component that was traditionally regarded as obstructing economic development and in need of transformation. Loss due to agriculture is driven by landowner desire to produce more from the land. Farmers drained wet areas using tiles and ditches and filled these low areas with trees cleared from woodland strips (figure 14.4).

"An epidemic of ditch-digging and land-booming set in. The marsh was gridironed with drainage canals, speckled with new fields and farmsteads" (Leopold 1949, 106). The one-two punch of water control and conversion of "reclaimed" lands to agriculture production has left us with wetland remnants, but these were not the only large-scale sources of wetland loss in Kentucky. Coal mining, in its various forms, has affected many of Kentucky's wetlands and continues to be one of several major sources of water pollution (Lindberg et al. 2011). The Western Kentucky Coalfields region had the highest concentration of historic wetlands, but today most of these wetlands no longer exist (Evans and Abernathy 2008). In the Appalachian region, contour mining and mountaintop-removal mining operations result in the filling of chemically and biologically functioning slope and headwater wetlands. These wetlands are important in maintaining downstream water quality, so their loss compounds the degradation of ecological services. Underground mining often creates acidic groundwater drainage that empties into streams and wetlands, thereby greatly limiting the ability of many organisms to survive (Ledin and Pedersen 1996). Forestry practices have also resulted in the loss and degradation of many of Kentucky's wetlands, and today just a few old-growth forest wetlands remain (figure 14.5). Roads can also be a source of wetland loss. Mining and forestry activities usually require road construction, and in mountain areas, public and private roads typically follow streams and rivers and thus affect wetlands.

Indirect Impacts and Wetland Degradation

Wetland loss is not always an abrupt disappearance of pristine habitat. Many wetlands succumb to anthropogenic disturbance through a process of gradual decline in quality and function. Often the end point is an alternative state that is stable but does not resemble the original natural condition. The hydrologic connectivity of wetlands to streams, rivers, lakes, and groundwater makes them vulnerable to changes in human land use that occur outside the wetland boundary. Activities that disconnect wetlands from surface water and groundwater (e.g., dams, dikes, and ditches) or alter groundwater flows (e.g., aquifer use, ditches, and drain lines) sever hydrologic connections and lead to wetland degradation and loss (Kusler, Mitsch, and Larsen 1994). For example, as sediment in stream channels erodes because of stream channelization or other alterations, over time the water table often drops to the bedrock and disconnects the stream from the floodplain and its wetlands. Additionally, forested wetlands are often cleared, and in the process the hydrology is modified such that ecological succession is impaired and a scrub-shrub or emergent wetland replaces the lost forest. Over time, the scrub-shrub wetland continues to degrade, sometimes through active changes such as excavation or damming to create open-water farm ponds.

14.3a, 14.3b, and 14.3c. Wetland drainage projects intended for (a) mosquito control, (b) agriculture, and (c) general-purpose drainage. Each image depicts drainage ditches excavated in Kentucky by the Works Progress Administration. All images are from the University of Kentucky Archives.

14.4a, 14.4b, and 14.4c. Images depicting farmers installing drain tiles and lines to allow farming in wetland areas of Powell County, Kentucky. Images provided by Thomas Biebighauser.

14.5. Old bald cypress trees in a wetland at Metropolis Lake State Nature Preserve, McCracken County. Photo taken by Tara Littlefield.

Invasive plant species also pose a threat to the ecological functions of wetlands, particularly in regard to biodiversity loss. Although many invasive species are exotic, introduced from other continents, some species are native but capable of rapidly spreading into new areas. For instance, native cattails (*Typha* spp.) can expand to dominate wetland communities and reduce biodiversity (Keddy 2010). Thus direct human impacts and other processes of global change can have incipient effects in which highly functioning wetlands slowly degrade to a point where they have little ecological and human value. The end point bears little resemblance to and almost none of the functions of the original natural wetland (LePage 2011).

Economics of Wetland Loss

The global human benefits provided by wetlands should be obvious; less apparent is their economic value for individuals and society. The average annual economic value of wetlands worldwide was valued at $4.9 trillion in the mid-1990s (and is certainly higher now), primarily because of their ecosystem services to provide flood abatement, water filtration, drinking water, and recreation (Costanza et al. 1997). Economic analysis reveals that the cost of floods, soil erosion, reduced carbon storage, and the loss of recreational resources from lost wetlands far exceeds the short-term gains of developing wetlands (Balmford et al. 2002).

Nevertheless, wetlands continue to be undervalued, viewed as impediments to progress, and degraded or destroyed. There is a strong connection between the economic engines of Kentucky and wetland loss, as well as a common view that wetlands impede farming and building on otherwise useful land. This is likely because of misconceptions by the general public and prevailing economic models that fail to capture the value of ecological services.

Although the Clean Water Act requires mitigation when wetlands are affected or destroyed, restored and constructed wetlands typically fall short of replacing natural wetland functions (Moreno-Mateos et al. 2012). We need better land-management practices. For example, best management practices for mining, forestry, and agriculture could be used and improved to include provisions for avoiding wetlands and incorporating suitable buffers around wetlands. This would retain wetland functions on the landscape, including the reduction of velocity, turbidity, and contaminant load of water reaching Kentucky streams and rivers. Such changes in management practice may add costs to business, but the societal benefits and long-term costs to taxpayers would make it economically sensible.

What Is Being Done about Wetland Loss?

It was not until 1977, when the Clean Water Act was passed, that wetlands received protection. Since that time, the U.S. Army Corps of Engineers (USACE) and the Kentucky Division of Water have regulated impacts on Kentucky's wetlands. The Corps is responsible for determining whether proposed impacts are allowed under Section 404 of the Clean Water Act (1977) and Section 10 of the Rivers and Harbors Act (1899) (USACE 2010). This legislation gives the Corps the ability to regulate and issue permits for the discharge of dredged or fill materials into navigable waters and other activities that may affect the capacity of waters to be navigable. The Kentucky Division of Water certifies that the federal permits do not violate water-quality standards, which are regulated through the Water Quality Certification Program (Section 401 of the Clean Water Act) (U.S. Environmental Protection Agency [USEPA] 2010). One important exemption from this regulation is for "normal farming and harvesting activities that are part of established, ongoing farming or forestry operations." For regulated activities, permit applicants must demonstrate that they have considered alternatives or made efforts to avoid or minimize the impact (http://www.epa.gov/owow/wetlands/facts /fact20.html). If the applicant has demonstrated that there are no alternatives and the impact cannot be avoided, then a permit may be granted. In an effort to offset wetland losses that are permitted, the Clean Water Act expresses a goal of "no net loss" of wetlands.

The goal of "no net loss" is to replace wetland function and ecological services lost to development. This is achieved through a process called compensatory mitigation. Compensatory mitigation operates by permitting wetland impacts that create "debits," which must be paid for with "mitigation credits." This can occur in one of four ways: (1) restoration of an existing wetland (or other aquatic resource), (2) enhancement of an existing site's functions, (3) establishment or creation of a new aquatic site, or (4) preservation of an existing site (USACE and USEPA 2008). There are three ways these different methods of compensatory mitigation can be achieved. In the past, the most common approach was for the permittee to directly mitigate and take responsibility for ensuring completion and assessing success. In such cases, the mitigation is completed at the impact site or at another location within the same watershed. The other two ways to fulfill compensatory mitigation are "mitigation banks" and "in-lieu fee" programs, both of which involve a third party taking responsibility for completing the mitigation at another location, but typically still within the same watershed, and ensuring compliance with performance standards. In 2008, the Final Mitigation Rule (USACE and USEPA 2008) was published, which established relative preference for the different approaches to compensatory mitigation, with wetland banks being most preferred, then in-lieu fee programs, and permittee-responsible mitigation as the least preferred option.

Mitigation banks and in-lieu fee programs differ in a number of ways. Mitigation banks are often operated by for-profit, private entities, and thus the financing for projects comes from private investors. In-lieu fee programs are generally operated by state or local governments (e.g., the Kentucky Department of Fish and Wildlife Resources) or nonprofit, nongovernmental organizations, and thus financing for these projects comes from money collected from permittees (USACE and USEPA 2008). For mitigation banks, specific milestones concerning project development and financial backing must be reached before a project is determined to be complete, whereas in-lieu fee programs begin projects only when money has been collected, often substantially after the permitted impact (USACE and USEPA 2008). The Kentucky in-lieu fee programs converted to a mitigation-bank-like structure in 2012 (http://www.lrl.usace.army.mil/Missions/Regulatory /Mitigation.aspx; Barbara Scott, Kentucky Division of Water, personal communication).

Why Are We Still Losing Wetlands and Their Critical Functions?

Although good progress has been made concerning no net loss of wetlands in the United States, there is still much that can be improved in function and quality. For example, much of the purported progress in wetland conservation is founded on recent increases in total acreage of wetlands across the United States; however, many created wetlands are merely farm ponds with minimal ecological function (Stedman and Dahl 2008). The capacity of a wetland to perform certain functions depends on the wetland type and the amount of impact it has sustained; therefore, it is important to consider these factors in setting appropriate mitigation ratios. Unfortunately, the established wetland-mitigation credit ratio is typically only at least 1:1. In Kentucky, the actual required ratio is usually 2:1, which means that for every acre of wetland affected, two acres of wetland must be mitigated regardless of the wetland's type or previous condition (USACE and USEPA 2008). This policy does not provide an incentive to create higher-quality wetlands because mitigation credits are determined on a per-acreage basis. Additionally, some wetland types such as depressional wetlands that are hydrologically isolated are not currently regulated under Section 404 of the Clean Water Act and so are not mitigated. Last, the success of compensatory mitigation projects in replacing the functions of wetlands lost is generally low (Moreno-Mateos et al. 2012). More research is needed to better understand how to restore and create wetlands to replace the functions of lost wetlands. This will be difficult in the current mitigation framework because mitigation requirements generally do not include funding for long-term monitoring or research to help improve the mitigation process. Until these problems are corrected, homogenizing the diversity of wetlands and their functions across Kentucky landscapes will continue (Brooks et al. 2005).

Moving Forward Despite the Challenges

One major hurdle in addressing this problem is that Kentucky has no established method to evaluate wetland condition and quality. To address this need, the Kentucky Division of Water, Eastern Kentucky University, and various state and federal agencies formed a partnership to develop a rapid assessment method for wetlands: the Kentucky Wetland Rapid Assessment Method (KY-WRAM). The method is now fully developed and published for use by the Kentucky Division of Water, other regulatory agencies, and environmental consultants to improve Kentucky's Section 401 Water Quality Certification Program, ambient monitoring of wetlands in Kentucky, and other ecological studies (see Guidugli-Cook et al. 2017). Incorporation of this method into the mitigation process allows for wetland quality to be assessed before impact and for the mitigation credit ratio to be increased above 2:1 for higher-quality wetlands. A future

goal is to modify KY-WRAM to create an assessment method for mitigation wetlands to evaluate and monitor their quality. Such information is vital to ensuring the replacement of functions, as well as quality, and not just lost wetland acreage.

Another means of enhancing wetland mitigation and water quality in the state is to encourage those who are designing stream-mitigation projects to include wetland restoration as part of their projects (Parola and Biebighauser 2011; Parola and Hansen 2011). This approach recognizes that wetlands historically occurred with streams and that their presence was integral to stream function and value (Biebighauser 2007). Wetlands on floodplains store surface water and attenuate flood surge; they also supply base flow to the stream during drier months through groundwater connections to the wetland and provide habitat for amphibians, fish, plants, and other organisms whose populations would not likely be established if only the stream were restored (Parola and Hansen 2011). This restoration philosophy is a shift from simply returning the stream to its natural geomorphology to an approach that focuses on reinstating a more complete, naturally functioning historic system. A weakness is that stream-mitigation projects receive credits for each foot of stream that meets established criteria, but current policies do not provide additional credits for including wetland construction in the broader design plans. Two exemplary projects that demonstrate the ecological value of incorporating wetlands into stream-mitigation projects are the Slabcamp Creek and Stonecoal Branch (Rowan County, Kentucky) and Dix River (Lincoln County, Kentucky) stream and wetland restoration projects (Parola and Biebighauser 2011).

Recommendations and Next Steps

Wetlands are critical components of Kentucky's landscape that serve to connect terrestrial and aquatic resources. They play a vital role in the cycling of water from air and land to waterways. Over 80 percent of Kentucky's wetlands have been lost, and without increased efforts for conservation, wetlands will continue to disappear; however, recent development of statewide partnerships and projects for wetland conservation gives reason for hope. For example, in 2008–2009 the state developed a framework for a planning process for wetland conservation (McSpirit et al. 2011). Such a planning process can establish conservation goals, the appropriate indicators and end points for achieving those goals, and guidance for developing research-based approaches to management that include adaptive-management feedback mechanisms. As an example of this approach, wetland ecologists from Eastern Kentucky University have been working with the U.S. Forest Service to enhance wetland construction techniques to provide better habitat for ephemeral pool-breeding amphibians (Denton and Richter 2013; Drayer and Richter 2016).

One critical need to help guide statewide planning is additional research on the actual distribution and abundance of wetlands within Kentucky. Although national-scale inventories and trend-monitoring programs provide a general picture of the distribution of wetlands in Kentucky, the data are out of date or not properly ground-truthed (Guidugli-Cook et al. 2017). Ultimately, stakeholders need to develop a new philosophy of wetland value to human communities. Essential will be conservation that incorporates adaptive changes to management as knowledge on physical and biological functions reveals new best practices for wetland conservation and replacement.

References (* Indicates Suggested General Readings)

Balmford, A., A. Bruner, P. Cooper, R. Costanza, S. Farber, R. Green, M. Jenkins, P. Jefferiss, V. Jessamy, J. Madden, K. Munro, N. Myers, S. Naeem, J. Paavola, M. Rayment, S. Rosendo, J. Roughgarden, K. Trumper, and R. Turner. 2002. Economic Reasons for Conserving Wild Nature. *Science* 297: 950–53.

Barry, J. M. 1997. *Rising Tide: The Great Mississippi Flood of 1927 and How It Changed America.* New York: Simon & Schuster.

*Biebighauser, T. R. 2007. *Wetland Drainage, Restoration, and Repair.* Lexington: University Press of Kentucky.

Brooks, R. P., D. H. Wardrop, C. A. Cole, and D. A. Campbell. 2005. Are We Purveyors of Wetland Homogeneity? A Model of Degradation and Restoration to Improve Wetland Mitigation Performance. In *Wetland Creation, Restoration and Conservation—The State of the Science,* edited by W. J. Mitsch, 331–40. Amsterdam: Elsevier.

Brown, D. B., and S. C. Richter. 2012. Meeting the Challenges to Preserving Kentucky's Biodiversity. *Sustain* 25: 22–33.

Callicott, J. B. 1987. *Companion to "A Sand County Almanac": Interpretive and Critical Essays.* Madison: University of Wisconsin Press.

Costanza, R., R. d'Arge, R. de Groot, S. Farber, M. Grasso, B. Hannon, K. Limburg, S. Naeem, R. V. O'Neill, J. Paruelo, R. G. Raskin, P. Sutton, and M. van den Belt. 1997. The Value of the World's Ecosystem Services and Natural Capital. *Nature* 387: 253–60.

Dahl, T. E. 1990. *Wetlands Losses in the United States, 1780s to 1980s.* Washington, DC: United States Fish and Wildlife Service; Jamestown, ND: Northern Prairie Wildlife Research Center.

Denton, R. D., and S. C. Richter. 2013. Amphibian Communities in Natural and Constructed Ridge Top Wetlands with Implications for Wetland Construction. *Journal of Wildlife Management* 77: 886–889.

Drayer, A. N., and S. C. Richter. 2016. Physical Wetland Characteristics Influence Amphibian Community Composition in Constructed Wetlands. *Ecological Engineering* 93: 166–174.

Evans, M., and G. Abernathy. 2008. *Presettlement Land Cover of Kentucky.* Frankfort: Kentucky State Nature Preserves Commission.

Guidugli-Cook, M., S. C. Richter, B. J. Scott, and D. R. Brown. 2017. Field-based Assessment of Wetland Condition, Wetland Extent, and the National Wetlands Inventory in Kentucky, USA. *Wetland Ecology and Management* DOI 10.1007/s11273-017-9533-3.

Jones, R. 2005. *Plant Life of Kentucky: An Illustrated Guide to the Vascular Flora.* Lexington: University Press of Kentucky.

*Keddy, P. 2010. *Wetland Ecology: Principles and Conservation.* Cambridge: Cambridge University Press.

Kentucky State Nature Preserves Commission (KSNPC). 2009. Natural Communities of Kentucky. Working draft. Frankfort, KY.

*Kusler, J. A., W. J. Mitsch, and J. S. Larsen. 1994. Wetlands. *Scientific American* 270: 64–70.

Ledin, M., and K. Pedersen. 1996. The Environmental Impact of Mine Wastes—Roles of Microorganisms and Their Significance in Treatment of Mine Wastes. *Earth-Science Reviews* 41: 67–108.

Leopold, A. 1949. *A Sand County Almanac.* Oxford: Oxford University Press.

LePage, B. A. 2011. Wetlands: A Multidisciplinary Perspective. In *Wetlands: Integrating Multidisciplinary Concepts,* edited by B. A. LePage, 3–26. Dordrecht: Springer.

Lindberg, T. T., E. S. Bernhardt, R. Bier, A. M. Helton, R. B. Merola, A. Vengosh, and R. T. Di Giulio. 2011. Cumulative Impacts of Mountaintop Mining on an Appalachian Watershed. *Proceedings of the National Academy of Sciences* 108: 20929–34.

*Luenn, N. 1994. *Squish! A Wetland Walk.* New York: Macmillan.

McSpirit, S., D. Brown, S. Scott, and J. Pulliam. 2010. Major Impacts and Challenges Facing Kentucky's Streams and Wetlands: A Summary of Agency, Other Expert, and Stakeholder Views. *Journal of the Kentucky Academy of Sciences* 71: 82–94.

*Mitsch, W. J., and J. G. Gosselink. 2007. *Wetlands.* Hoboken, NJ: John Wiley and Sons.

*Mitsch, W. J., J. G. Gosselink, C. J. Anderson, and L. Zhang. 2009. *Wetland Ecosystems.* New York: John Wiley and Sons.

Moreno-Mateos, D., M. E. Power, F. A. Comin, and R. Yockteng. 2012. Structural and Functional Loss in Restored Wetland Ecosystems. *PLOS Biology* 10: e1001247.

Mushet, D. M., A. J. K. Calhoun, L. C. Alexander, M. J. Cohen, E. S. DeKeyser, L. Fowler, C. R. Lane, M. W. Lang, M. C. Rains, and S. C. Walls. 2015. Geographically Isolated Wetlands: Rethinking a Misnomer. *Wetlands* 35: 423–431.

Parola, A. C., Jr., and T. R. Biebighauser. 2011. The Stream Institute, University of Louisville's Stream and Wetland Restoration Program. *Sustain* 22: 8–13.

Parola, A. C., Jr., and C. Hansen. 2011. Reestablishing Groundwater and Surface Water Connections in Stream Restoration. *Sustain* 22: 2–7.

Stedman, S., and T. E. Dahl. 2008. *Status and Trends of Wetlands in the Coastal Watersheds of the Eastern United States, 1998 to 2004.* Washington, DC: National Oceanic and Atmospheric Administration, National Marine Fisheries Service, and United States Fish and Wildlife Service.

United States Army Corps of Engineers (USACE). 2010. *Interim Regional Supplement to the Corps of Engineers Wetland Delineation Manual: Eastern Mountains and Piedmont Region.* Edited by J. S. Wakeley, R. W. Lichvar, C. V. Noble, and J. F. Berkowitz. ERDC/EL TR-10-9. Vicksburg, MS: U.S. Army Engineer Research and Development Center.

United States Army Corps of Engineers (USACE) and United States Environmental Protection Agency (USEPA). 2008. Rules and Regulations: Part II: Department of Defense, Department of the Army, Corps of Engineers 33 CFR Parts 325 and 332 Environmental Protection Agency 40 CFR Part 230 Compensatory Mitigation for Losses of Aquatic Resources; Final Rule. *Federal Register* 73: 19594–19705.

United States Environmental Protection Agency (USEPA). 2010. *Clean Water Act Section 401 Water Quality Certification: A Water Quality Protection Tool for States and Tribes.* Washington, DC: USEPA, Office of Wetlands, Oceans, and Watersheds.

Wellborn, G. A., D. K. Skelly, and E. E. Werner. 1996. Mechanisms Creating Community Structure across a Freshwater Habitat Gradient. *Annual Review of Ecology and Systematics* 27: 337–63.

Woods, A. J., J. M. Omernik, W. H. Martin, G. J. Pond, W. M. Andrews, S. M. Call, J. A. Comstock, and D. D. Taylor. 2002. *Ecoregions of Kentucky* (color poster with map, descriptive text, summary tables, and photographs). Reston, VA: U.S. Geological Survey (map scale 1:1,000,000).

Chapter Fifteen

Reconnecting through Stream Restoration

Carmen T. Agouridis

> Water is the most critical resource issue of our lifetime and our children's lifetime. The health of our waters is the principal measure of how we live on the land.
> —Luna Leopold

Streams are complex and changing systems consisting principally of an active channel and an adjacent floodplain, both of which are influenced by the landscape context in which they exist. These flowing water bodies support a myriad of functions, including the movement of water and sediment, provision of aquatic and terrestrial habitats, and nutrient cycling, as well as societal functions such as transportation and recreation. When streams are fulfilling these functions, they are considered healthy (FISRWG 1998). Nevertheless, not all streams are healthy. Anthropogenic activities such as urbanization, agriculture, or resource extraction can diminish, impair, or destroy the structure and functions of stream systems. Excess anthropogenic discharges or sediment loads can cause a stream to adjust its width, depth, slope, bed-material composition, and flow pattern (Lane 1955). When these forms of degradation occur, a stream-restoration intervention may be required in order to bring the stream back to a healthy functioning state (NRC 1992; Sarr 2002). A restoration can have impacts on not only the geomorphic, chemical, and biological characteristics of a stream but also the human individual and community relationship to it. This chapter examines how the process of stream restoration can improve stream health and function, as well as how it can reconnect communities to streams.

From a technical perspective, stream restoration involves the conversion of an unstable, altered, or degraded stream as closely as possible to its predisturbance or reference condition while considering present and future upstream and overall watershed uses. A reference condition refers to a section of stream that is morphologically stable, not necessarily one that is pristine. This reference condition serves as a template for the design of the restored stream (Rosgen 1998; White 2001). The restoration process consists of restoring the stream's dimension or cross section, pattern, and profile to achieve dynamic equilibrium (Harrelson, Rawlins, and Potyondy 1994; Rosgen 1994; Hey 2006). Dynamic equilibrium represents the stream's stable condition for the present environment, meaning that the processes of degradation and aggradation are not excessive but are in balance (Leopold, Wolman, and Miller 1964). Typically, the primary goal of a stream-restoration project is to restore flow and sediment-transport regimes (Hey 2006) with the intention that the chemical and biological or habitat-related components and functions will recover over time (Lakly and McArthur 2000). In some projects, several or all of the objectives are met quickly; in others, recovery is slow, particularly if in-stream water quality is poor (Selvakumar, O'Connor, and Struck 2010).

The stream-system-restoration process is complex. Conceptually, restoration projects consist of six major factors that contribute to their complexity: morphology (dimension, pattern, and profile), floodplain connection, in-stream structures, stream-bank stabilization, riparian-corridor vegetation, and both in-stream and riparian-zone habitat enhancement (FISRWG 1998; NRCS 2007). From this list alone, it is apparent that stream restoration requires integrating multiple knowledge domains,

15.1. Team efforts are essential in successful stream-restoration projects because their complexity requires knowledge and skills in a large number of areas.

all of which are extremely rarely found in one person, to carry a project from conception through completion. Interaction among several disciplines in the fields of engineering and life sciences and, depending on the project, education and social sciences is a must (figure 15.1). Technical knowledge, as well as partnerships working together, is essential for a project that is both sound and embraced by the community. The old adage that there is no "I" in "team" is never more apparent than in successful stream-restoration projects.

The Millcreek Elementary (School) Stream and Wetlands Restoration Project and Outdoor Classroom in Lexington, Kentucky (Fayette County), is a prime example of developing partnerships across disciplines and entities to restore a degraded urban stream reach. Before restoration, the stream was incised with steep, eroding banks. Gabion baskets, which are wire-mesh baskets filled with rocks and placed along stream banks to stop erosion, were failing (figure 15.2). Stream erosion not only was threatening utilities but also was creating a safety hazard. Children were routinely warned to stay away from the creek while they were playing outside for fear that they would fall into it (figure 15.3). These concerns gave rise to an opportunity to use a stream-restoration process to reestablish a stream channel that would be safer on multiple levels. The stream-restoration process has changed how the stream functions and how

it is perceived by the children and adults who use it. It has now become an asset to the school and the community rather than something to be feared in the landscape.

This project design was led by the University of Louisville Stream Institute with project partners from several different groups: the University of Kentucky's Tracy Farmer Institute for Sustainability and the Environment led the education initiative, the Kentucky Department of Fish and Wildlife Resources provided funding for the project, and Ridgewater LLC and EcoGro constructed and vegetated the project. The project involved the creation of a lower, more accessible floodplain and in-stream and riparian habitats, water-quality-improvement features, and educational structures. Hundreds of yards of soil were excavated to create access to a new floodplain at a lower elevation. The lower stream banks, as seen in figure 15.4, now allow waters to exit the stream and spill out onto the floodplain during storms. Greater floodplain access means that storm water places much less stress on the stream banks, and hence much less streambank erosion occurs. Also created were a wide variety of in-stream and riparian habitats, such as long riffles, deep pools, large woody habitat structures, and riparian wetlands. These features provide aquatic macroinvertebrates, fish, salamanders, frogs, and the like with good places to live and thrive. In addition to the riparian wetlands, which help improve water quality, carbon in the form of wood chips and small branches was mixed with the floodplain soils to help promote beneficial microbial processes, such as denitrification, or the removal of nitrogen from the water. Millcreek, like many urban streams, receives runoff with high concentrations of nutrients. These nutrients promote eutrophication or high levels of algal and plant growth, which in turn depletes dissolved oxygen levels in the water. If dissolved oxygen levels drop too much, all but the most tolerant aquatic species will die. Last, educational structures, such as a small upland wetland, mulched walking paths, and large rock structures allowing students to access key stream features such as pools, were added to the stream. The project resulted in the restoration of approximately 625 feet of stream and riparian zone and the creation of about 2,200 square feet of wetlands.

15.2. Hard-engineering measures to stop stream-bank erosion were failing, further threatening utilities, as well as making the stream uninviting for the community. Photo courtesy of Eric Dawalt of Ridgewater LLC.

15.3. Steep banks on Millcreek were adjacent to a children's playground, which made for conditions that were perceived to be unsafe. Photo courtesy of Eric Dawalt of Ridgewater LLC.

In addition, stream restoration at this site not only offered a means of restoring the stream to a healthier and better-functioning state but also reconnected the school and community to this local water resource. The restored site serves as an outdoor classroom where students, who may have few if any other opportunities, can connect with nature. To a child, streams offer adventure with the opportunity to discover new aquatic insects living under rocks or fish swimming in pools (figure 15.5). Exploration and play become the foundation for offering lessons in subjects such as mathematics, chemistry, and biology. "Prior to our stream restoration, the stream was not utilized much for educational purposes," said Kelli Faulkner, a science lab teacher at Millcreek Elementary.

15.4. The Millcreek project restored approximately 625 feet of stream to a stable state by reducing the height of the stream banks and creating habitat features such as pools and riffles in the stream and wetlands in the riparian zone. Photo courtesy of Carmen Agouridis of the University of Kentucky.

15.5. Millcreek, which was once avoided by students and teachers, is now eagerly used in science lessons as an extended classroom. Photo courtesy of Matt Barton of the University of Kentucky.

"Now with the restoration in place, we have implemented four days ('Stream Days') throughout the school year where students are immersed in activities centered on our stream and wetland area. We have the kids in the stream conducting macroinvertebrate and water-quality tests during each Stream Day along with an additional activity." Faulkner noted that with the success of "Stream Days" in helping kids learn through hands-on education, more teachers are using the site for lessons at other times of the school year as well.

15.6. The Cane Run Watershed Council has helped forge strong partnerships among stakeholders by serving as a forum for identifying problems and collectively developing solutions. Photo courtesy of Steve Patton of the University of Kentucky.

15.7. The Cane Run Watershed Festival offered community members of all ages a chance to learn about their watershed and ways to improve its health. Photo courtesy of Carmen Agouridis of the University of Kentucky.

Although over 37,000 stream-restoration projects had been completed nationwide as of 2007 (Sudduth, Meyer, and Bernhardt 2007), the practice of stream restoration is still in its youth. A number of stream-restoration projects have been designed and implemented, restoring stream functions in many, but not all, cases. There is still much to learn not just about geomorphic, chemical, and biological functions but also about policy and social implications. Understanding the technical pieces, such as the hydrology, hydraulics, and sediment transport, is an important aspect, but there must also be an understanding about how what might seem like unrelated stream-system puzzle pieces help in the shaping of personal and community perspectives.

In addition to the science, understanding how streams fit into communities and how communities function around them is critical for healthy stream-system function. Projects cannot end at the property or project line but instead must seek to bring in partners from the larger watershed because it is the cumulative effect of all our actions in a watershed that affects a stream's functions. The Cane Run Watershed Assessment and Restoration Project at the University of Kentucky is one such endeavor to improve community awareness and education about watershed and stream issues in addition to the technical aspects of stream restoration. Although the Cane Run watershed is primarily in Lexington, it serves as the main drinking-water source for the city of Georgetown, located just to the north. The Kentucky Division of Water has identified the Cane Run watershed as impaired because of high sedimentation/siltation levels, pathogens, and nutrient/organic enrichment from both point and nonpoint sources (Kentucky Division of Water 2011).

The University of Kentucky is leading a two-phase project funded by the U.S. Environmental Protection Agency through the 319(h) program to complete a watershed-based plan and begin its implementation. However, landowner and community buy-in is essential for the development and implementation of this plan. Therefore, education and outreach strategies that focus on involving stakeholders through a watershed council help bring together different interests that result in powerful partnerships to assess current watershed conditions and develop restoration action plans (figure 15.6). Partners reflect a community spectrum and include federal, state, and local governments, private landowners, private businesses, and nonprofit organizations. Educational and outreach activities are targeted at project partners, watershed residents, watershed professionals, students, and visitors (figure 15.7). By building awareness, stewardship, and cooperation, the project serves as a catalyst to promote watershed- and stream-restoration efforts at federal, state, and local levels.

Stream systems are powerful forces that appeal to our human nature on a basic level. People often have a con-nection to a stream, from the backyard stream that we explored in our childhood to one that flows by our home or one we drive or bicycle over on our way to work. Although the practice is relatively young, stream restoration holds much promise not just in reestablishing stream functions but also in reconnecting people and communities to streams.

References

FISRWG (Federal Interagency Stream Restoration Working Group; fifteen agencies of the U.S. government). 1998. *Stream Corridor Restoration: Principles, Processes, and Practices.* GPO Item No. 0120-A; SuDocs No. A 57.6/2:EN 3/PT.653. http://www.nrcs.usda.gov/Internet/FSE_DOCUMENTS/stelprdb1044574.pdf (accessed October 15, 2015).

Harrelson, C. C., C. L. Rawlins, and J. P. Potyondy. 1994. *Stream Channel Reference Sites: An Illustrated Guide to Field Techniques.* USDA Forest Service Rocky Mountain Forest and Range Experiment Station General Technical Report RM245, 67. http://www.stream.fs.fed.us/publications/PDFs/RM245E.PDF (accessed October 15, 2015).

Hey, R. D. 2006. Fluvial Geomorphological Methodology for Natural Stable Channel Design. *Journal of the American Water Resources Association* 42: 357–74.

KDOW (Kentucky Division of Water). 2011. *2010 Integrated Report to Congress on the Condition of Water Resources in Kentucky.* Volume II. 303(d) List of Surface Waters. Frankfort, KY: Kentucky Energy and Environment Cabinet, Department for Environmental Protection, Division of Water.

Lakly, M. B., and J. V. McArthur. 2000. Macroinvertebrate Recovery of a Post-thermal Stream: Habitat Structure and Biotic Function. *Ecological Engineering* 15: S87–S100.

Lane, E. W. 1955. The Importance of Fluvial Morphology in Hydraulic Engineering. *Proceedings of the American Society of Civil Engineers* 81: 1–17.

Leopold, L. B., M. G. Wolman, and J. P. Miller. 1964. *Fluvial Processes in Geomorphology.* New York: Dover Publications.

NRC (National Research Council). 1992. *Restoration of Aquatic Ecosystems: Science, Technology, and Public Policy.* Washington, DC: National Academy Press.

NRCS (National Resources Conservation Service). 2007. National Engineering Handbook-Part 654 Stream Restoration Design. U.S. Department of Agriculture. http://directives.sc.egov.usda.gov/viewerFS.aspx?hid=21433.

Rosgen, D. L. 1994. A Classification of Natural Rivers. *Catena* 22: 169–99.

———. 1998. The Reference Reach—A Blueprint for Natural Channel Design. ASCE conference on river restoration, Denver, CO, March. Reston: ASCE. http://www.wildland hydrology.com/assets/The_Reference_Reach_II.pdf (accessed October 15, 2015).

Sarr, D. A. 2002. Riparian Livestock Exclosure Research in the Western United States: A Critique and Some Recommendations. *Environmental Management* 30 (4): 516–26.

Selvakumar, A., T. P. O'Connor, and S. D. Struck. 2010. Role of Stream Restoration on Improving Benthic Macroinvertebrates and In-Stream Water Quality in an Urban Watershed: Case Study. *Journal of Environmental Engineering* 136 (1): 127–39.

Sudduth, E. B., J. L. Meyer, and E. S. Bernhardt. 2007. Stream Restoration Practices in the Southeastern United States. *Restoration Ecology* 15 (3): 573–83.

White, K. E. 2001. *Regional Curve Development and Selection of a Reference Reach in the Non-urban, Lowland Sections of the Piedmont Physiographic Province, Pennsylvania and Maryland.* U.S. Geological Survey Water-Resources Investigations Report 01-4146. http://pa.water.usgs.gov/reports /wrir01-4146.pdf (accessed October 15, 2015).

Recommended Further Reading

FISRWG (Federal Interagency Stream Restoration Working Group; fifteen agencies of the U.S. government). 1998. *Stream Corridor Restoration: Principles, Processes, and Practices.* GPO Item No. 0120-A; SuDocs No. A 57.6/2:EN 3/PT.653. http://www.nrcs.usda.gov/Internet/FSE_DOCUMENTS /stelprdb1044574.pdf (accessed October 15, 2015).

Hunter, C. J. 1991. *Better Trout Habitat: A Guide to Stream Restoration and Management.* Washington, DC: Island Press.

Izaak Walton League of America. 2006. *A Handbook for Stream Enhancement and Stewardship.* 2nd ed. Blacksburg, VA: McDonald and Woodward; Gaithersburg, MD: Izaak Walton League of America.

Leopold, L. B. 1994. *A View of the River.* Cambridge, MA: Harvard University Press.

Leopold, L. B., M. G. Wolman, and J. P. Miller. 1964. *Fluvial Processes in Geomorphology.* New York: Dover Publications.

Riley, A. L. 1998. *Restoring Streams in Cities: A Guide for Planners, Policymakers, and Citizens.* Washington, DC: Island Press.

Rosgen, D. L. 1996. *Applied River Morphology.* Pagosa Springs, CO: Wildland Hydrology.

CHAPTER SIXTEEN

Urban Water Management

Responding to Federal Regulation

Jack Schieffer

Urban areas face many challenges in managing water quantity and quality, particularly for sanitary and stormwater systems. Increasing urbanization places greater demands on the water supply and infrastructure, while simultaneously age, antiquated design, and lack of maintenance can lead to sanitary sewer overflows into homes and streets, as well as discharges of untreated sewage into surface waters. Proliferation of impervious surfaces in densely populated areas also contributes to flooding and contaminated stormwater runoff. Water pollution is regulated under the Federal Water Pollution Control Act of 1948 and its subsequent amendments (1972, 1977, and 1987), commonly referred to as the Clean Water Act (CWA). Under the CWA, local governments are charged with protecting water quality. Violations of this law have resulted in Kentucky's three largest urban centers entering into consent decrees with the U.S. Environmental Protection Agency (EPA): Louisville in 2005, northern Kentucky (i.e., the Cincinnati area) in 2007, and Lexington in 2008. These consent decrees are court-sanctioned settlements of the EPA's lawsuits against each of these cities, and they specify fines and codify the steps that the cities will take to overhaul their water infrastructure and achieve the mandated levels of water quality (e.g., usable for aquatic habitat or recreation). This chapter compares and contrasts the experiences of these three Kentucky communities and their consent decrees: the nature of the problems they face, the policies they have implemented, and their successes (or lack thereof) in achieving compliance.

All three consent decrees include standard approaches taken by numerous U.S. cities under agreements with the EPA. These typical responses include sewer user and stormwater fees collected from property owners to fund programs to repair and upgrade entire systems. However, these three consent decrees also illustrate newer types of measures, such as funding for various environmental projects (e.g., stream restoration), increased use of green infrastructure, and informational and educational programs to promote water-conscious behavior among residents. Of particular note is the northern Kentucky consent decree because it is one of the first in the nation to incorporate a watershed-based approach. This approach encourages consideration of water-impairment sources other than sewer overflows, thereby allowing the local government more flexibility in addressing water-management issues. Additionally, this flexibility is expected to allow northern Kentucky to meet water-quality targets at a lower cost, compared with Lexington and Louisville.

This chapter focuses on three cities' experiences that illustrate the ways in which urbanization often leads to the degradation of water quality. These examples also provide an instructive look at the state of water-related law and policy in the United States and the difficulties that society faces in balancing the needs of a growing population with the protection of important resources such as water. Fortunately, these stories also offer hope in the form of important lessons learned and innovative approaches for managing urban water challenges as more people live in urban and urbanizing areas.

Balancing Urbanization and Water-Quality Protection

Cities provide many benefits to people: employment and business opportunities, top-flight restaurants, cultural events, and shopping and recreational amenities, among others. In contrast to these positive aspects, urban areas also present substantial threats to water quality. As populations and urbanization increase, the pressure on water resources also grows. For example, in 2010, about 82 percent of the U.S. population lived in urban areas. From 1982 to 2007, the amount of land classified as large urban development in the forty-eight contiguous states almost doubled, from 45 million acres to 81.8 million acres. In Kentucky, urban areas account for approximately 58 percent of the total population, and the strong urbanization trend parallels that of the nation.[1]

Increasing use of water resources, the discharge of waste into waterways, and physical change in the landscape of urban areas all lead to a variety of water-related problems. Collectively, these factors contribute to a general pattern often called urban stream syndrome. Streams near urban centers develop substantial differences from their counterparts in undeveloped areas in terms of hydrology, stream-channel size and shape, and water quality. Urban streams typically have more frequent and intense high-volume flows and flooding because the impervious surfaces in cities cause more runoff to travel rapidly overland and via storm sewers to the receiving waters rather than infiltrating more slowly as groundwater. Development changes the size and shape of streambeds and disturbs the surrounding vegetation and habitat. In addition, urban streams are typically characterized by contamination related to roads and automobiles, construction, wastewater flows, and similar sources.[2]

Two of the most important causes of urban water degradation are sanitary or combined sewer overflows and stormwater runoff. Municipal discharges of sewage (including overflows) contribute to pollution in approximately 14.3 percent of impaired rivers and streams in the United States, while runoff affects about 9.2 percent of those waters.[3]

Sewer overflows are often associated with wet-weather events (i.e., heavy precipitation) when stormwater combines with wastewater flows to exceed the capacity of the system. During overflows, undertreated or untreated sewage discharges into surface waters and can back up into streets and buildings, creating a public health hazard. In a combined sewer system (CSS), wastewater generated by homes and businesses and storm runoff are collected and then carried in the same pipes to treatment facilities before their ultimate discharge into waterways. CSSs represent an older sewer-system design, built primarily before the early twentieth century and before wastewater treatment became more common. In Louisville, for example, the use of underground sewers dates to the mid-1800s, and CSSs still constitute approximately 20 percent of the city's sewer infrastructure. CSSs are concentrated in the Northeast and Great Lakes regions of the United States, since these areas were settled earlier in the nation's history. In these systems, overflows are designed to occur at designated outfalls into streams and rivers, with the stormwater diluting the sewage. However, high volumes of water can still exceed the system's capacity and lead to overflows at unintended locations. In addition, although the discharge is diluted, it still contains untreated sewage that contaminates waterways.[4]

Newer sanitary sewer systems (SSSs) are designed to carry wastewater separately from stormwater runoff, which is collected in municipal separate storm-sewer systems (often abbreviated MS4). In principle, this design should insulate a sanitary sewer from wet-weather flows and mitigate the overflow problems. In practice, wet-weather overflows can still occur in an SSS because storm runoff and groundwater enter the system in various ways. For example, cracked pipes allow water that has soaked into the ground to enter the system. Some households illegally connect their gutter downspouts or basement sump pumps to the sanitary sewers. These higher flows during wet weather can overwhelm the system and lead to overflows.

Both types of sewer systems can also suffer from inadequate maintenance, pipe leaks, pump failures, power outages, and blockages, which can cause overflows in dry weather as well as wet. At the time of its 2005 consent decree, Louisville's average overflows were over 500 million gallons annually of untreated sewage from its SSS and over 4.5 billion gallons annually of combined sewage and stormwater from its CSS. In northern Kentucky, sanitary sewer overflows averaged 82 million gallons annually, and combined sewer overflows averaged over 850 million gallons annually. To put these numbers in perspective, a million gallons would result in an American-rules football field covered almost three feet deep in sewage.[5]

Even in the absence of overflows, stormwater runoff carried by MS4s often degrades water resources. In urban areas, much of the ground is covered by impervious surfaces, such as roads, sidewalks, parking lots, and buildings, and the open ground that remains is often more heavily compacted than the soil in undeveloped areas. Precipitation tends to collect and flow across these surfaces rather than soaking into the ground. As the water flows, it picks up a variety of contaminants, including sediment, lawn chemicals, fluids and particulates from automobiles, and debris. The MS4s carry these pollutants to the receiving waterways without any intervening treatment. The typical purpose of traditional storm sewers is to move large volumes of water away from developed areas as quickly as possible to prevent flooding in those areas. However, this functional mind-set is often at odds with other needs, such as the protection of water quality and ecological attributes of the watershed.

In summary, urban stream syndrome and related overflow and runoff issues create several problems. Sewage discharges and contaminated runoff pollute watersheds, and this pollution affects public health, water-based recreational opportunities, and drinking-water supplies, as well as causing ecological damage. Hydrological and streambed changes, such as flooding and erosion, can damage property and reduce real-estate values. Water policy and regulation attempt to balance these impacts with accommodation of a growing urban population. However, the regulatory framework established in the 1970s has at times struggled to keep pace as new types of issues emerge.[6]

The Regulatory Framework

Around the turn of the twenty-first century, the U.S. government initiated a wave of lawsuits against cities that were in violation of the Clean Water Act (CWA), the federal law that regulates water pollution and protects the quality of the nation's water resources. Among the defendants were the three largest urban areas in Kentucky—Louisville, Lexington, and northern Kentucky (i.e., the Cincinnati area south of the Ohio River)—in separate legal actions. All three areas are currently operating under consent decrees, which are court-approved settlements with the enforcing agencies that require massive overhauls of sewer systems and other steps to improve watershed quality over several years.

The Environmental Protection Agency (EPA) is the federal agency charged with developing regulations under the CWA and monitoring and enforcing compliance. However, state governments also play a substantial role because the EPA delegates many administrative functions to state agencies. In Kentucky, the Energy and Environment Cabinet (formerly the Environmental and Public Protection Cabinet) and its divisions are responsible for implementing federal environmental laws and regulations.

Initially, regulation under the CWA focused on controlling waste flows from point sources such as industrial point sources and municipal sewage-treatment plants. These pollution sources are required to obtain permits from the EPA under the National Pollutant Discharge Elimination System (NPDES). The permits place limits on the types and amounts of pollutants that sources can emit, as well as requiring specific technologies in certain cases (e.g., sewage-treatment plants). Exceeding these emission limits or discharging pollutants without a permit constitutes a violation of the CWA. On balance, the NPDES program has been relatively successful in reducing water pollution from these traditional end-of-pipe discharges by point sources.

However, current regulation has been less successful in controlling sewer overflows and stormwater runoff, partly because of the geographic nature of these problems. Traditional point-source discharges are characterized by fairly predictable and consistent effluent flows with known pollutants. Because the discharges typically originate from identifiable locations, they are easily monitored for compliance with permits. In contrast, overflows and runoff are more geographically dispersed and unpredictable because they are tied closely to weather events. The types and concentrations of contaminants that these discharges contain can change from one event to the next, as can the volume of water involved and the locations of the discharges. These properties impede the management of overflows and stormwater, so the regulatory framework designed with traditional point sources in mind tends to fit poorly with these two important causes of urban water problems. However, policy measures have begun to evolve to address these issues.[7]

In 1994, the EPA issued its Combined Sewer Overflow Control Policy, which was formally endorsed by Congress in 2001. This policy applies to CSSs and requires adherence to nine minimum controls that cover proper maintenance of the sewer system, prohibition of dry-weather combined sewer overflows (CSOs), reduction in wet-weather CSOs and their impacts, and similar objectives. In addition, CSS operators must develop and implement long-term control plans to minimize CSOs to protect water quality. Although a comparable program specifically for controlling sanitary sewer overflows (SSOs) does not exist, SSOs are considered discharges without permits and are thus illegal under the CWA.[8]

With regard to storm runoff, the EPA initially considered treating storm sewers as NPDES point sources in 1973 but rejected this approach and instead exempted those systems from regulation. In *Natural Resources Defense Council v. Costle* (1977), a federal appellate court overturned this exemption and others, ruling that the CWA required the EPA to regulate stormwater. In response, the EPA created a set of rules and permit requirements for MS4s and industrial and commercial activities. However, these new stormwater regulations generated considerable controversy and were challenged in lawsuits by both trade associations and environmental groups.[9]

In 1987, Congress acted to settle the issue of whether stormwater should be regulated by passing the Water Quality Act. This legislation amended the CWA and explicitly required the EPA to develop a comprehensive program for stormwater management. Phase I of the new program took effect in November 1990 and applied NPDES permit requirements to large municipalities (populations greater than 100,000) and industrial sites

(construction runoff or storm-sewer discharge). Phase II took effect in 1999 and extended the NPDES permitting requirements to smaller municipalities.[10]

The stormwater program has resulted in a huge increase in permitting by the EPA. As of 2008, over 500,000 stormwater permits had been issued, as compared with approximately 100,000 non-stormwater (i.e., wastewater) NPDES permits. This proliferation of permits has put a strain on the EPA's resources for administering programs and monitoring regulatory compliance.[11]

Stormwater permits generally require the permit holders—such as MS4s, industrial facilities with their own stormwater systems, and construction sites—to develop plans for managing stormwater and implement best management practices (BMPs) to control runoff. These BMPs may include infrastructure design and operation, structural features such as bioswales or detention ponds, "good housekeeping" practices (e.g., safeguarding construction materials from exposure to weather) to limit the contamination of stormwater, and similar measures. Although the EPA provides some guidance, the stormwater sources and state agencies are left with a great deal of latitude in determining appropriate BMPs. In addition, monitoring requirements tend to be weak relative to those for traditional point sources.[12]

Starting in the late 1990s, a lack of progress in implementing CSO policy and correcting SSOs led the EPA to file a number of lawsuits against municipalities across the nation that were in violation of their NPDES permits. Once legal action has been initiated in such a case, negotiations between the EPA and the municipality usually result in a consent decree—such as those in Louisville, northern Kentucky, and Lexington—to resolve the conflict. In many ways, a consent decree is similar to an out-of-court settlement between private parties. After the parties negotiate, the defendant agrees to take certain actions over a specified period in return for the plaintiff dropping the lawsuit. However, since the EPA represents the public interest in enforcing regulation, consent decrees must also be approved by the court. The court supervises the completion of obligations under the decree, and a defendant who does not abide by the terms can be found in contempt of court and face additional punishment.

Three Consent Decrees in Kentucky

The legal actions in Louisville, northern Kentucky, and Lexington exemplify some of the common patterns and trends in this approach to water policy. These cases are generally settled via consent decrees specifying the remedial actions that the municipalities will take in exchange for avoiding the full penalties possible under the CWA. However, the three cases in Kentucky also showcase some differences in structure and approach to resolving the issues.

Louisville

The Metropolitan Sewer District (MSD) was created in 1946 to take over Louisville's sewer system and expand its operations in Jefferson County. It currently operates six major regional wastewater-treatment plants (WWTPs), as well as fourteen smaller facilities, and 286 pumping stations. As in other relatively old cities, combined sewers still constitute a significant portion of Louisville's system, over 600 miles of the approximately 3,200 miles of sewer overall. In February 2004, the Commonwealth of Kentucky sued the MSD, a unit of the Louisville–Jefferson County Metro government, for violations of the CWA. The EPA later joined the suit on behalf of the plaintiff.[13]

The MSD and the state and federal agencies reached an agreement on the terms of a consent decree, which was approved by the U.S. District Court (Western District of Kentucky) in August 2005. The largest requirement under the consent decree is a major overhaul of the city's sewer infrastructure. Intended to bring the sewer systems into regulatory compliance and mitigate overflows, this program encompasses a variety of projects, such as identification and assessment of overflow locations and damaged infrastructure, repair and capital improvement, separation of combined sanitary and storm sewers at certain locations, improved control systems, and installation of backup power supplies. The MSD has named this program Project WIN (Waterway Improvements Now), and the initial estimated budget was $500 million (later revised to $850 million), with projects scheduled through 2024. As of the MSD's 2010 annual report, 22 percent of the Project WIN projects had been completed, were under way, or had been bid.[14]

The consent decree also stipulated a $1 million penalty, as well as requiring several supplemental projects at an estimated cost of $2.25 million. Supplemental projects represent activities that the municipality would not otherwise undertake, but that would contribute to a broad objective of watershed and environmental protection. In contrast, the infrastructure overhaul described earlier reflects efforts that are necessary to avoid continued violations of the CWA.

One supplemental project is to provide funding for educational and community outreach activities promoting environmental awareness, with emphasis on watershed-related education. A second project involves public health screenings and education for residents close to heavily industrialized areas. A third supplemental project sponsors one-time stream-restoration activities in Jefferson County locations affected by MSD's discharges.

The Louisville consent decree reflects the conventional regulatory approach to urban water issues. It emphasizes sewer overflows, and the primary solution is to upgrade the so-called gray infrastructure: pipes, pumps, and treatment facilities. However, the MSD has also undertaken some "green infrastructure" initiatives related to its consent-decree obligations. The term "green infrastructure" describes the use of natural systems (e.g., undeveloped green spaces, strategically placed vegetation) to protect a community from flooding, improve water quality, reduce erosion, and provide similar benefits. With regard to managing stormwater and sewer overflows, green spaces and permeable pavement can reduce the amount of water runoff entering sewer systems during wet weather.[15]

Detention ponds, rain gardens, permeable pavement, and similar measures all improve the capacity of the landscape to absorb precipitation, allowing it to infiltrate into groundwater rather than traveling overland or via storm sewers as runoff. The gray infrastructure is better able to handle this lower volume of runoff, so that fewer overflows and similar problems occur. While conventional mitigation measures (i.e., sewer upgrades) can be very expensive and require many years to implement, green infrastructure projects can be relatively quick and inexpensive in many cases. For example, the MSD has implemented a $1.5 million green infrastructure project on the University of Louisville campus. The project includes landscaping (e.g., rain gardens, bioswales), porous pavement, infiltration pits in parking lots, and rooftop vegetation. These elements all increase the volume of water absorbed by the ground rather than carried away in storm sewers, which the MSD hopes will save tens of millions of dollars in costs versus more conventional approaches. In addition to lower costs, green infrastructure is often touted as providing other environmental benefits, such as improved air quality and more aesthetically pleasing landscapes. Although the use of green infrastructure was not explicitly mentioned in the consent decree, the MSD and the EPA agreed to use pilot projects to assess the approach's effectiveness.[16]

Northern Kentucky Sanitation District No. 1

In October 2005, Kentucky's Environmental and Public Protection Cabinet filed a CWA lawsuit against Sanitation District No. 1 of northern Kentucky (SD1), the utility operating sewers and treatment facilities in several Kentucky counties (Boone, Kenton, and Campbell) within the Cincinnati metropolitan area. The system currently encompasses approximately 1,600 miles of sewers, including both separate and combined sewers, and SD1 operates two major WWTPs and 142 pumping stations.

In April 2007, the U.S. District Court (Eastern District of Kentucky) approved the consent decree negotiated by the parties. Like the decree in Louisville, the northern Kentucky consent decree mandates a thorough evaluation of the gray infrastructure, remedial upgrades and repairs, and improved maintenance and monitoring in the future. The estimated cost of this component of the consent decree is over $880 million. In addition, SD1 paid a fine of almost $500,000 and was required to complete several supplemental projects. One project involved extending sewer service to areas with known defective septic tanks and straight pipes that discharge raw sewage. Other projects involve land conservancy, improving the monitoring of water quality, public education programs about water quality, and funding watershed restoration.

A particularly innovative feature of the northern Kentucky consent decree is the adoption of a watershed-based approach. Under this management framework, which was proposed by SD1 in the consent-decree negotiations, the CSOs and SSOs in a watershed are considered together with other point and nonpoint sources in the watershed-management area. This provides SD1 with a broader array of tools and mitigation strategies to achieve the overall goal of improving water quality in the watershed. The agency can use tools such as green infrastructure, as discussed earlier in this chapter, to mitigate overflow by reducing runoff volume in lieu of upgrading gray infrastructure to handle higher volumes. It can also use watershed controls that would reduce pollution from other point sources (e.g., industrial waste) or nonpoint sources (e.g., agriculture) as a substitute for reducing pollution by eliminating overflows.

These substitutions require evidence that they will be as effective in reducing pollution as conventional approaches, but they offer increased flexibility and the potential for lower costs in meeting watershed-improvement goals. Some of the alternative measures would also offer additional environmental benefits, such wildlife habitat provided by installing riparian buffers alongside streams.

Lexington

The EPA, joined by Kentucky's Environmental and Public Protection Cabinet, filed a lawsuit against the Lexington-Fayette Urban County Government (LFUCG) in November 2006. The LFUCG operates a sanitary sewer system with over 1,400 miles of sewers, two WWTPs (after closing the Blue Sky plant in accordance with the consent decree), and 81 pumping stations.

The parties agreed to the terms of the consent decree in March 2008, although court approval was delayed because of concerns about the $425,000 fine to be paid by LFUCG. During the public comments period for the proposed decree, Lexington residents argued that the money would be better spent on addressing local water problems than paid as a penalty. The district court judge agreed and ordered the EPA and the LFUCG to renegotiate the terms of the consent decree. When the EPA appealed the judge's decision, the Sixth Circuit Court of

Appeals overruled the lower court. The original terms of the consent decree, including the fine, were approved and reinstated in January 2011.

The consent decree includes the typical requirement to evaluate and improve the sewer infrastructure, with a budget of $290 million. Lexington does not have a combined sewer system (although improper connections do exist), so the emphasis is on identifying and eliminating SSOs. The LFUCG must also engage in several supplemental projects with a budget of $2.7 million. One example is the Coldstream Park Stream Corridor Restoration and Reservation project ($750,000 construction budget), involving the construction of wetlands, planting of riparian buffers, and other measures to improve a degraded stream in northern Lexington. The project will reduce flooding and pollution loads in the stream, as well as provide additional benefits in the form of wildlife habitat, recreational opportunities, and community education. The LFUCG is also funding a number of small-scale grants to schools, local businesses, and neighborhood associations to improve stormwater-related infrastructure, restore streams, and provide education.

A distinctive feature of the Lexington consent decree is its emphasis on stormwater. The original lawsuit complaint included allegations that the LFUCG's operation of its storm-sewer systems violated regulations under the Clean Water Act. This stands in contrast to other cases' focus on CSS and SSS problems, and to some degree violations of WWTP permits, as the only transgressions motivating legal action. The consent decree mandates actions specifically aimed at mitigating stormwater problems. The LFUCG's Storm Water Quality Management Plan requires improved monitoring and inspection, stricter ordinances on construction sites and other sources of runoff contamination, and funding for stormwater-management programs. Since the decree was approved, the LFUCG has reported progress in implementing its terms. Numerous maintenance and upgrade projects for the sanitary sewer system have been completed, including the shutdown of the Blue Sky treatment plant and the consolidation of its service area with another plant, and others have begun. Mandated inspections of the storm-sewer system have been implemented and are planned to continue. Although much work remains to be done over the decade-plus time frame of the consent decree, the city government has taken some much-needed steps.

Meeting the Challenges Ahead

Although regulation and control of sewer overflows, stormwater runoff, and other urban water problems have proven difficult, there has been progress on this front. The overarching policy framework is recognizing the importance of stormwater runoff and considering a variety of approaches other than conventional gray infrastructure solutions to urban water degradation. Alternative approaches, including green infrastructure, low-impact development, watershed-level management, and others, have substantial potential for increasing water policy's effectiveness and lowering its costs. This broader scope for addressing urban water issues can be seen both in the regulations themselves, such as the development of CSO and stormwater programs by the EPA, and in policy instruments such as consent decrees that allow for more flexible approaches.

In some ways, these shifts in policy reflect a change in focus to the root causes of urban water problems. Although sewer overflows may be the most visible manifestation of inadequate water management, the fundamental issues are land-use choices, the ways in which we conduct urban development, and how we educate the public about reducing people's environmental impacts in their daily lives. As urbanization continues, finding cost-effective ways to manage wastes and runoff will become even more important. If we act wisely, however, incorporating elements of stormwater control into the design and planning of urban development should prove more effective and less costly than retrofitting measures on already-developed land.

Notes

1. National population statistics are taken from Central Intelligence Agency, *The World Factbook* (Washington, DC: Central Intelligence Agency, 2010), https://www.cia.gov/library/publications/the-world-factbook/index.html (accessed April 23, 2014). Land-use statistics are taken from U.S. Department of Agriculture, *Summary Report: 2007 National Resources*

Inventory (Washington, DC: Natural Resources Conservation Service, 2009), http://www.nrcs.usda.gov/technical/NRI/2007/2007_NRI_Summary.pdf (accessed April 23, 2014). Kentucky population statistics are taken from U.S. Department of Agriculture, *State Fact Sheet: Kentucky* (Washington, DC: Economic Research Service, 2014), http://www.ers.usda.gov/data-products/state-fact-sheets.aspx (accessed April 23, 2014). The urbanization trends are described by Bill Estep, "Two Kentuckys: Cities Grow While Rural Areas Decline, Census Shows," *Lexington Herald-Leader,* March 18, 2011, http://www.kentucky.com/2011/03/18/1674369/kentuckys-urban-areas-growing.html (accessed April 23, 2014).

2. Christopher J. Walsh, Allison H. Roy, Jack W. Feminella, Peter D. Cottingham, Peter M. Groffman, and Raymond P. Morgan II, "The Urban Stream Syndrome: Current Knowledge and the Search for a Cure," *Journal of the North American Benthological Society* 24, no. 3 (2005): 706–23; Center for Watershed Protection, *The Impacts of Impervious Cover on Aquatic Systems* (Ellicott City, MD: Center for Watershed Protection, 2003).

3. U.S. Environmental Protection Agency, *National Water Quality Inventory: Report to Congress, 2004 Reporting Cycle,* EPA 841-R-08-001 (Washington, DC: U.S. Environmental Protection Agency, Office of Water, 2009), 12, http://water.epa.gov/lawsregs/guidance/cwa/305b/2004report_index.cfm (accessed April 23, 2014).

4. U.S. Environmental Protection Agency, *Report to Congress: Impacts and Control of CSOs and SSOs,* EPA 833-R-04-001 (Washington, DC: U.S. Environmental Protection Agency, Office of Water, 2004), http://cfpub.epa.gov/npdes/cso/cpolicy_report2004.cfm (accessed April 23, 2014); U.S. Environmental Protection Agency, "Clean Water Act Agreement Announced with Louisville and Jefferson County," press release, April 25, 2005, http://yosemite.epa.gov/opa/admpress.nsf/b1ab9f485b098972852562e7004dc686/d099dfec771c4cc785256fee00750c86!OpenDocument (accessed April 23, 2014).

5. U.S. Environmental Protection Agency, "Clean Water Act Agreement Announced with Louisville and Jefferson County"; U.S. Environmental Protection Agency, "Clean Water Act Agreement Announced with the Sanitation District No. 1 of Northern Kentucky," press release, October 7, 2005, http://yosemite.epa.gov/opa/admpress.nsf/e51aa292bac25b0b85257359003d925f/95c56084461c38f7852570d00058e310!OpenDocument (accessed April 23, 2014).

6. Quan Truong, "Quiet Gahanna Creek Now 'Raging Torrent' Threatening Subdivision," *Columbus Dispatch,* February 21, 2012; Jerald R. Barnard, "Externalities from Urban Growth: The Case of Increased Storm Runoff and Flooding," *Land Economics* 54, no. 3 (1978): 298–315.

7. Dave Owen, "Urbanization, Water Quality, and the Regulated Landscape," *University of Colorado Law Review* 82 (2011): 431–504.

8. U.S. Environmental Protection Agency, "Combined Sewer Overflow (CSO) Control Policy," *Federal Register* 59, no. 75, April 19, 1994, http://www.epa.gov/npdes/pubs/owm0111.pdf (accessed April 23, 2014).

9. *Natural Resources Defense Council v. Costle*, 568 F.2d 1369 (D.C. Cir. 1977); National Research Council, *Urban Stormwater Management in the United States* (Washington, DC: National Academies Press, 2008), 42–43, 56–57, http://www.epa.gov/npdes/pubs/nrc_stormwaterreport.pdf (accessed April 23, 2014); *Natural Resources Defense Council v. Environmental Protection Agency*, 673 F.2d 392 (D.C. Cir. 1980); Theodore L. Garrett, "*NRDC v. EPA*: The D.C. Circuit's Long-Awaited Decision in the NPDES Permit Rules Litigation," *Environmental Law Reporter* 19 (1989): 10223–29.

10. U.S. Environmental Protection Agency, *Overview of the Storm Water Program*, EPA 833-R-96-008. (Washington, DC: U.S. Environmental Protection Agency, Office of Water, 1996), http://www.epa.gov/npdes/pubs/owm0195.pdf (accessed April 23, 2014).

11. National Research Council, *Urban Stormwater Management in the United States,* 29–30, 42, 56.

12. Ibid., 84–102.

13. Louisville and Jefferson County Metropolitan Sewer District (MSD), *2010 Annual Report* (2010), 2, http://www.msdlouky.org/aboutmsd/pdfs/msd10ar.pdf (accessed April 23, 2014).

14. Ibid., 4.

15. Anna Brittain, "Green Infrastructure: Investing in Nature to Build Safer Communities," *Resources* 183 (2013): 20–21. For more information, see the U.S. EPA's and Nature Conservancy's websites on the topic of green infrastructure, listed under "Further Resources" at the end of this chapter.

16. Natural Resources Defense Council, *Rooftops to Rivers: Green Strategies for Controlling Storm Water and Combined Sewer Overflows* (New York: Natural Resources Defense Council, 2006), 7–10; James Bruggers, "MSD Hopes to Save Money, Property by 'Greening' University of Louisville," *Louisville Courier-Journal*, July 3, 2011, http://www.courier-journal.com/article/20110703/NEWS01/307040017/MSD-hopes-save-money-property-by-greening-University-Louisville (accessed April 23, 2014).

Additional Resources

Lexington-Fayette Urban County Government's website regarding its consent decree. http://www.lexingtonky.gov/index.aspx?page=840 (accessed October 15, 2015).

Louisville and Jefferson County Metropolitan Sewer District's website on Project WIN. http://msdprojectwin.org/ (accessed October 15, 2015).

Natural Resources Defense Council. 2001. *Stormwater Strategies: Community Responses to Runoff Pollution.* http://www

.nrdc.org/water/pollution/storm/stoinx.asp (accessed October 15, 2015).

Nature Conservancy. Building the Case for Green Infrastructure. http://www.nature.org/about-us/working-with-companies/companies-we-work-with/building-a-case-for-green-infrastructure.xml (accessed October 15, 2015).

Sanitation District No. 1's (Northern Kentucky) Federal Court Order (Consent Decree). http://www.sd1.org/AboutSD1/FederalCourtOrder.aspx (accessed October 15, 2015).

U.S. Environmental Protection Agency. Civil Cases and Settlements. http://www2.epa.gov/enforcement/sanitation-district-no-1-northern-kentucky-clean-water-settlement (accessed October 15, 2015).

———. Green Infrastructure. http://water.epa.gov/infrastructure/greeninfrastructure/index.cfm (accessed October 15, 2015).

———. Managing Urban Runoff. http://water.epa.gov/polwaste/nps/urban.cfm (accessed October 15, 2015).

———. Stormwater Program. http://cfpub.epa.gov/npdes/home.cfm?program_id=6 (accessed October 15, 2015).

———. Urban Waters. http://www2.epa.gov/urbanwaters (accessed October 15, 2015).

CHAPTER SEVENTEEN

Looking to the Past, Designing and Demonstrating Alternative Methods Today to Address Urban Stormwater Challenges

Jason Hale

Water is becoming recognized more as a valuable resource to communities today and is being addressed less as a nuisance. This holds true especially for stormwater. The way many Kentucky communities have urbanized for the past sixty years or so has unfortunately often had negative impacts on our water resources. Storm-sewer systems were designed and implemented under a mindset that viewed rainwater and subsequent stormwater runoff as waste or a nuisance and therefore sought to remove it as quickly as possible from the site or landscape. Consequently, this efficient removal of stormwater also provided for the efficient transport of urban pollutants into the nation's waterways without significant treatment.

Urban stormwater runoff is identified by many as a resource that, with proper management, can improve the quality of life for its citizens by providing a place where natural processes can be appreciated. This chapter focuses specifically on large and small communities by demonstrating a variety of projects that have explored and implemented low-impact development strategies by design professionals around the Commonwealth of Kentucky having landscape architects or engineers on their team. These projects have been designed to manage stormwater runoff by either storing and reducing runoff volumes or treating the water so that cleaner water is returned to our streams and waterways. Particular emphasis is placed on alternative material used in contrast to traditional material usage and approaches, as well as the challenges associated with changing ingrained mind-sets more familiar with the conventional materi-

als and methods of addressing stormwater runoff. This is a story about how people are seeing the necessity to change perspectives on water resources and how the implementation occurs.

First, some background and context are needed to set the stage for this chapter. Many effective techniques exist to manage stormwater runoff. Some are tried-and-true methods that have been used in the construction industry for decades, while other newer methods that offer alternatives for stormwater management continue to be developed. These alternative approaches and materials are becoming readily available to the construction industry, and some have been tested extensively and installed to great success. Many materials without supporting data or a history of successful implementation have properties that resemble those of materials that have been thoroughly tested. These products seem to have appeared with the intent to capitalize on the popularity of emerging industry trends and should be scrutinized and carefully considered before being installed. It should be understood that alternative construction materials have their place and will not serve as a substitute for conventional materials. The intent of this chapter is not to advocate for the complete replacement of conventional approaches and materials with alternative approaches and materials; conventional materials and methods have a purpose, but how we address urban runoff needs to evolve.

It is critically important in considering changes to hydraulic systems to understand how existing conditions perform currently on sites while putting them in the context of the naturally occurring systems before improvements are proposed. Understanding how water moves through a landscape minimally affected by anthropogenic influences is essential in considering alternatives. In addition, understanding how preindustrial cultures manage water can be a source of inspiration in addressing modern dilemmas. For example, Machu Picchu is a development in Peru believed to have been constructed in the fifteenth century that used simple but intricate drainage networks that allowed for the construction of a village on the steep mountainside. Without carefully considering how the runoff was generated from rooftops and through permeable terraced platforms, the development could not have been constructed and certainly would not have remained on the steep slopes of the Andes Mountains without eroding down the sheer mountain slopes.

Implementing alternative materials requires a thorough analysis of each site-specific design solution in the up- and downstream context. These materials should be considered as part of a larger, interconnected system when one is evaluating what each material offers as an improvement to stormwater management with each application. The recommendation or decision for the particular use of stormwater-management material should be left to specialized design professionals who can evaluate landscape conditions, compare approaches and materials, and determine the most appropriate application since the design professional will ultimately accept the responsibility associated with both successes and failures.

A community member or water resources professional must be aware that opposition to using alternative approaches or development material exists in some municipalities, as well as among some design professionals. In my experience, I have found that this opposition typically exists either because these materials are perceived to be too new for the decision makers and implementers and not tried-and-true or, more often, because individuals are not familiar with the recommended and typically superior alternative approaches and materials.

Before urban development occurs, water-quantity calculations are now typically required as part of the proposed construction document set. These calculations compare stormwater runoff volumes under existing and postdevelopment conditions in likely storm events in the region. The postdevelopment calculations reveal how the proposed site will respond to site conditions after construction. In order to address the anticipated hydrologic system change, a number of different strategies known as best management practices (BMPs) can be used. In most cases, the hydrograph of urbanized areas is altered from the preurbanization state by reducing base flow and increasing the amount of water, as well as shortening the time during which the water moves through a landscape.

Several different types of alternative stormwater-management methods are part of this chapter, including rain gardens, several permeable pavement materials, and subsurface detention structures. Some or all of these may be familiar to readers, but the following will serve as a brief review so that terminology is used consistently. I will not be very technical; more information is provided in the resources listed at the end of this chapter. For starters, rain gardens, or rainwater gardens, offer an uncomplicated approach to stormwater management in a relatively small area. A representative residential rain garden may capture and treat a small volume of water directed from a roof downspout into an area typically sized to hold the first flush or a half-inch rainfall. Rain-garden installation has been popular with the public in the past few years because the gardens offer individuals an easy way to manage runoff generated from their homes while providing a personal enhancement to the property and the community. This small form of stormwater management is minimal but can effectively address urban runoff in at least three important ways. First, the quantity of water captured from the home's roof is kept from entering the public storm-sewer system, so the volume is reduced. Second, the runoff directed to a storm system is slowed down, and the energy that would otherwise disrupt a stream and accelerate bank erosion is reduced. Third, the runoff is also treated within the rain garden's amended soil. Much research has been performed that indicates that pollution in the form of nutrients (nitrogen and phosphorus) and toxic metals (copper, zinc, and lead) carried from paved surfaces of roads and parking lots is removed from stormwater runoff.[1] Rain gardens are a basic method of stormwater management intended to mimic larger natural systems, like wetlands, where water is stored and pollutants are broken down biologically within the soil, but they are effective in reducing runoff volumes, slowing the velocities of flows, and removing pollutants such as nutrients and toxic metals within the first flush.

Rain gardens are a model method describing how stormwater storage approaches, much like traditional retention ponds, affect runoff after rainstorms. Rain gardens are recognized by some municipalities as a stormwater-management device and are credited with improving stormwater quality when improvement measures are required, but they are not the only way to work with water to improve quality. A similar type of facility used for the conveyance of stormwater is a bioretention swale, a bioinfiltration swale, or simply a bioswale. Rather than just quickly directing runoff through a piped network to the storm-sewer system, this approach can offer comparable treatment to rain gardens and increase infiltration typically through a sand filter with a subdrain. One recent example of a bioinfiltration swale that was proposed for construction is located at the University of Kentucky Coldstream Research Campus (figures 17.1 and 17.2). An existing drainage channel that carries runoff toward Cane Run was forced nearly 90 degrees before flowing beneath a bridge located along McGrathiana Parkway. Rather than closely mimic a natural flow path, this abrupt turn in the channel resulted in severe erosive forces from the runoff generated from an approximately sixty-five-acre watershed. In addition to widening the bottom of the channelized flow path as one solution to distribute the water and reduce erosive energy, a more gradual curve was designed within the channel, as well as a bioinfiltration swale defined by a low retaining wall. This wall acts as a dam for the bioswale and has several spillways of varied width that will slow the velocity of the runoff during large rains but allow water to pass through at controlled elevations. Smaller rains ranging up to a half inch will be captured and will slowly infiltrate into a subdrain system directed beyond the dam farther downstream. In combination, these elements provide benefits in both improved water quality and increased water-quantity storage capacity.

Cisterns are also a BMP focused on reducing stormwater volume. Typically, cisterns are used to collect rainwater runoff to be harvested. Rainwater harvesting is the collection, storage, and use of rain as a resource rather than a nuisance. The water is frequently screened or filtered, so this results in an improvement to water quality as well. Cisterns can be closed for rainwater-harvesting purposes or remain open so the captured runoff is allowed to slowly infiltrate rather than enter the storm-sewer system through open drainage-conveyance channels or drain structures. Cisterns to store potable water have been used as long as civilization has attempted to collect and store water, so this practice should not be considered new, although renewed interest is occurring in particular for use in landscape watering. Some state agencies are beginning to realize the limits of existing freshwater resources. Colorado, a state that is often considered progressive in its developmental practices, had placed limits on rainwater harvesting. Before passing a law in 2009 making residential collection legal, Colorado prohibited private rainwater-collection systems because they reduced flows to streams belonging to owners with water rights to streams.

In larger residential and commercial or industrial projects, another similar water-saving method that is becoming more popular is the use of structural stormwater storage facilities with storage pipes, cells, or chambers. These are becoming more common in areas where real estate is limited and expensive and the luxury of building a surface detention or retention basin is not a realistic solution. Along with underground solutions, using rooftops is also another strategy to intercept rain as it begins the process of moving through the landscape. Vegetative roofs, or green roofs, are another BMP that can greatly reduce stormwater volumes. Green roofs are generally recognized as improving water quality, although some research suggests that runoff from young vegetated roofs may result in elevated rates of total suspended solids and certain nutrients, especially phosphorus, upon installation.[2] This may be considered a temporary negative attribute, but the benefits associated with stormwater storage capacity (among other positive qualities, such as increased thermal insulation values, aesthetics, and removal of other pollutants) can

17.1. Plan image, UK Coldstream Research Campus Bioinfiltration Swale. CDP Engineers.

17.2. Conceptual Section A. CDP Engineers.

outweigh the negatives. Vegetative roofs in the form of sod roofs have been used in Finland and Scandinavia since the Middle Ages. This roofing type originated through use of readily available material and was not intended as a stormwater BMP, but traditional sod roofs greatly reduced roof runoff. The sod allows rainwater to infiltrate rather than shed directly onto the ground, like the majority of industrial roof materials. Another example of a built project that combines multiple strategies in its stormwater management is a modified "green" roof application that is not vegetated and is not as much a roof as it is a plaza or park. An example of this kind of project is the celebrated Park Güell in Barcelona, Spain, designed by the noted architect and landscape designer Antoni Gaudí and originally built from 1900 to 1914. This park is in a Mediterranean climate where rain is uncommon during summer months. Park Güell is a creative space that, among other features, uses a crushed aggregate and sand paving surface of a plaza. As the summer season leads to fall, thunderstorms are a regular occurrence, and the plaza is designed to filter rainwater, which is then directed to an underdrain system. This filtered water is stored beneath grade in a cleverly concealed cistern for use in several fountains. One fountain is an ornate dragon sculpture decorated with brightly colored mosaic tiles that was designed to become animated once water emerges from its mouth after rainwater has percolated through this hidden drainage system. This design has been in place since the early 1900s and, with ongoing maintenance, has continued to function as originally intended.

Several types of permeable paving technology are not brand new but are becoming more recognized and associated with modern stormwater-management benefits. In the history of paving surfaces, they are actually new to the United States. Porous concrete was first identified and used for its drainage properties in England in the 1970s[3] and then was introduced in Florida to meet specific stormwater requirements. Compared with the use of traditional brick or stone paving, which originated in ancient Rome on the famed Appian Way (one of the first constructed transport systems), porous pavements are beginning to be considered as a viable paving option in the United States. Advancements have

been made in alternative paving methods since ancient Rome, and established research supports how permeable materials are highly capable of changing the hydrograph after storms. Studies have shown that some properly constructed porous pavement systems have infiltration rates of 670 to 900 inches of stormwater per hour,[4] which is more rainwater than any storm can produce. Some systems were even capable of infiltrating more than 4 inches per hour eight years after being installed,[5] which means that they respond more similarly to lawn conditions than conventional pavements. The amount of water infiltrated through the porous pavement is limited by the subbase conditions, which are designed with the intention of either storing a specific volume of water in a closed system or diverting the water away through a piped network. Of course, to function properly, the pavement must also be regularly maintained and kept free of debris and fine sediment that could potentially clog the system and prohibit infiltration.

With advanced technology, porous paving systems have become more refined, so installations are unique from site to site. This variability results in a greater probability of error due to faulty design or inappropriate installation, making the materials more susceptible to failure. The appropriate paving type varies and depends on a site-specific application (this is where designers come in—to select the most appropriate material within the context and use of the site). Their recognition as a stormwater-quality BMP should also be accepted because of their effective removal of pollutants like total suspended solids (sediment) and trace metals that result from vehicular use.[6] Regardless, when they are considered as an alternative to traditional and largely impervious paving materials, reluctance is encountered from individuals not familiar with the various pervious paving types, who become more distracted by potential risks associated with the materials rather than how the benefits can be used to their advantage. The built environment is constructed largely of traditional paving materials, and those are what municipalities, designers, and the construction industries are familiar and comfortable with installing. Ultimately there is not a single paving material, porous or nonporous, that is a replacement for another; each material has both positive and negative attributes to consider before a decision can be made. However, if people and communities limit themselves only to conventional or the most common materials, they are not using all the available tools in the toolbox in developing a design solution, and the benefits offered from using porous paving materials are no longer available.

Here is an example of how to combine multiple strategies to address stormwater on an urban environment site. Porous pavements are extremely effective at reducing stormwater volumes when they are designed in combination with manufactured detention units designed to store stormwater underground. When the surface is porous and allows for infiltration of stormwater beneath grade, materials and products are available to store this water with the intention of harvesting it for future use. Attention is returning to rainwater capture to help ensure adequate water supplies. Systems can range from a simple barrel placed beneath a gutter downspout in a residential application to a more complicated arrangement of tanks with electronic pumps and controls. Many examples can be cited where captured stormwater is used for irrigation purposes intended to offset and reduce the use of potable water from municipal treatment utilities. A different use has been planned for a proposed pump station located in Jeffersontown near Louisville.

The Grand Avenue Pump Station has been designed for the Louisville Metropolitan Sewer District so 60 percent of the roof is directed toward a permeable paving surface. Here, roof runoff can be infiltrated through a permeable system rather than be diverted into the closest stream, Chenoweth Run (figure 17.3). Beneath the permeable surface, a network of arched storage chambers collects the runoff in a system contained by an impermeable liner. One must first understand the purpose of the pump station to understand the intended use for the captured water. In general, the pump station has large concrete storage basins below grade that fill with solid waste during periods of peak flow (figure 17.4). Once this waste is pumped and the levels can be lowered, solid residue remains along the bottom and sides of the basin, which need to be flushed clean. The captured runoff stored beneath the pavement is used to refill aluminum tipping

buckets that flush this residue from the below-grade containment system. There is a backup water line for filling the tipping buckets when the volume of stored water has been emptied. But the roof area and the conventional pavement, as well as the permeable pavement's surface area, result in nearly one-third of an acre that is directed to the storage area. For example, a one-inch rainfall over this area will generate nearly 8,800 gallons of water, which is enough to fill the tipping buckets more than three times without using treated water. However, the entire storage system is capable of storing much more water—actually, enough to fill the tipping buckets more than two hundred times before being drawn down completely. This not only helps reduce runoff that would otherwise be diverted to streams but also limits the use of potable water that would only be flushed down the sanitary sewer system for additional treatment.

In general, the use of stormwater BMPs offers a tremendous benefit to property owners and developers interested in maximizing their developable area. Case studies can be cited where a portion of property, otherwise required for stormwater storage in the form of a conventional detention pond if traditional materials had been used, was reduced in size, if not eliminated entirely, because the amount of runoff was reduced through the use of various BMPs and permeable materials. One example in Versailles, Kentucky, is Equestrian Park, a proposed commercial development where all paving surfaces within the public right-of-way are constructed of permeable interlocking concrete pavers. The developer and designers worked closely with the City of Versailles to design a stormwater system that reduced the need for stormwater structures and piping and reduced the size of the detention area required, which reserved a greater developable area within the project limits.

Alternatively, stormwater BMPs can be used to retrofit project sites where trouble areas exist because of the use of traditional materials. One example is Sunset Avenue in Richmond, Kentucky. Several residents on this historic street close to downtown cited flooding problems because the existing infrastructure was either failing or was undersized to accommodate the current amount of stormwater runoff directed to the storm-

17.3. Grand Avenue Pump Station illustration. CDP Engineers.

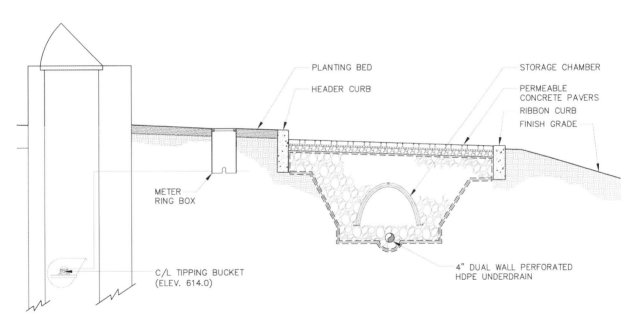

17.4. Cross-section detail through storage chambers. CDP Engineers.

sewer system (figure 17.5). A conventional solution (figure 17.6) to correct the problem would be to enlarge the current pipes, but when pipes are increased in size and directed downstream to smaller pipes, those pipes need to be replaced as well because failure to do so risks creating another problem with the potential for flooding during large rains. This solution would result in a great expense to the city of Richmond through the purchase of easements on private property, in addition to the construction material and labor required. Therefore, another mind-set and approach were needed. A more site-specific design solution proposed the use of permeable paving

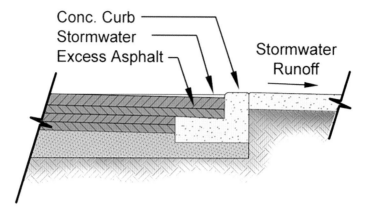

17.5. Section of curb and gutter: preconstruction conditions. CDP Engineers.

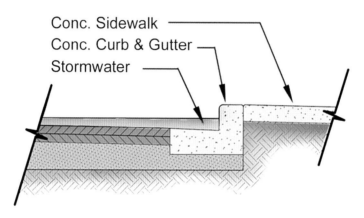

17.6. Section of curb and gutter: conventional design solution. CDP Engineers.

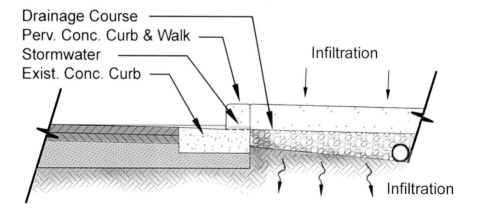

17.7. Section of curb and gutter: porous concrete solution. CDP Engineers.

17.8. Georgetown Fire Station parking lot. Image Courtesy of the City of Georgetown.

materials within the sidewalk along one side of the street (figure 17.7). This alternative design solution resulted in fewer disturbances to property downstream from the immediate problem area and was more cost effective for the city when compared with the conventional design solution.

A special project that was designed to demonstrate a variety of site-scale BMP solutions in addressing stormwater is located in Georgetown, Kentucky. The Georgetown Fire Station Parking Lot Demonstration Project was designed and constructed partly with federal grant dollars made available through the federal Clean Water Act, Section 319(h) (figure 17.8). This project site was likely to become a conventional, impervious asphalt parking area for Fire Station #3 and the adjacent recreational park in Georgetown. However, grant funding was awarded to this project because it demonstrated the ability, through an application process, to reduce nonpoint-source pollutants and improve water quality through the use of several BMPs intended to be installed on the site. Critically important was the large public involvement component, where professionals, public works officials, and the general public were invited to attend educational field days focused on seven different stormwater BMPs and the benefits they can offer to water quality. The intent was for the community to take the knowledge and experience gained from witnessing the BMPs installed so they might be considered viable, if not preferred, material alternatives for other sites. The implemented BMPs included sediment- and erosion-control devices, three types of permeable paving materials (porous concrete, pervious asphalt, and permeable interlocking concrete pavers), subsurface detention units, rain gardens, and a rain barrel. The site also features multiple interpretive sign panels designed to orient visitors interested in taking a self-guided tour of the site so they can educate themselves about the benefits to water quality offered by the BMP on display.

Even if we set aside generally improved aesthetics for the moment, the use of alternative materials should be encouraged within the built environment because of the benefits offered to stormwater management. If people and communities are not already in a mind-set of seeing water as a resource rather than a nuisance, they are stuck in a late twentieth-century way of thinking about dominating and controlling water as opposed to working with it. The BMPs briefly described here demonstrate some of the options to consider in combination with traditional material usage in new and retrofitted urban contexts. The challenge associated with changing ingrained mind-sets is typically more difficult than developing unique site solutions for the water. With more frequent installation and continued use, and with stormwater becoming a major factor in urbanization at site and watershed scales, alternative materials and methods and, more important, twenty-first-century thinking are needed so that they are the tools of choice in the toolbox for people and communities.

Notes

The contents of this chapter reflect project work while the author was employed by CDP Engineers, Lexington, Kentucky.

1. United States Environmental Protection Agency, *Low Impact Development (LID): A Literature Review,* EPA-841-B-00-005 (October 2000), 18–22, http://water.epa.gov/polwaste /green/upload/lid.pdf (accessed October 15, 2015).

2. Charles Glass and ETEC, *Green Roof Water Quality and Quantity Monitoring,* 2007, 11–17, http://www.asla.org/uploaded Files/CMS/Green_Roof/Green_Roof_Water_Monitoring _Report.pdf (accessed October 15, 2015).

3. D. P. Maynard, *A Fine No-Fines Road,* Concrete Construction, Publication C70016 (Aberdeen Group, 1970).

4. Bruce K. Ferguson, *Porous Pavements* (Boca Raton, FL: CRC Press, 2005), 124.

5. Ibid.

6. United States Environmental Protection Agency, *Low Impact Development (LID).*

Additional Resources

Center for Watershed Protection. http://www.cwp.org/ (accessed September 18, 2016).

Green Roofs for Healthy Cities. http://www.greenroofs.org (accessed September 18, 2016).

Low Impact Development Center. http://www.lowimpact development.org/ (accessed September 18, 2016).

National Conference of State Legislatures. State Rainwater | Graywater Harvesting Laws and Legislation. http://www .ncsl.org/issues-research/env-res/rainwater-harvesting .aspx (accessed September 18, 2016).

National NEMO (Nonpoint Education for Municipal Officials) Network. http://nemonet.uconn.edu/ (accessed September 18, 2016).

U.S. Environmental Protection Agency. Water: Low Impact Development. Last updated September 1, 2016. http:// water.epa.gov/polwaste/green/ (accessed September 18, 2016).

Chapter Eighteen

Who Pooped and How It Was Found

One Scientist's Story of Source Tracking Fecal Bacteria in an Inner Bluegrass Watershed

Tricia Coakley

As a child, I loved to walk around the old family farm in Green County, Kentucky, with my dad and granddad. As we walked, they would tell me stories of all the funny, and sometimes scary, things that had happened on that farm long ago. I wished that I could have been there to experience the hard work, long days, and near-fatal accidents. I wanted to be the one catching chickens and chasing rats away from the hams curing in the shed. I wanted to see the old rusty tractor that tried to kill people. I especially wanted to make soap out of lard in the big iron kettle and then use it to wash my clothes and hang them out to dry in the sun. The romance of farm living is thick in a city kid, and one day of walking the farm and listening to stories would yield at least a month of daydreaming about the good old days. I remember rolling up my pants legs and wading through the creek one day to cool off. As I walked along in my state of farm bliss, I thought, "Even the mud in the creek is better on a farm." It was the softest, silkiest mud I could imagine. There was no worry about cutting your foot on a rock in this creek. No sir, this creek was padded for comfort. When I asked my granddad why the mud was so soft, he simply pointed to a herd of cattle upstream from us. Just at that moment, one cow raised her tail and provided me with a timely demonstration of her soft-mud-contribution efforts. For a moment, I wondered whether this meant that we should not be walking barefoot in the creek. My wise, old granddad, however, calmly walked ahead with no concerns, so I decided to trust the poop as well and considered it a blessing as I squished my toes

into the creek bottom. I have learned a lot about poop since that day on the farm, most important, that the majority of the pollution in creeks cannot be so easily traced visually to its source, and that my instinct to avoid the poop was well founded. This is a story about the endeavors of an environmental scientist in Kentucky to clean up the polluted streams of her community, and the lessons learned along the way about the power of community involvement.

First, I will provide a little background information. The Kentucky Environmental Quality Commission reported in 1995 that fecal coliform bacteria were one of the top two pollutants affecting Kentucky's waters. Coliform bacteria are ubiquitous in the environment and are expected to occur in healthy streams. The subcategory fecal coliform bacteria, including *E. coli,* comes from the feces of humans and animals and is measured as an indicator of the severity of fecal pollution in the environment. Although most *E. coli* bacteria will not cause illness in humans, if they are found in large quantities in a stream, they indicate the potential for finding other more serious fecally associated pathogens that can cause illnesses, like hepatitis A and giardiasis (also known among travelers as Montezuma's revenge).

The Kentucky Division of Water recommends a maximum limit for *E. coli* bacteria in streams used for primary-contact recreation to protect human health and prevent gastrointestinal illness. This limit is 240 colony-forming units of *E. coli* bacteria per 100 milliliters of water. To get a sense of the severity of the problem, the average *E. coli* concentration in the streams of the Wolf Run watershed of Lexington during the summer months of 2010 was more than fifty times the maximum allowance. Sanitary sewer and storm-sewer overflows, known to occur in cities with outdated sewage infrastructure, contribute vast amounts of *E. coli* to the surrounding waters. The Environmental Protection Agency (EPA) has started to take action against many cities, including Lexington (see chapter 16 by Schieffer on consent decrees in this book), for chronic sewage discharge violations. Kentucky cities under consent decrees with the EPA are required to spend a combined total of more than $1.6 billion to resolve the problems, and Lexington alone will have to spend over $300 million to bring its infrastructure into compliance with federal law.

If this type of pollution is causing so many problems, why are we not fixing it quickly? What is the holdup? To effectively remove or reduce fecal bacteria entering waterways, we must first determine the locations of the pollution sources. For years, there has been a series of finger-pointing statements between municipalities and farmers, but there were no data available to show the explicit pollution source. Unlike the cow in Granddad's creek, most fecal pollution sources are difficult to determine, and *E. coli* is an insufficient indicator for the job because it comes from all animal sources. *E. coli* can also persist in the stream for some time, depending on stream chemistry and temperature, and travels with the stream flow away from its origin. *Bacteroides* bacteria, like *E. coli*, are found in the intestines and feces of humans and animals. Unlike *E. coli,* however, *Bacteroides* bacteria do not persist well in the environment because they are strict anaerobes and cannot survive the increased oxygen levels found outside the host organism's intestines. Because of this characteristic, *Bacteroides* has been proposed as the ideal fecal indicator organism for watershed research. This indicator still comes from humans and animals alike, however, and we need to differentiate between host sources to successfully resolve the associated pollution problems.

Modern laboratory techniques, using fecal bacterial DNA sequences found in samples collected from potential polluters like humans, cows, chickens, and other animals, have provided a template for comparison with DNA sequences found in polluted streams. These techniques have led to the formation of a new field of environmental forensics, known as molecular fecal source tracking. It is now possible to provide data concerning the sources of fecal pollution in streams in a manner similar to the way we determine the identity of an alleged criminal by skin or hair found at a crime scene. The DNA sequence from the *Bacteroides* bacteria has been shown to have regions that are specific to the animal host of its origin (human, cow, and so on). This is good news because molecular biologists around the world have begun developing DNA biomarkers specific to these unique regions of the *Bacteroides* genome, and the biomarkers can be used to differentiate between fecal sources contributing to the pollution found in waterways.

All this background information came together in a very real water-pollution problem in the Wolf Run watershed of Lexington. As an employee of the Environmental Research and Training Lab and a graduate student in the Earth and Environmental Sciences Department at the University of Kentucky in 2008, I was challenged with the task of putting these new molecular fecal source markers to the test. A nonprofit watershed-protection organization, Friends of Wolf Run, came with a request to perform a study using molecular source tracking techniques to locate broken, leaking, or cross-connected sanitary sewer lines. The city had already identified several storm-sewer overflows within the area, but they did not explain the consistently high *E. coli* signal throughout the watershed. The storm-sewer system should carry only storm water, but the high *E. coli* concentrations found in many creek samples indicated that they were very likely being derived from more concentrated sources, like sanitary sewer lines. The goal was to use the DNA biomarker concentrations from samples throughout the watershed to provide a mapped spatial distribution of human fecal intensity, which could lead to likely location(s) of sewage inputs from cross connections between the storm-sewer and sanitary sewer systems or from sanitary sewer leaks that the city had not found previously.

To begin the project, positive and negative controls were needed to ensure that the results would be valid. For a study of fecal source identity, control samples come in the form of freshly formed fecal donations from as many species as possible. The data from these controls were necessary to ensure that the human-specific biomarker we chose for our study could indeed come only from humans. I have spent many years of my life in the lab, and suffice it to say, I am the stereotypical lab nerd. It is true that I have an arguably unhealthy fondness for pipettes, test tubes, petri dishes, and charts. Knowing this about me, as well as the farm romance of my youth, one can imagine how excited I was to step outside familiar territory and locate positive controls.

My carefully collected lab notes indicate that cows are apathetic, otters are elusive, horses are mischievous, and llamas . . . well, llamas are just plain mean. I am proud to report that I successfully obtained thirty positive control samples from various animals, including humans (the only species requiring bribery), but I will picture llamas as schoolyard bullies for the rest of my days. Once the lab techniques had been tested with the collected fecal control samples, we were ready to sample the watershed.

Friends of Wolf Run volunteers, along with students and faculty at Bluegrass Community and Technical College, assembled a large team and collected samples from twenty locations throughout the watershed nearly simultaneously. This provided a snapshot in time of the water quality at multiple locations. The volunteer efforts were essential to the success of the study because a single person could not have done this simultaneous sampling alone. All the samples were transported in coolers of ice to the lab, where the DNA was extracted and frozen.

Each sample was then analyzed for a DNA biomarker of human fecal pollution. As hypothesized, the concentration of the marker was greater at some locations than others, and we were able to define hot spots of human fecal pollution that warranted further investigation. The technique allowed a scientific approach to identify potential locations of sewage leaks within a large watershed with several miles of streams and a complex system of storm-water and sanitary sewer lines.

We found one such hot spot at a storm-sewer overflow near the intersection of Lafayette Parkway and Rosemont Garden. Although the city was aware of this overflow, it did not appear to be a greater threat to human health than any of the other storm-sewer overflows on the basis of the *E. coli* signals obtained. When the location was viewed with respect to the DNA marker for human feces, however, it really stood out among the crowd and indicated a much greater contribution to the total fecal load of the stream than could be explained by our understanding of the storm-sewer system. The results led us to suspect a broken or leaking sanitary sewer line nearby, or possibly a cross connection between the storm-sewer and sanitary sewer systems.

These data were of interest to the city government employees charged with meeting the EPA's requirements and led them to begin searching the area around this hot spot in a more detailed fashion. While the city employees were walking the path of the storm-sewer lines leading to the overflow location and looking into manholes, they observed a sudden gush of water flowing into the storm sewer. The day was dry, with no storm runoff, which could mean only one thing: they had found a cross connection with the sanitary sewer. Further investigation of historical records for that sewer line revealed that the effluent plumbing from an adjacent building, including the toilets, had been inadvertently connected to the storm-sewer line rather than the sanitary sewer line during construction. We knew then that the building was a likely source of the fecal pollution found at the storm-sewer overflow identified as a hot spot farther downstream. The city, empowered with this information, rerouted the unintentional cross connection so that the sewage from the building flowed only into the sanitary sewer system.

After the correction of the cross connection, we collected additional samples from the original hot-spot stream location at the storm-sewer overflow. The DNA biomarker results from these samples indicated a substantially reduced human fecal signal, confirming that the cross connection had indeed been the pollution source.

Stepping back from the science-driven part of this story, I point out that there is a bigger and in some ways a more important point to take away, one that comes back to the importance of the community in addressing environmental challenges. When there was a big project to do on the old family farm, my granddad would call on neighbors from around the community to pitch in because they all understood that some things just could not be accomplished alone. As a society, we need to channel the wisdom of our grandparents and pull together teams of people with different perspectives and skills. Using one's data to make a difference in the real world is every lab nerd's dream, and I am one lucky lab nerd. I could have spent the rest of my life generating similar data, feeling compelled by its value, sharing it with other scientists, and garnering new funding to continue the cycle, all without cleaning up one single fecal hot spot. This can happen in scientific research, but it does not have to be the case. As a collaboration of stakeholders, including a very dedicated group of citizens, a well-organized nonprofit organization, an incentivized city government, and scientific researchers with innovative techniques, we made real improvements in the environment by putting scientific data to work for the community. After all, isn't that what it is all about?

Additional Resources

Hagedorn, Charles, Anicet R. Blanch, and Valerie J. Harwood. 2011. *Microbial Source Tracking: Methods, Applications, and Case Studies.* New York: Springer.

Kentucky Environmental Quality Commission. *State of Kentucky's Environment: 1994 Status Report.* 1995. Accessed October 15, 2015, http://www.e-archives.ky.gov/pubs/Natural/environment/state_of_ky_envir94.pdf.

Kentucky Legislative Research Commission. *401 KAR 10:031 Surface Water Standards. Kentucky Administrative Regulations.* 2008. Accessed October 15, 2015, http://www.lrc.ky.gov/kar/401/010/031.htm.

United States Environmental Protection Agency. Office of Research and Development. 2005. *Microbial Source Tracking Guide Document.* EPA/600-R-05-064. June. Accessed October 15, 2015, http://www.sfbayjv.org/tools/EPAMicrobialSourceTrackingGuideDocument_June2005.pdf.

CHAPTER NINETEEN

The Watershed Atlas Project

A Different Way to Look at the Commonwealth

Brian D. Lee, Corey L. Wilson, and Angela Schörgendorfer

The Watershed Atlas Project represents an approach that uses existing publicly available geospatial data to visualize landscape indicator aspects of multiple watersheds simultaneously. Initial grounding for this work can be found in Jones et al. (1997), and it has gone through multiple iterations over the past decade as development has continued. By viewing the landscape from a watershed perspective, this atlas is intended to present an understanding of land use and management decisions and insight into how those decisions affect waterways and water. The project used the hydrologic unit code–14 (HUC-14) watersheds as the fundamental unit of analysis (Kentucky Geological Survey). In this atlas, each of the subwatersheds was characterized using data available from the Kentucky Geography Network in the Environmental Systems Research Institute's ArcGIS environment.

The atlas is divided into six sections with the following themes: geographic introduction, geomorphic, human, vegetative, riparian, and specialty indicators; these are followed by a condensed representation of the assessment in a data table. An indicator glossary is included at the end of each section, providing source information and a description of each analysis. The geographic introduction section of data provides base information to introduce the basin and its subwatersheds and waterways in the context of notable landmarks. The second section provides a geomorphic watershed characterization that focuses on data attributes such as size, elevation, and terrain. Section three contains data about the human-modified aspects of the watersheds, such as impervious cover, development amount, and roadways in relationship to water resources. Section four details the vegetative land cover of the watersheds; specific indicators focus on percentage of agriculture and forest, as well as the relationship of crop agriculture on land slopes of 3 percent and greater. In addition, this section characterizes changes in land cover that occurred between 2001 and 2006 as measured by satellite-based data. Section five focuses on stream- and riparian-area characteristics, and section six focuses on specialty indicators, such as Kentucky Pollutant Discharge Elimination System points, package plants, outfalls, and regulated dams.

Building on specific indicators as mapped in this atlas, a method was sought to group subwatersheds by human-induced and geomorphic characteristics. A statistical approach known as cluster analysis with complete linkage was used to determine thirteen clusters using twelve of the more than one hundred indicators derived during this project. The indicators used in the statistical cluster analysis were determined after a bivariate correlation analysis was performed using all variables to identify highly related indicators. The data and geospatial methods were subjectively balanced by the researchers to reduce the number of indicators in order to use them in the statistical cluster analysis. It is important to remember that these are simply descriptive clusters and are exploratory rather than inferential. Further linking these clusters to real water-quality and quantity data is the next step in the evolution of this project. Some limited studies have been undertaken, but

Subwatershed Size

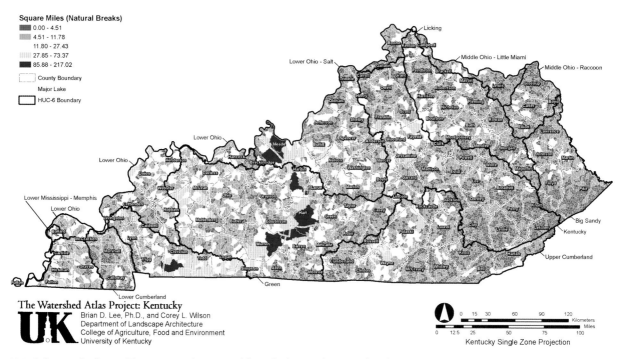

19.1. Subwatershed size. The size was determined by calculating the size of each subwatershed (hydrologic unit code-14) once projected in Kentucky Single Zone Projection.

Elevation Standard Deviation–Topographic Roughness

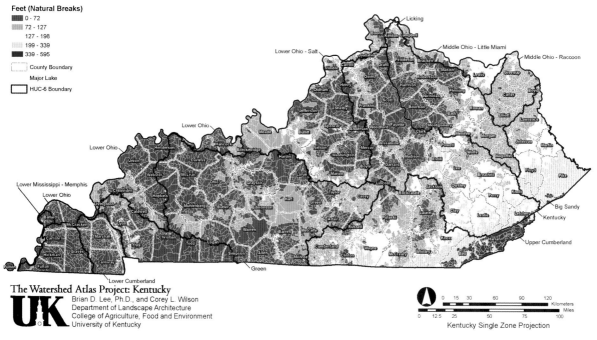

19.2. Elevation standard deviation—topographic roughness. Elevation standard deviation is calculated as the standard deviation of topographic elevations in the HUC-14 subwatershed, derived from Shuttle Radar Topography Mission (SRTM) 2000 data. This indicator is a measure of subwatershed topographic roughness.

Sinkhole Density

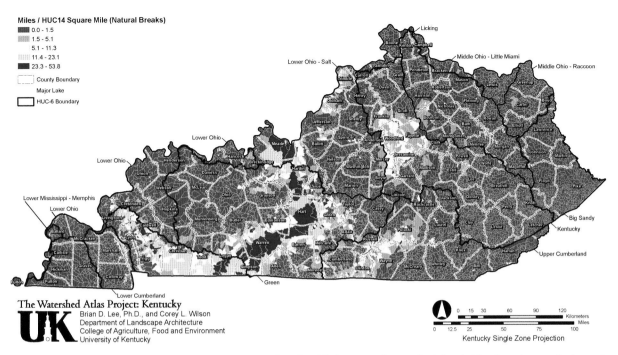

Miles / HUC14 Square Mile (Natural Breaks)
- 0.0 - 1.5
- 1.5 - 5.1
- 5.1 - 11.3
- 11.4 - 23.1
- 23.3 - 53.8
- County Boundary
- Major Lake
- HUC-6 Boundary

The Watershed Atlas Project: Kentucky
Brian D. Lee, Ph.D., and Corey L. Wilson
Department of Landscape Architecture
College of Agriculture, Food and Environment
University of Kentucky

Kentucky Single Zone Projection

19.3. Sinkhole density. Sinkhole density is the square miles of sinkholes in a subwatershed divided by the subwatershed area in square miles. This indicator is a measure of karst geologic features.

Topographic Change (Cut)

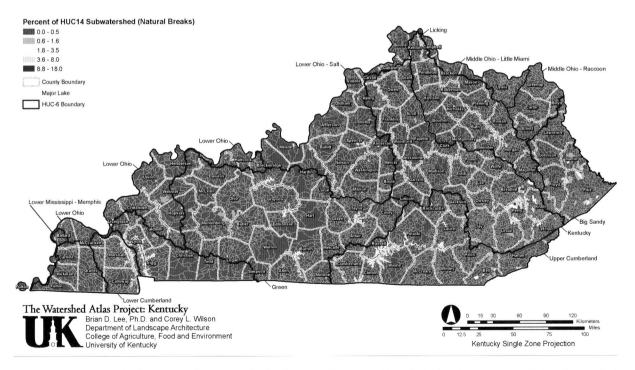

Percent of HUC14 Subwatershed (Natural Breaks)
- 0.0 - 0.5
- 0.6 - 1.6
- 1.8 - 3.5
- 3.6 - 8.0
- 8.8 - 18.0
- County Boundary
- Major Lake
- HUC-6 Boundary

The Watershed Atlas Project: Kentucky
Brian D. Lee, Ph.D. and Corey L. Wilson
Department of Landscape Architecture
College of Agriculture, Food and Environment
University of Kentucky

Kentucky Single Zone Projection

19.4. Topographic cut(s). The topographic change (cut) indicator is determined by calculating the percentage of the subwatershed changed according to USGS Topographic Change data identified as "cut." A higher percentage of change is considered an indicator of a more impacted watershed.

Flow Accumulation Standard Deviation

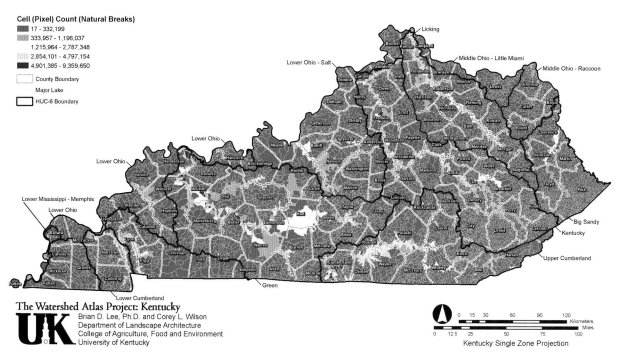

19.5. Flow accumulation standard deviation. Flow accumulation standard deviation is the standard deviation of the accumulation of all subwatershed cells that flow into a downslope cell in the SRTM 2000 data. The indicator was calculated using a flow accumulation function with the statewide SRTM 2000 data before extraction for subwatershed analysis.

2006 Imperviousness

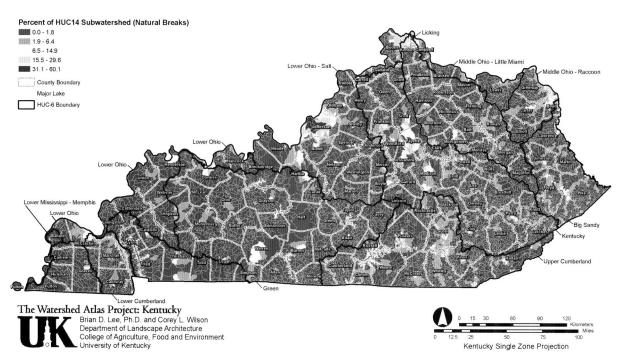

19.6. Imperviousness, 2006. Imperviousness is calculated as the subwatershed percentage covered by impervious surfaces according to the Multi-Resolution Land Characteristics Consortium (MRLC)–National Land Cover Database (NLCD) 2006 Percent Imperviousness data. A lower percentage of impervious surface is considered an indicator of a less impacted watershed.

2001 – 2006 Impervious Cover Change

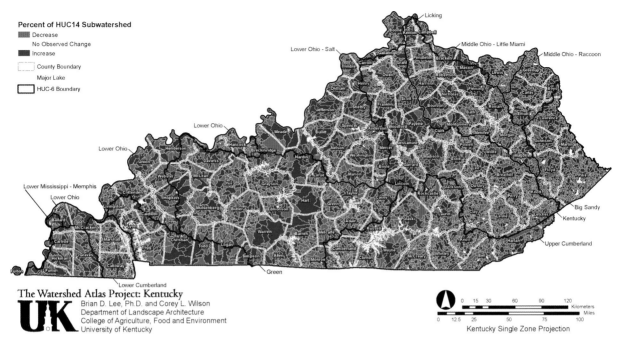

Percent of HUC14 Subwatershed
- Decrease
- No Observed Change
- Increase
- County Boundary
- Major Lake
- HUC-6 Boundary

The Watershed Atlas Project: Kentucky
UK Brian D. Lee, Ph.D. and Corey L. Wilson
Department of Landscape Architecture
College of Agriculture, Food and Environment
University of Kentucky

Kentucky Single Zone Projection

19.7. Impervious cover change, 2001–6. Imperviousness change is calculated as the percentage change in imperviousness from 2001 to 2006 based on the National Land Cover Data (NLCD) 2001 (Version 2) and the 2006 National Land Cover Database on a subwatershed unit of analysis. The percentage indicates how much more the subwatershed has been made impervious during the period.

2006 Forest Cover

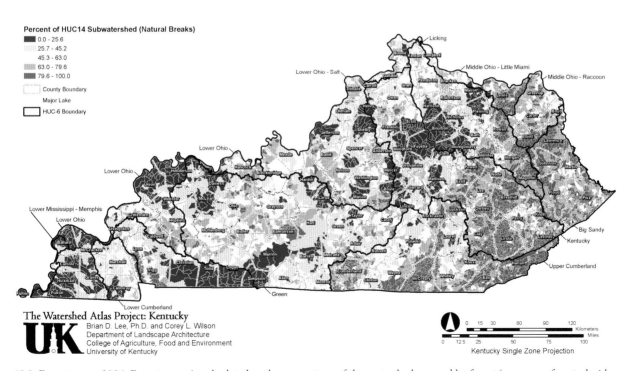

Percent of HUC14 Subwatershed (Natural Breaks)
- 0.0 - 25.6
- 25.7 - 45.2
- 45.3 - 63.0
- 63.0 - 79.6
- 79.6 - 100.0
- County Boundary
- Major Lake
- HUC-6 Boundary

The Watershed Atlas Project: Kentucky
UK Brian D. Lee, Ph.D. and Corey L. Wilson
Department of Landscape Architecture
College of Agriculture, Food and Environment
University of Kentucky

Kentucky Single Zone Projection

19.8. Forest cover, 2006. Forest cover is calculated as the percentage of the watershed covered by forest (evergreen forest, deciduous forest, and mixed forest) according to the respective 2006 NLCD data set. A higher percentage of forest cover is considered an indicator of a less impacted watershed.

2001 – 2006 Forest Cover Change

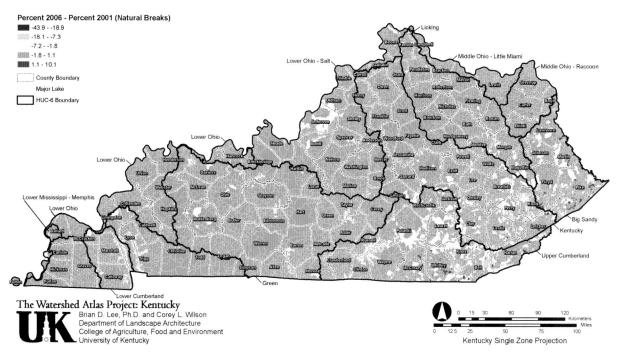

19.9. Forest cover change, 2001–6. Forest cover (evergreen forest, deciduous forest, and mixed forest) change is calculated as the percentage change in forest land cover from 2001 to 2006 based on the NLCD 2001 (Version 2) and the 2006 NLCD. The percentage indicates how much the forest cover of the subwatershed has been reduced during the period.

2006 Crops on Slopes (3% or greater)

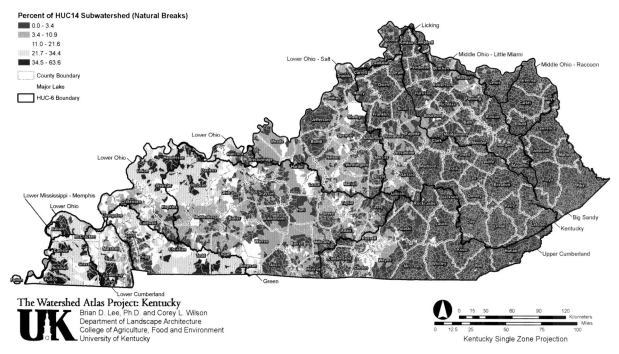

19.10. Crops on slopes (3 percent or greater), 2006. Crops on slopes are calculated as the percentage of the watershed with crops on slopes greater than or equal to 3 percent (Jones et al. 1997). The respective 2006 NLCD data (Cultivated Crops) and the derived slope map from SRTM were combined to identify cells that met the slope and cultivated-crop conditions. A lower percentage of crops on slopes is considered an indicator of a less impacted watershed.

Stream Density

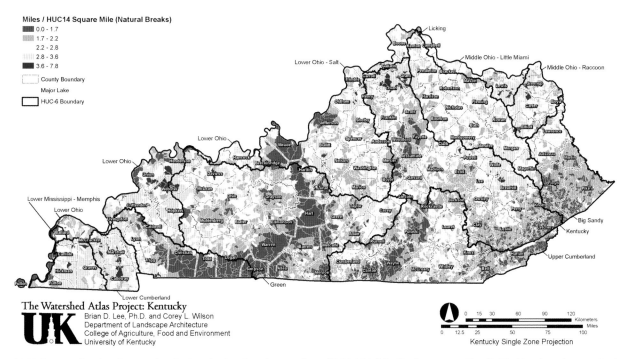

19.11. Stream density. Stream density is calculated as the number of 1:24,000 (blueline) stream miles divided by the subwatershed area in square miles.

Wetlands

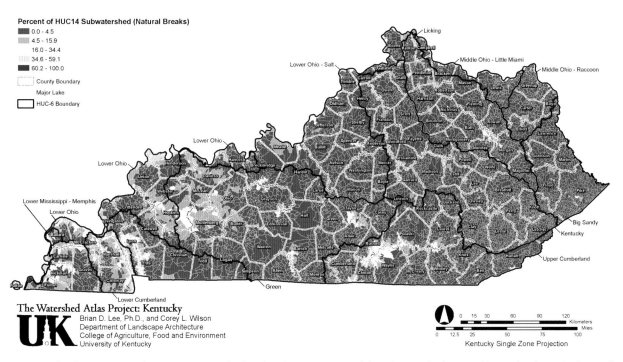

19.12. Wetlands. The wetlands percentage is calculated as the percentage of the subwatershed covered by wetlands on the basis of the National Wetlands Inventory.

Subwatershed Clusters

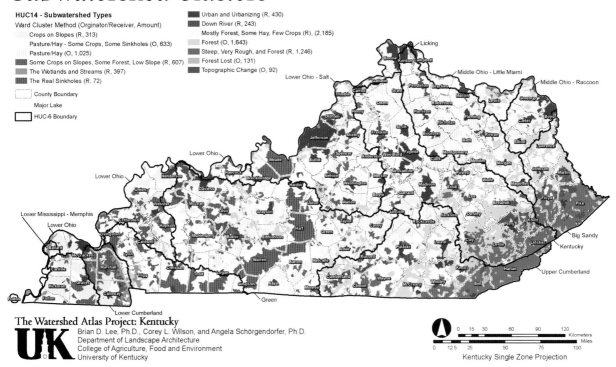

19.13. Subwatershed clusters. The thirteen subwatershed clusters or types were determined using the twelve previous indicators by a statistical method. The clusters are useful in identifying subwatersheds that have similar characteristics but may not be geographically close to one another.

unfortunately, at this time the resources are not available to undertake this effort in a scientifically robust way. It is believed that this clustering approach will allow for common planning and implementation development on a geographic scale that will be meaningful to people working from field to river-basin scales.

An overarching benefit of characterizing the landscape from a watershed-atlas perspective is the ability to recognize that the impacts of human influences on the natural environment do not necessarily observe political boundaries. A watershed-based approach to making land-management decisions takes into account that although cities, counties, and states may appear to be distinct entities, they are connected by ecological features and processes. There is an inherent argument in this perspective that land decisions should be made with these

characterizations in mind. This atlas can be used as a tool to identify which landscape characteristics are potentially relevant as a guide for future prioritization and management decisions, which may influence waterway quality.

The twelve indicators used resulted in the characterization of thirteen watershed types (clusters) representing over nine thousand HUC-14 subwatersheds in the Commonwealth of Kentucky. The indicators described different geomorphic and human-induced vegetative cover and hydrologic conditions, as well as forest and impervious change from 2001 to 2006. Figures 19.1 to 19.13 depict each of the indicators used in the resulting cluster analysis. Each extended caption describes the data and approach for each HUC-14 analysis unit.

References

Jones, K. Bruce, Kurt H. Riitters, James D. Wickham, Roger D. Tankersley Jr., Robert V. O'Neill, Deborah J. Chaloud, Elizabeth R. Smith, and Anne C. Neale. 1997. *An Ecological Assessment of the United States Mid-Atlantic Region: A Landscape Atlas.* USEPA/600/R-97/130. Washington, DC:

U.S. Environmental Protection Agency. http://archive.epa.gov/emap/archive-emap/web/html/atlas.html (accessed October 18, 2015).

Kentucky Geological Survey. *14 Digit Hydrologic Units* (statewide, shapefile). Lexington, KY. http://kgs.uky.edu/kgsweb/download/rivers/huc14.zip (accessed October 18, 2015).

CHAPTER TWENTY

Reinventing Water-Landscape Monitoring and Management in the Age of Geoenabled Environmental Sensor Webs and Social Networks

Demetrio P. Zourarakis

Kentucky's human communities and waters intersect in space and time, and the intensity of this interaction has been accelerating because of population changes, anthropogenic and natural disturbances, and changing socioeconomic drivers. At the same time, rapid technological innovation, social-network mobility, and interconnectedness offer an opportunity to satisfy the converging interests of environmental-awareness and resource-conservation initiatives.

Water and Humans Shaping Kentucky's Physiography

Kentucky's a rural character and physiography determine that human settlement and activity across the commonwealth share its ubiquitous waterways and water bodies (U.S. Census Bureau 2016a; U.S. Census Bureau 2016b) (figures 20.1 and 20.2). This familiarity provided by our proximity to water, however, can act in two opposing ways. On the one hand, the experiential value of water as an indispensable element in a sustainable quality of life is self-evident. On the other, the great risk is that this value can be negated when degradation of the resource can become imperceptible as our perceptual adjustments to small changes occur over time (Brody 2004).

Through governmental and citizen actions, communities have long recognized their stake in the fate of the waters of the commonwealth by showing increasing sensi-

tivity to improving water quality and quantity in Kentucky. The Watershed Management Framework implemented by the Kentucky Department of Environmental Protection has been in existence since the late 1990s (Kentucky Division of Water 2016a). Grassroots-dependent efforts, such as the Water Watch and Watershed Watch programs, have been active in Kentucky for about the same time (Kentucky Division of Water 2016b; Kentucky Division of Water 2016c). Nongovernmental organizations have implemented similar efforts, such as the Sierra Club National Water Sentinel Program (Sierra Club 2013).

Available to people with access to the Internet through the use of browsers, online mapping tools have continued to proliferate and now include essential and authoritatively maintained data such as the National Hydrography Dataset (U.S. Geological Survey 2016a; U.S Geological Survey 2016b). These viewers provide dynamic contextualization, in addition to displaying the data against a backdrop of periodically acquired imagery. Mapping applications allow for on-the-fly extraction of geospatial information essential to model water behavior (U.S. Geological Survey 2016c). A weakness is that they lack the synchrony of real-time data feeds such as the ones we often see, for example, in weather-related mapping sites such as the Kentucky Geonet–Kentucky Mesonet collaboration or the U.S. Geological Survey (USGS) Stream Gage Station system (Kentucky Geography Network 2016a; U.S. Geological Survey 2016d).

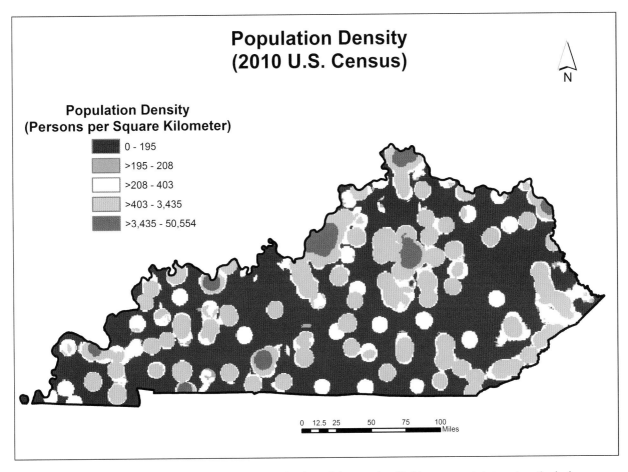

20.1. Spatial distribution of population density in Kentucky; kernel density classified by geometric interval method; class ranges are feature values (2010 Census Data 2014; 2010 Census TIGER/Line Shapefiles 2014).

The increased ability to generate or use real-time environmental data is becoming a valuable commodity for people and communities. Feed-forward information loops, such as advance forecasts of episodic meteorological events like storms and flash floods or environmental impacts (e.g., chemical spills), are now routinely procured and provided via mobile devices. The convergence of multiple and seemingly disparate technological and social network revolutions has implications for how water-resource issues can be addressed in the future. This chapter attempts to provide a better understanding of how multiple and mutually interacting innovations and developments are changing the way people view water in their landscapes and their roles in relation to

it—a view that has implications for us as a community and provides a glimpse at future possibilities.

The Digital Landscape: Sensor Webs and Social Networks

Advances in sensor capabilities, mobile platform design, and software and communication technology set the stage for the provision of both authoritative and volunteered georeferenced information. Enabled by social networks, ubiquitous volunteered geographic information has led to the realization of a virtual community of neo-geographers (Goodchild 2008; Hudson-Smith et al. 2009). In the current context, the challenge before us is twofold:

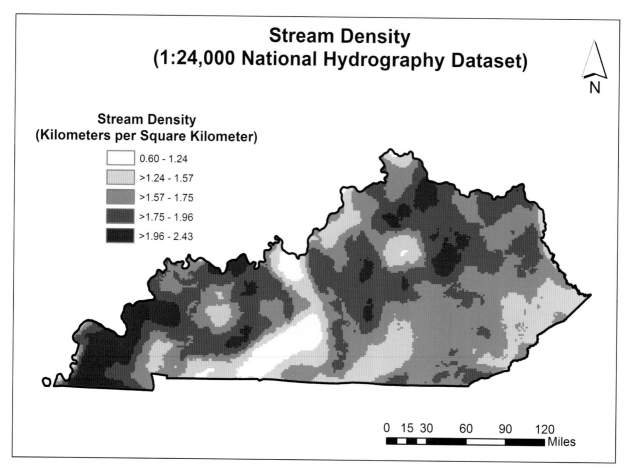

**Stream Density
(1:24,000 National Hydrography Dataset)**

N

**Stream Density
(Kilometers per Square Kilometer)**

- 0.60 - 1.24
- >1.24 - 1.57
- >1.57 - 1.75
- >1.75 - 1.96
- >1.96 - 2.43

0 15 30   60   90   120
Miles

20.2. Spatial distribution of stream density in Kentucky; kernel density classified by natural breaks method; class ranges are feature values (National Hydrography Dataset 2014; High Resolution—National Hydrography Dataset File Geodatabase 2014).

1. To enable monitoring of landscape-human interactions in a timely manner while maintaining information accuracy standards in order
2. to better explain, predict, and communicate results of these interactions with the appropriate tools while engaging communities in safeguarding and conserving valuable water resources.

A sign of the times, the social networks in existence now number in the hundreds, with tens of millions of visits per day for the top three or four most popular sites (Prescott 2010; Van Grove 2012; Webster 2010). Thanks to software-hardware integration and the development of specialty applications, portable, wireless devices have become location aware and able to collect multiple types of environmental data, thereby enabling people to become a sensor node. Geopositioning or geolocation capabilities are thus enabled through the integration of telecommunication and navigation—local or global, stationary (e.g., cellular phone towers) or mobile (e.g., satellites) systems. With digital imagery acquired at shorter intervals to update a photorealistic base map, the advent of online viewers brought digital image interpretation and analysis within almost everyone's reach (Google 2016; Microsoft 2016). Mapping applications, also known as "apps," developed to run on these faster, smaller, and

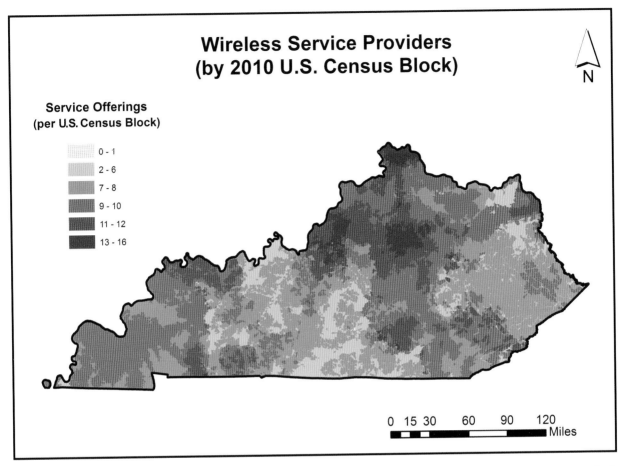

20.3. Spatial distribution of number of wireless service providers per 2010 census block; classified by manual intervals method; class ranges are feature values (2010 Census Data 2014; 2010 Census TIGER/Line Shapefiles 2014; December 31, 2010, National Broadband Map Datasets 2014).

mobile computing platforms are already pervasive (Kentucky Geography Network 2016b).

The ability to map the mobile landscape will become more commonplace as ease of access to geodatabases containing information on device locations and communication traffic improves (Massachusetts Institute of Technology, 2006). In recent years, a denser statewide broadband coverage in Kentucky has become a reality (figure 20.3), but this is not to undervalue those communities that are still in unserved or underserved locations because of unfavorable topographies (Broadband KY 2016; National Telecommunications and Information Administration 2016).

The Human Aspect: Water-Quality and Quantity Data Collection with Sensor-Augmented Mobile Devices

Both point- and nonpoint-source water pollution can translate into impaired designations and listings of water bodies (Kentucky Division of Water 2016d). Short- and long-term effects on water-body health indicators, as well as the watersheds they influence, are more readily visualized. Resources exist and the timing is right for an environmentally and spatially aware, empowered community, a sort of roaming, human-assisted sensor web, to be developed into a critical resource in helping communicate monitoring data and information in real time. Internet-based tools exist for raising environmental awareness through social-media networks and digital badges encompassing many different demographic groups, such as the K–12 community (U.S. Department of Education, 2016). Mobile fluidity is added to these interactions, targeting a larger community on the landscape and effectively increasing the ubiquity and human capacity of this network. The power of volunteered geospatially and temporally referenced data can be further capitalized on to shape water across landscapes, people, and Kentucky communities (Ye et al. 2011; Zook and Graham 2007).

The likely continued development of mobile device variety and numbers has energized the development of portable, miniaturized systems endowed with special capabilities. Further developments can include but need not be limited to telemetry/telematics (making measurements at a distance and transmitting them) and photogrammetry (deriving dimensional measures from images captured). Mobile devices such as smart phones, tablets, notebooks, notepads, and laptops, for example, have begun to be interfaced with attachments such as laser ranging devices, useful for triangulation and positioning, and with sensors that can measure physical-chemical parameters such as pH, temperature, solute concentrations, sediment/turbidity, flow velocity, and water depth (Chen and Steele 2012; Karol 2013). Spectral sensors, such as spectrophotoradiometers, LiDAR (light detection and ranging), and radar plug-ins, and apps to run them are increasingly possible, making these devices responsive to fluctuations in some environmental signals. An integration of multiple existing monitoring networks providing real-time or near-real-time feeds can conceivably be used to further explore these ideas (Inletkeeper 2016; Kentucky Division of Water 2016b; Kentucky Division of Water 2016c; Kentucky Geography Network 2016c; Kentucky Mesonet 2016)

Although it can be argued that computational literacy does not necessarily follow proficiency in social-networking and texting skills, the information content exchanged can grow beyond being just social. Just as simultaneous membership in social networks can be an advantage in the exchange of digital images, it can also be an advantage in the exchange of standardized environ-

mental feeds. Information transfer between systems in the sensor web has been codified in the form of languages and standards (OGC 2016). Text messages or feeds or spatially enabled environmental information can be exchanged. Once they are armed with sensor data collected, these social networks can gain and retain environmental and spatial identities while being able to communally tap into cloud-based models and expert systems to extrapolate spatial and temporal trends detected by an expansive, decentralized virtual network of position-centric information collection devices (Montella and Foster 2010).

The Geoenabled Citizen Scientist and Citizen Sensor: Environmental Information Exchange and New Automated Sensing and Interfacing Modalities

These technological capabilities add to the role and capability of the citizen scientist while posing new societal challenges and opportunities (Liu et al. 2011). Examples of engagement of communities in identifying and monitoring species frequency, diversity, and distribution already exist (Discover Life 2016). Several information-added products can be distributed to communities and societal segments for consumption that are alerts and advisories about water bodies or watersheds that can be triggered on the basis of deviations from baseline or predefined acceptable criteria that can also include geospatial representations, such as on-screen cartographic products. As a national precedent, the U.S. Environmental Protection Agency's WaterSentinel initiative, developed in partnership with drinking-water utilities and other stakeholders as a response to Homeland Security Presidential Directive HSPD-9, is an example of such an online contamination-incident warning system (U.S. Environmental Protection Agency 2005; U.S. Environmental Protection Agency 2010).

It is realistic that time lapses between data collection and actions will be shortened as rates of transfer and sharing of geospatially attributed environmental information accelerate (Liu et al. 2008a). The data and interpretive layers resulting from model output will be moved through this network, creating feed-forward loops and feedback loops that will in essence not be much different from those we have been perfecting and using in meteorological forecasts and observations for many decades now. The difference will be that the sensor-web information will come from ground or water level and will be field verified through proximal sensing systems rather than captured primarily from space-based orbiting sensors (U.S. National Oceanic and Atmospheric Administration 2016). The technology to produce an on-the-fly assessment of a digital record (e.g., graphic or numeric) acquired from a specific sample location by relaying information to and retrieving information from a cloud-based knowledge-based repository, automated or human mediated, is within reach (Barry Tonning, e-mail message to author, July 13, 2013).

The amalgamation of social networks and eyes in the sky mounted on pilotless craft, known as "drones," contributing to collective and almost synchronous knowledge of both water and landscape status is quickly emerging (Kahn 2012). Examples of automated, autonomous systems such as surfboard drones and unmanned or uninhabited aerial vehicles (UAVs), or unmanned aerial systems (UASs), as they are also called, as means of collecting information are increasing exponentially (Carmon 2010; Hardy 2011; Shurkin 2012; United Nations Environment Programme 2013; U.S. National Aeronautics and Space Administration Agency 2016; Whitehead and Hugenholtz 2014; Whitehead et al. 2014). As rules for airspace use by UASs evolve, new forms of surveillance, such as the use of miniaturized, almost insect-sized, nano- or micro-UASs, will continue to expand the frontiers of telemetry, monitoring, and real-time analytics (Anderson 2010; Hennigan 2011; Klotz 2012; Murphy 2011).

In order to track spatiotemporal trajectories of biota and streams, on-the-ground or in-system monitoring technology exists that at the same time provides telemetry of indicators and metrics for both internal and external environments. Microtagging with radio frequency ID, both active and passive, geospatially enabled or not, is becoming more possible for environmental applications (CISCO 2014; GS1 2014; Yoder 2011). Similarly, Quick Response (QR) codes are finding use in the dissemination of ecologic information to the population (Gonzalez 2011; KAYWA QR Code 2016; Osocio 2010). Pending legislation affecting the use and release to the environment

of nanoparticle-size taggants, development of mobile, miniaturized versions of these QR codes and other multidimensional barcodes by using nanotechnology is entirely feasible. These new microtag varieties, similar to those in existence for pharmaceuticals, can be used to convey information on effluents and nonpoint sources of pollution (ip.com 2007).

As more sophisticated sensors and platforms are incorporated into our daily panoply of digital gear, the notions of "sensor fusion" and perhaps also of "platform fusion" arise. The human as platform carrying her or his own sensorial array is synergized with the added sensor array her or his mobile device platform carries. Thus the minimum requirement for a citizen scientist is first to be a citizen sensor. Fulfillment of this prerequisite enables collection of scientifically scrutinized, standardized data, generation of information for hypothesis testing experimentally or observationally, and formulation of valid inferences. The degree to which this evolution from citizen sensor to citizen scientist takes place may be determined by the volume, quality, and complexity of sensor-array data and information input that person receives from the environment (Citizen Sense Project 2015; Citizen Sensor: DIY Pollution Monitoring 2015; COST 2016; European Commission 2016 Institute for Advanced Architecture of Catalonia 2016).

Through crowd sourcing, the collective sensor-data input resulting from individual and potentially diverse sensor- and platform-fusion configurations can be transmitted, stored, and mined for data or exploited at different levels for n-dimensional visualization and display. The information thus procured and provisioned can find its way into clouds, both private and public, and from there can be communicated in near-real-time fashion to decision makers (Acrobotics 2016).

Social Media, Mobile Digital Landscapes, Sensor Webs, and the Internet of Things: The Ultimate Fusion Environment

A recent characterization of social media as a focus group encapsulates a vision for a future in which stakeholder status reaches new levels of both collective awareness and potential for positive action (Sussman 2012).

By its own nature and somewhat in contrast to science, information technology moves along at a frenzied pace, an embodiment of faster rates of hardware and software obsolescence (Sandborn 2007). Expectations for a holistic assessment of waters based on the latest technology may prove elusive, however, because the main indicators of water quality are based more on scientific inference than on technological advancement.

Catastrophic events, either human induced or natural, such as a chemical spill or dam breach, can be telegraphed to a multidimensional and multifaceted network of users. This creates the opportunity for involvement, concerted remediation, and actions by people and communities. Existing examples of similar links between rainfall upstream and flood advisories or emergency notifications exist today. Densification of data-collection points makes it possible to extend this to mesoscale and microscale geographies if political will and resources are available.

Potentially, the idea of the existing network of USGS stream gauges can be expanded to ever-growing capabilities, with added dimensions and multiple perspectives for improved display and geovisualizations (U.S. Geological Survey 2016d). Technologies derived from existing implementations of time-lapse videography, such as those based on webcam monitoring for security purposes, can be implemented to outfit every water intake and major outfall in Kentucky (Nichols 2010). Basic sensors can be added to the monitoring node in order to improve plant operations and collect more geospatially dense data about water parameters. Concepts and algorithms applied to and derived from the field of change detection currently used on ecocams can be implemented to trigger alerts based on changes in the parameters sensed and fed into the sensor web (McDowell 2015; Virginia ECOCAM 2016).

The development of virtual sensors continues at a rapid pace, stimulated by ever-growing cyberinfrastructures (Kabadayi and Julien 2007; Sensor Web Alliance 2010). Virtual environments in which a blend of real and virtual sensor data is streamed into and overlain on the real object of a watershed are now possible. An example of these ideas is the amalgam of augmented reality and sensors, allowing for fusion of data and information on the go (Kofman, 2015; The Glass Explorer Program 2014). As new forms of digital observatories emerge, the notion of "virtual watershed tours" may take on new dimensions (Myers et al. 2008; Ohio State University Extension 2016). Current watershed- and ecosystem-management approaches are undoubtedly affected by the possibility of deploying both real and virtual sensors inside a fusion environment (Izurieta et al. 2011; Liu et al. 2008b; Tim 2006; X10 Irrigation 2016). Technologies such as the cloud and the "Internet of Things Ecosystem" concepts are coming of age, bringing increased automated interdependence and interoperability to the interaction between humans and water-centered activities and systems (Higginbotham 2011; Texas Instruments 2016).

Social networks and sensor webs are engaging in a dialogue or intercourse as our communal experience progresses (MacManus 2009). The Facebook and Twitter paradigms easily allow us to represent the life of a water body or watershed (Facebook 2016; Twitter 2016). The persona of a friend—an avatar and attendant digital badges—and the data and stream or watershed health state they represent can be translated into an analog representation, such as human language, in ways very similar to how a friend in a social network communicates how she or he is feeling at any given time. These technologies can be used to learn and give an augmented voice to the waters of Kentucky while learning about a richer story that is integrally linked to people.

Predictable futures are very hard to find, particularly in technology. In the context of a rapidly changing technological and social landscape, what are we to expect? We all use, share, enjoy, and frankly depend on the commonwealth's waters and the landscapes they traverse and shape; we all participate in a communion of sorts that comes from sharing spaces and places. Interpenetrating technological and social capabilities are creating a new level of knowledge and awareness. More than just an overlap, this anastomosis-like connection between social networks and sensor webs is now able to add potentiated value to environmental parameters monitored at particular places (e.g., a water feature or resource) and times. Thus our current culture and society have the chance to become richer.

References and Additional Resources

Acrobotic Industries. 2016. "The Smart Citizen Kit: Crowdsource Environmental Monitoring." https://www.kickstarter.com/projects/acrobotic/the-smart-citizen-kit-crowdsourced-environmental-m (accessed October 6, 2016).

Anderson, Mickie. 2010. "UF Team's Work Pays Off with Unmanned-Flight System That Captures Valuable Data." University of Florida News. http://news.ufl.edu/archive/2010/10/uf-teams-work-pays-off-with-unmanned-flight-system-that-captures-valuable-data.php (accessed October 6, 2016).

Broadband KY. 2016. "Kentucky Broadband Mapping Viewer v 2.0." http://www.bakerbb.com/kybroadbandmapping/ (accessed October 6, 2016).

Brody, Samuel D. 2004. "Does Location Matter? Measuring Environmental Perceptions of Creeks in Two San Antonio Watersheds." *Environment and Behavior* 36: 229–50.

Carmon, Irin. 2010. "Arad Metering Technologies Conserves Water via Battery-Operated Drones." Fast Company. https://www.fastcompany.com/1676952/arad-metering-technologies-conserves-water-battery-operated-drones (accessed October 6, 2016).

Chen, Ian, and Jim Steele. 2012. "What Do Smartphone Sensors sense All Day?" EDN NETWORK. http://edn.com/design/systems-design/4403623/What-do-smartphone-sensors-sense-all-day- (accessed October 6, 2016).

CISCO. 2014. "RFID Tag Considerations." http://www.cisco.com/c/en/us/td/docs/solutions/Enterprise/Mobility/WiFiLBS-DG/wifich6.html (accessed October 6, 2016).

Citizen Sense Project. 2015. "Citizen Sense." http://www.citizensense.net/about/ (accessed October 6, 2016).

"Citizen Sensor: DIY Pollution Monitoring." 2016. http://www.citizensensor.cc/ (accessed October 6, 2016).

COST - European Cooperation in Science and Technology. 2016. "Mapping and the Citizen Sensor." http://www.citizensensor-cost.eu/ (accessed October 6, 2016).

Discover Life. 2016. "Discover Life." http://www.discoverlife.org/ (accessed October 6, 2016).

European Commission. 2016. "Citizen Sensing and Environmental Practice: Assessing Participatory Engagements with Environments through Sensor Technologies." CORDIS (Community Research and Development Information Service). http://cordis.europa.eu/project/rcn/106442_en.html (accessed October 6, 2016).

Facebook. 2016. http://www.facebook.com/ (accessed October 6, 2016).

Gonzalez, Carlos. 2011. "How QR Codes Can Create a Social Transmedia Environment." Gamifixation. Digital Media Strategies. http://gamifixation.wordpress.com/2011/03/11/how-qr-codes-can-create-a-social-transmedia-environment/ (accessed October 6, 2016).

"The Glass Explorer Program." 2014. Google. http://www.google.com/glass/start/ (accessed October 6, 2016).

Goodchild, Michael F. 2008. "Citizens as Sensors: Web 2.0 and the World of Volunteered Geography." Minghi Distinguished Lecture, University of South Carolina, February 2008. http://geog.ucsb.edu/~good/presentations/columbia.pdf (accessed October 6, 2016).

Google. 2016. "Google Maps." https://maps.google.com/maps?hl=en (accessed October 6, 2016).

GS1. 2014. "EPCglobal—RFID." http://www.gs1.org/epcglobal (accessed October 6, 2016).

Hardy, Quentin. 2011. "Catching a Wave, and Measuring It." *New York Times—Business Day,* November 1. http://www.nytimes.com/2011/11/02/business/wave-glider-a-floating-robot-seeks-to-network-the-oceans.html?_r=1&hpw (accessed October 6, 2016).

Hennigan, W. J. 2011. "Idea of Civilians Using Drone Aircraft May Soon Fly with FAA." *Los Angeles Times,* November 27. http://articles.latimes.com/2011/nov/27/business/la-fi-drones-for-profit-20111127 (accessed October 6, 2016).

Higginbotham, Stacey. 2011. "Behind the Sensor Web Lies a Cloud." http://gigaom.com/2011/11/26/behind-the-sensor-web-lies-a-cloud/ (accessed October 6, 2016).

Hudson-Smith, Andrew, Andrew Crooks, Maurizio Gibin, Richard Milton, and Michael Batty. 2009. "NeoGeography and Web 2.0: Concepts, Tools and Applications." *Journal of Location Based Services* 3 (2): 118–45. http://www.css.gmu.edu/andrew/pubs/LBSneogeographyPaper.pdf (accessed October 6, 2016).

InletKeeper. 2016. "Stream Tempearature Monitoring Network." https://inletkeeper.org/healthy-habitat/stream-temperature-monitoring-network (accessed October 6, 2016).

Institute for Advanced Architecture of Catalonia. 2016. "Smart Citizen." Intelligent Cities. Fab Lab Barcelona. http://iaac.net/research-projects/intelligent-cities/smart-citizen/ (accessed October 6, 2016).

ip.com. 2007. "Micro Tag Tracers for Downhole Detection of Water Influx and Its Source Location in the Well." 2007. ip.com. http://ip.com/IPCOM/000149707 (accessed October 6, 2016).

Izurieta, Clemente, Sean Cleveland, Ivan Judson, Pol Llovet, Geoffrey Poole, Brian McGlynn, Lucy Marshall, Wyatt Cross, Gwen Jacobs, Barbara Kucera, David White, F. Richard Hauer, and Jack Stanford. 2011. "A Cyber-Infrastructure for a Virtual Observatory and Ecological Informatics System—VOEIS." Creative Commons Attribution 3.0. Unported License. http://www.cs.montana.edu/izurieta/pubs/eim2011.pdf (accessed October 6, 2016).

Kabadayi, Sanem, and Christine Julien. 2007. "Remotely Deployed Virtual Sensors." Center for Excellence in Distributed Global Environments (UTEDGE). Department of Electrical and Computer Engineering. University of Texas at Austin. http://mpc.ece.utexas.edu/Papers/TR-UTEDGE-2007-010.pdf (accessed October 6, 2016).

Kahn, Matthew. 2012. "Social Networking and Drones Can Save the Earth." *Christian Science Monitor—Green Economics,* February 25. http://www.csmonitor.com/Business/Green-Economics/2012/0225/Social-networking-and-drones-can-save-the-earth (accessed October 6, 2016).

Karol, Lawrence. 2013. "How Your Smartphone Can Tell You Whether You're Breathing Dirty Air." http://www.lawrencekarol.com/2013/06/11/how-your-smartphone-can-tell-you-whether-youre-breathing-dirty-air/ (accessed October 6, 2016).

"KAYWA QR Code." 2016. http://qrcode.kaywa.com/ (accessed October 6, 2016).

Kentucky Division of Water. 2016a. "Watershed Management Framework." http://water.ky.gov/watershed/Pages/WatershedManagementFramework.aspx (accessed October 6, 2016).

Kentucky Division of Water. 2016b. "Kentucky Water Watch." http://water.ky.gov/ww/Pages/default.aspx (accessed October 6, 2016).

Kentucky Division of Water. 2016c. Kentucky Watershed Watch Program. "Watershed Watch in Kentucky." http://kywater.org/watch/ (accessed October 6, 2016).

Kentucky Division of Water. 2016d. "2012 303(d) Reports—Total Maximum Daily Load Program." http://water.ky.gov/waterquality/Pages/303dList.aspx (accessed October 6, 2016).

Kentucky Geography Network. 2016a. "Kentucky Weather Mapping: With Live Updates from the Kentucky Mesonet." http://kygeonet.ky.gov/kyweather/ (accessed October 6, 2016).

Kentucky Geography Network. 2016b. "KyGovMaps." http://kygeonet.ky.gov/govmaps (accessed October 6, 2014).

Kentucky Geography Network. 2016c. "The Commonwealth Map—Kentucky's Base Map." http://kygeonet.ky.gov/tcm/ (accessed October 6, 2016).

Kentucky Mesonet. 2016. "The Commonwealth's Official Source for Weather and Climate Data." http://www.kymesonet.org/ (accessed October 6, 2016).

Klotz, Irene. 2012. "U.S. Opening Up Airspace to Use of Drones." Science on NBCNEWS.com. http://www.nbcnews.com/id/46499162/ns/technology_and_science-science/#.V-sZw_krL0P (accessed October 6, 2016).

Kofman, Ava. 2015. "Dueling Realities." *The Atlantic*, June 9. http://www.theatlantic.com/technology/archive/2015/06/dueling-realities/395126/ (accessed October 6, 2016).

Liu, Yong, David J. Hill, Tarek Abdelzaher, Jin Heo, Jaesik Choi, Barbara Minsker, and David Fazio. 2008a. "Virtual Sensor-Powered Spatiotemporal Aggregation and Transformation: A Case Study Analyzing Near-Real-Time NEXRAD and Precipitation Gage Data in a Digital Watershed." Paper presented at Environmental Information Management Conference, Albuquerque, New Mexico, September 10. https://eim.ecoinformatics.org/eim2008/session1-sensors/5-liu-eim2008.ppt (accessed October 6, 2016).

Liu, Yong, Luigi Marini, Rob Kooper, Alejandro Rodriguez, David J. Hill, James D. Myers, and Barbara S. Minsker. 2008b. "Virtual Sensors in a Web 2.0 Virtual Watershed." Paper presented at eScience. 4th IEEE Fourth International Conference on e-Science.

Liu, Yong, Pratch Piyawongwisal, Sahil Handa, Liang Yu, Yian Xu, and Arjmand Samuel. 2011. "Going beyond Citizen Data Collection with Mapster: A Mobile+Cloud Citizen Science Experiment." Paper presented at IEEE eScience 2011, International Workshop on Computing for Citizen Science, Stockholm, Sweden, December 5–8.

MacManus, Richard. 2009. "When Sensors and Social Networks Mix." *Readwrite.* http://readwrite.com/2009/04/08/when_sensors_and_social_networks_mix#awesm=~odbbI581AsYdCS (accessed October 6, 2016).

Massachusetts Institute of Technology. 2006. "mobile landscape|graz in real time." SENSEable City Lab. http://senseable.mit.edu/graz/ (accessed October 6, 2016).

McDowell, William H. 2015. "NEON and STREON: opportunities and challenges for the aquatic sciences." *Freshwater Science 34* (1): 386–91.

Microsoft. 2016. "Bing Maps." http://www.bing.com/maps/ (accessed October 6, 2016).

Montella, Rafaelle, and Ian Foster. 2010. "Using Hybrid Grid/Cloud Computing Technologies for Environmental Data Elastic Storage, Processing, and Provisioning." In *Handbook of Cloud Computing*, edited by Borko Furht, and Armando Escalante, 595–618. New York: USA Springer Science+Business Media.

Murphy, Jack. 2011. "Nano-UAVs Snapped Up by the UK MoD." KIT UP! Military.com http://kitup.military.com/2011/11/nano-uavs-snapped-uk-mod.html (accessed October 6, 2016).

Myers, James, Luigi Marini, Rob Kooper, Terry McLaren, Robert E. McGrath, Joe Futrelle, Peter Bajcsy, Andrew Collier, Yong Liu, and Shawn Hampton. 2008. "A Digital Synthesis Framework for Virtual Observatories." National Center for Supercomputing Applications, University of Illinois at Urbana-Champaign. http://www.nesc.ac.uk/talks/ahm2008/1192.pdf (accessed October 6, 2016).

National Telecommunications and Information Administration. 2016. "Dec. 31, 2010 National Broadband Map Data sets." http://www2.ntia.doc.gov/Dec-2010-datasets (accessed October 6, 2016).

Nichols, Russell. 2010. "Neighborhood Watch Goes Digital in Ramsey County, Minn." *Government Technology,* September 20. http://www.govtech.com/public-safety/Ramsey -County-Minn-Neighborhood-eWatch.html (accessed October 6, 2016).

OGC. 2016. "Why Is OGC Involved in Sensor Webs?" http:// www.opengeospatial.org/domain/swe (accessed October 6, 2016).

Ohio State University Extension. 2016. "Ohio Watershed Network—Watershed Tour." http://ohiowatersheds.osu.edu /resources/watershed-tour (accessed October 6, 2016).

Osocio. 2010. "The Big Wild QR codes." http://osocio.org /message/the-big-wild-qr-codes/ (accessed October 6, 2016).

Prescott, LeeAnn. 2010. "54% of US Internet Users on Facebook, 27% on MySpace." VentureBeat/Social. http://venture beat.com/2010/02/10/54-of-us-internet-users-on-facebook -27-on-myspace/ (accessed October 6, 2016).

Sandborn, Peter. 2007. "Software Obsolescence—Complicating the Part and Technology Obsolescence Management Problem." *IEEE Transactions on Components and Packaging Tech.* 30 (4): 886–88. http://www.enme.umd.edu/ESCML/Papers /IEEE_SoftwareObs.pdf (accessed October 6, 2016).

Sensor Web Alliance. 2010. "GEOSS Sensor Web Workshop Report." http://www.ogcnetwork.net/system/files/GEOSS _Sensor_Web_Workshop_2010_Report.pdf (accessed October 6, 2016).

Shurkin, Joel N. 2012. "Surfboard-Sized Drones Crossing Pacific to Monitor Sea Surface." *NEWS Science—US News & World Report,* January 30. http://www.usnews.com/science /articles/2012/01/30/surfboard-sized-drones-crossing -pacific-to-monitor-sea-surface (accessed October 6, 2016).

Sierra Club. 2013. Kentucky Water Sentinels Summer Activity Report. https://content.sierraclub.org/grassrootsnetwork /team-news/2013/09/kentucky-water-sentinels-summer -activity-report (accessed October 6, 2016).

Sussman, Nadia. 2012. "Video: Social Media as Focus Group." *New York Times,* July 31. http://bits.blogs.nytimes.com /2012/07/31/video-social-media-as-focus-group/?_r=0 (accessed October 6, 2016).

Texas Instruments. 2016. "Internet of Things. Connect More." http://www.ti.com/ww/en/internet_of_things/index.html ?DCMP=ConnectMore&HQS=connectmore (accessed October 6, 2016).

Tim, U. Sunday. 2006. "Watershed Management and the Emerging Environmental Cyberinfrastructure." *Watershed Update* 4 (3): 1–9. http://www.awra.org/committees/techcom /watershed/pdfs/0403WU.pdf (accessed October 6, 2016).

Twitter. 2016. http://twitter.com/ (accessed October 6, 2016).

United Nations Environment Programme. 2013. "UNEP Global Environmental Alert Service (GEAS)." http://www .unep.org/pdf/UNEP-GEAS_MAY_2013.pdf (accessed October 6, 2016).

U.S. Census Bureau. 2016a. "2010 Census Data". http://www .census.gov/2010census/data/ (accessed October 6, 2016).

U.S. Census Bureau. 2016b. "2010 Census TIGER/Line Shapefiles." ftp://ftp2.census.gov/geo/pvs/tiger2010st/21_Kentucky /21/ (accessed October 6, 2016).

U.S. Department of Education. 2016. "Future Ready Learning. 2016 National Education Technology Plan." http://tech.ed .gov/files/2015/12/NETP16.pdf (accessed October 6, 2016).

U.S. Environmental Protection Agency. 2005. "WaterSentinel System Architecture Draft, Version 1.0." EPA 817-D-05-003. https://www.hsdl.org/?view&did=34214 (accessed October 6, 2016).

U.S. Environmental Protection Agency. 2010. "Sensor Network Design for Drinking Water Contamination Warning Systems." EPA/600/R-09/141. http://cfpub.epa.gov/si/si _public_file_download.cfm?p_download_id=498251 (accessed October 6, 2016).

U.S. Geological Survey. 2016a. "National Hydrography Dataset." http://nhd.usgs.gov/ (accessed October 6, 2016).

U.S. Geological Survey. 2016b. Kentucky Water Science Center. "Water Availability Tool for Environmental Resources." http://ky.water.usgs.gov/projects/waterbudget/index.html (accessed October 6, 2016).

U.S. Geological Survey. 2016c. "Streamstats in Kentucky." http://streamstats.usgs.gov/kentucky.html (accessed October 6, 2016).

U.S. Geological Survey. 2016d. "USGS Current Water Data for the Nation." NWIS: Web Interface. http://waterdata.usgs .gov/nwis/rt (accessed October 6, 2016).

U.S. National Aeronautics and Space Administration Agency. 2016. "The Uninhabited Aerial Vehicle (UAV) Coffee Project." http://www.nasa.gov/centers/ames/research/factsheets /FS-020901ARC.html (accessed October 6, 2016).

U.S. National Oceanic and Atmospheric Administration. 2016. "Eastern U.S.—Infrared Channel 2 Loop." NOAA Satellite and Information Service. http://www.ssd.noaa.gov/goes/east /eaus/flash-ir2.html (accessed October 6, 2016).

Van Grove, Jennifer. 2012. "Every 60 Seconds in Social Media (Infographic)." VentureBeat/Social. http://venturebeat.com /2012/02/25/60-seconds-social-media/ (accessed October 6, 2016).

"Virginia ECOCAM." 2016. https://ecocam.evsc.virginia.edu /drupal/ (accessed October 6, 2016).

Webster, Tom. 2010. "The Social Habit—Frequent Social Networkers." Edison Research. http://www.edisonresearch.com /the_social_habit_frequent_social_networkers_in _america/ (accessed October 6, 2016).

Whitehead, Ken, and Chris H. Hugenholtz. 2014. "Remote Sensing of the Environment with Small Unmanned Aircraft Systems (UASs), Part 1: A Review of Progress and Challenges." *Journal of Unmanned Vehicle Systems* 2 (3): 69–85.

Whitehead, Ken, Chris H. Hugenholtz, Stephen Myshak, Owen Brown, Adam LeClair, Aaron Tamminga, Thomas E. Barchyn, Brian Moorman, and Brett Eaton. 2014. "Remote Sensing of the Environment with Small Unmanned Aircraft Systems (UASs), Part 2: Scientific and Commercial Applications." *Journal of Unmanned Vehicle Systems* 2 (3): 86–102.

X10 Irrigation. 2016. "Virtual Rain Sensor." http://www.virtualrainsensor.com/About.aspx (accessed October 6, 2016).

Ye, Mao, Krzysztof Janowicz, Wang-Chien Lee, and Christoph Mülligann. 2011. "What You Are Is When You Are: The Temporal Dimension of Feature Types in Location-Based Social Networks." https://helios.geog.ucsb.edu/~jano/acmgis 2011tsi.pdf (accessed October 6, 2016).

Yoder, Steve. 2011. "Reduce Counterfeit Risk with Process-Friendly Authentication Technologies." http://www.colorcon .com/about/newsletter/eNewsletter-APR-2011 (accessed October 6, 2016).

Zook, Matthew A., and Mark Graham. 2007. "Mapping Digi-Place: Geocoded Internet Data and the Representation of Place." *Environment and Planning B: Planning and Design* 34: 466–82. http://www.geospace.co.uk/files/b33111.pdf (accessed October 6, 2016).

CHAPTER TWENTY-ONE

Water as the Context for Community-Based Science Projects

Teaching the Next Generation

Carol Hanley and Kelly Taylor

How can a trip to a Kentucky stream or pond be used to enhance a child's understanding of science? Can that same trip be used to encourage the child to pursue a career in science? Science education professionals are concerned about current trends in test scores and the number of students entering science careers nationwide (Glod 2008). The National Assessment Governing Board released findings in May 2012 on science mastery in the National Assessment of Educational Progress (NAEP). The average eighth-grade score rose in 2009 but is still well below proficiency. The NAEP tested a nationally representative sample of 122,000 students in the eighth grade from 7,290 public and private schools. "I'm disappointed," said Gerry Wheeler, the interim executive director of the National Science Teachers Association, in an interview. "Two points is certainly nothing to cheer about. If these kids can't do better in science, our nation is in trouble" (Sparks 2012).

Although the percentages of high school graduates who have taken science courses increased between 1990 and 2009 (Aud et al. 2013, 121), Mel Schiavelli discusses the difficulties employers have in finding enough qualified workers to fill science, technology, engineering, and math (STEM) jobs (Schiavelli 2011). Science educators who have researched and implemented professional development programs have a sense that things could possibly be different and believe that changes could be implemented in the way science is taught, referred to as pedagogy, which would help motivate and interest students in STEM careers.

The future does not have to be so dim if we ask, What are the current motivational practices employed in K–12 natural resource or environmental education? Educators at the University of Kentucky (UK) College of Agriculture, Food and Environment's (CAFE) Environmental and Natural Resources Initiative consider these questions as we use real-life contexts to teach science to Kentucky's K–12 students. The central question becomes, Is it possible to use project-based learning in conjunction with community-based issues to interest students in science topics? The work we have been doing has attempted to shed light on these questions and, at the same time, expose K–12 students and their teachers to conceptually rich, multisensory experiences with specific water-related themes. This chapter is the story of what has been done and shows promise for involving young people of today in science careers for the future.

Setting the Context

In 2009, fifteen-year-old students from around the world took part in the Program for International Student Assessment (PISA). This international test of reading, math, and science measures how well students from more than seventy countries are prepared to meet future academic challenges. The test showed that students in the United States were average performers. The scores of U.S. students showed no statistically significant difference from the Organization for Economic Cooperation and Development (OECD) average. Countries that performed near the average, at the same level as the United States, included Hungary, the Czech Republic, Norway, Denmark, and France (OECD 2010, 151). Arne Duncan, U.S. secretary of education, discussed his concerns in 2010 alongside officials from OECD (Duncan 2010). "The most encouraging finding from PISA is that our average science score is up. In 2006, American 15-year-old students had below-average skills in scientific literacy, compared to their OECD peers. Today, U.S. students have improved enough to become average performers in science among OECD nations, earning 17th place in the OECD rankings. Still, that's not much to celebrate. Being average in science is a mantle of mediocrity—and especially in a knowledge economy where scientific literacy is so central to sustaining innovation and international competitiveness."

Science educators are concerned because an understanding of science and technology is important to a young person's preparedness for life in modern society. A scientifically literate populace is crucial in a democratic society where determining public policy on issues of science and technology has an effect on everyone's life. Students whose proficiency in science is limited will find it difficult to participate in society at a time when science and technology play a large role in daily life. Those students capable of advanced scientific work have the potential to become part of a group of future innovators who will boost their countries' technological and innovative capacities in science-related industries (OECD 2010, 137).

Mediocre international science test results are only one indicator that the United States is behind; we see evidence of poor performance on domestic measures of academic performance. For example, in 2012, only 31 percent of high school graduate students of the fifty states and the District of Columbia who took the American College Testing (ACT) assessment met or exceeded the ACT College Readiness Benchmark in science of 24 (ACT 2013). In Kentucky, that number was just 22 percent, which means that approximately one in five Kentucky students has a 50 percent chance of obtaining a B or higher or about a 75 percent chance of obtaining a C or higher in a science credit-bearing college course such as biology (ACT 2013).

In addition to the lackluster performance of U.S. students on national and international assessments, there are also important data on high school graduation. An analysis of 2011 data by Editorial Projects in Education showed that from 2000 to 2010, the national graduation rate increased by 7.9 percentage points to 74.7 percent in 2010, a gain of less than one point per year on average. During this period, graduation rates rose in forty-six states. Kentucky's rate rose 13.5 percent in that period to 77.2 percent in 2010 (Editorial Projects in Education Research Center 2013, 6). Unfortunately, 1 million public high school students in the class of 2013 will fail to graduate with a diploma across the nation. That amounts to a loss of more than 5,500 students from the U.S. graduation pipeline every school day, or one student every thirty-one seconds. In Kentucky, this amounts to a projected 11,457 nongraduates in 2012–13, or 64 students lost each day (Editorial Projects in Education Research Center 2013, 8).

Dropout research shows that there are individual and institutional factors that predict which students might drop out of school (Rumberger and Lim 2008). One of the most important factors is student engagement, which includes students' active involvement in academic work. Rumberger and Arellano (2007) found that in schools where students described their classes as being more interesting and challenging, there was a very small but statistically significant effect on the odds of students graduating from California high schools. Therefore, it appears that pedagogy matters in ways that go beyond simply how well students learn content.

Although direct correlations between dropout rates and science courses or teaching are not available, it seems likely that there is a relationship to pedagogy. "The reason kids drop out in Kentucky, and I suspect that this is the case nationwide, is not because they're falling behind," stated Lisa Gross, director of communications at the Kentucky Department of Education, in an interview with Claudio Sanchez on National Public Radio in January 2012. "It's because they don't see a connection between what they're learning in high school and what their lives are going to be like as adults" (Sanchez 2012).

In addition, some insight can be drawn from data collected as students participated in state-mandated testing known as the Commonwealth Accountability Testing System (CATS). In 2006, the Kentucky Department of Education required students to answer questions related to science pedagogy as part of the state testing system. The fourth-grade students were asked how often they were able to do experiments or investigations about things in which they were interested. A total of 53 percent of those students answered "never" or "sometimes but not every week." The seventh- and eleventh-grade students were asked the question in a slightly different manner, how often they designed and conducted scientific investigations about things in which they were interested. A full 73 percent of the seventh-grade students and 70 percent of the eleventh-grade students responded "never" or "sometimes but not every week" (Kentucky Department of Education 2006).

This disaffection with science is not limited to Kentucky. Cleaves (2005) interviewed year-nine (high school freshmen) students from the United Kingdom who planned to continue studying STEM fields. Many reported that they found their science experiences boring and that they did not enjoy secondary science. This view is corroborated by other researchers (e.g., Osborne, Simon, and Collins 2003) who showed that although students have positive attitudes about science as an endeavor, they do not have the same attitudes about the science they experienced in the classroom.

Increasing Student Engagement

Traditional Western education is based on a novice-expert relationship associated with the transmission of knowledge in decontextualized situations (Morgan 2009). In action-oriented learning, the lesson or transmission of knowledge is predicated on resolution or amelioration of an issue. It occurs best in real-world or community-based situations. Students must sense that the learning is relevant to their lives and feel that they play an active role in addressing an issue through inquiry, problem solving, and critical reflection (Morgan 2009).

How do we increase student engagement in science coursework? When do we admit that students are not engaged with the flat diagrams on the worksheet that call to be labeled? Why would students care that the

questions at the back of the chapter must be answered by first period Monday morning? Will those uninspiring and static exercises lead to the next generation of undergraduate students and professionals who will address critical science-based issues, including the challenges of water resources in Kentucky and beyond? There is a more effective way of teaching that fosters student engagement and enhances student motivation.

We know the importance of hands-on, minds-on, and experiential education across multiple fields, including science. However, we must also acknowledge the barriers, both real and perceived, that exist within the public school classroom, which disallow the use of new and exciting pedagogies that can employ a more learner-centered and community-based approach to science education.

How can we introduce teachers to pedagogy that motivates students to study natural resources and the environment? After all, the ultimate goal of our environmental education efforts is the sustainable use of natural resources, which "requires an intimate knowledge of biological and physical realities. Moreover, it is children and their acquisition of that knowledge that make subsistence possible over generations" (Young 2002, 855).

Educators in geography and other fields put great emphasis on a pedagogical strategy they call place-based education. Place-based education uses issues or questions from the community to teach concepts from across the curriculum. It emphasizes hands-on, real-world learning to increase academic achievement and helps students develop stronger ties to their community, enhances students' appreciation for the natural world, and creates a heightened commitment to community service (Sobel 2004). In place-based education, the content is specific to the place, and the learning is inherently interdisciplinary and experiential (Boyer 2006). Place-based learning gives a sense of belonging and strengthens students' connections to others and to the place in which they live (Borgelt et al. 2009). The local environment provides an obvious and immediate connection with real-life experience and opportunities for authentic and meaningful teaching and learning experiences. It also provides an avenue for students to connect their local experiences to wider ideas, issues, and experiences (Lane et al. 2005).

Our Local Science Education Story

For over a decade, we have focused on changing the story of science education in real, local, and meaningful ways. Although water is not the only resource we use to focus our science education efforts, it is an important one. What follows is a demonstation of our approach to science-based educational efforts and how water resources have been central to our most recent work in this area.

Each community-based investigation begins with an essential question to drive it. Essential questions give teaching and learning a focus and ultimately a purpose. For example, questions for water investigations have included "How does development in our community impact our water?" and "What are the sources of water pollution in our community?" Such questions were used by each teacher to construct an instructional plan, which included classroom learning activities, field trips, and expert guest presenters' visits. The content for every community-based project has been carefully aligned with Kentucky's Program of Studies and Kentucky Core Content for Assessment, Kentucky's curriculum and assessment standards. Future programs will align with the Next Generation Science Standards, which were formally adopted by the Kentucky Department of Education in the summer of 2013.

Our community-based projects are a form of project-based learning (PBL). PBL is a learner-centered strategy that provides students with authentic tasks connected to their personal interests (Grant 2011). The approach used and described here adheres to commonly cited characteristics of PBL, including an essential question and the production of one or more artifacts as representations of learning (Adderley 1975; Blumenfeld et al. 1991). The projects serve as artifacts or representations of the learners' solution of the essential question (Rieber 2004). PBL is an approach to teaching that is based on constructivist theories of learning (Harel and Papert 1991; Kafai and Resnick 1996). This approach engages learners in content through the process of constructing artifacts, which are critical to the learning goals (Prince and Felder 2006; Williams van Rooij 2009). Project-based learning favors in-depth investigations over memorization of broad content knowledge (Harris and Katz 2001).

Our work on community-based projects, our version of PBL and place-based education, at UK's CAFE began in 2004 with an agricultural problem related to horses called Mare Reproductive Loss Syndrome or MRLS important to central Kentucky. The project was funded by the United States Department of Agriculture and implemented in central Kentucky schools. We developed and implemented a community-based natural resource program because we wanted students and teachers to act as scientists and use the tools of scientists as they identified and explored important issues within their communities. The students' work mirrored the work of the UK researchers; some of those researchers visited the schools and worked alongside the teachers and students, for example. Furthermore, it was our contention that subject matter is often separated from interesting and important contexts, resulting in sterile instructional materials and models and disengaged students. Therefore, our instructional philosophy resembles that of Barab et al. (1998) because we feel that many decontextualized learning environments produce situations where "learning becomes the memorization of seemingly abstract, self-contained entities, not useful tools for understanding and interacting with the world" (Barab et al. 1998, 15). The learning environment and the learning tasks used should be authentic and reflect the complexity of the environment where learners are expected to be able to function after training (Savery and Duffy 1995).

More recently, the focus of our place-based instructional approach has been water resources. In 2008, we received funding from National Science Foundation (NSF) to create a program called the Information Technology through Community-Based Natural Resources Program for Students and Teachers, which was designed to increase students' and teachers' use of technology in science courses. Teachers and students from across central and eastern Kentucky used geospatial technologies, including geographic information systems (GISs) and global positioning systems (GPSs), to study local environmental issues. Many schools focused on water-related questions. Over the course of the project, approximately three thousand students from elementary through high school took part in community-based natural resource investigations that focused on water-quality issues. The students asked questions such as "What is a watershed?" "How clean is the water in my community?" "How can water quality be measured?" and "How can I live as a responsible and informed citizen in my watershed?"

All students took at least one field trip to visit a stream or pond in their community. These proved to be highly engaging and motivating experiences. The instructional objectives for the field trips included cognitive objectives, such as biological indicators of water quality and water chemistry; process-skills objectives, such as measuring, observing, and collecting data; and affective objectives, such as appreciating the beauty of the stream and its natural surroundings. John Amos Comenius, as far back as the seventeenth century, understood the importance of the affective domain. He discussed the learning process, which involved the moral and emotional aspects of the relationship between student and teacher. "Without interest, learning seldom takes place, or if it does, it cannot rise above the level of rote memory" (Woodhouse 2001, 1). It was remarkable to learn that many of the students had their first one-on-one interaction with a stream or a stream organism such as a crayfish or tadpole.

Under the guidance of CAFE staff, students learned key science concepts in a rich, local context with which they could personally identify. For example, students carefully and critically examined the flow of matter and energy through an ecosystem, a key science concept, by looking at food-web relationships that exist in a typical Kentucky stream flowing in their community. Before this experience, the majority of the students seemed unaware of the existence of such creatures as the macroinvertebrates that inhabit nearly every stream and pond in the state. They would impatiently crowd around a small plastic container that held samples of the day's catch and would ask questions about the characteristics or habits of cranefly larvae, dragonfly nymphs, or snails. Before this experience, their only formal science interaction with streams or macroinvertebrates had been with a worksheet—and it is rare that students crowd around a worksheet. Through this type of experience, the team addressed the project's cognitive objectives, described next.

Students conducted stream surveys of macroinvertebrates as a biological measure of water quality. They also conducted chemical analyses of collected water samples for evidence in a physical sense. These data were used in the classroom as students engaged in critical-thinking exercises related to their school-developed and community-focused essential questions. Using geospatial technology, including GPS units and GISs, students constructed maps of their research areas that were used for presentations to others about the work and the conclusions reached. These activities helped students gain science process skills and thus addressed pragmatic skills objectives.

The community field experiences also fostered other less obvious but profound student effects. Amid the sounds of splashing water, excited squeals, and laughter, one could overhear comments from the students such as "This is awesome!" "Can we come here again tomorrow?" and "I never knew there was so much cool stuff in a creek." Educators have long recognized the value of what is referred to as the affective domain of instruction (Littledyke 2008; Kuwahara 2013). In simple terms, this is merely an overt recognition of the fact that educational experiences have an emotional dimension in addition to the cognitive one. It is clear that many of the students were affected emotionally as well as intellectually, and that this deepened their understanding of water in their environment.

As one would imagine, the teachers were gratified to see their students so actively engaged with learning. On numerous occasions, the teachers reported to UK staff that the enthusiasm generated at the creek or pond continued into the classroom and made students much more eager to learn and apply their learning to consideration of local environmental or water-related issues. One young man who appeared especially disinterested in mathematics was finally motivated to learn addition and subtraction when the tasks were applied to the numbers of macroinvertebrates he found in the creek.

Reyhner (2010) and Trent and Reyhner (2002) suggest that students have trouble finding meaning in a decontextualized, one-size-fits-all curriculum and instruction that does not relate to their communities. Through place- and community-based curriculum and instruction, teachers can provide students with a relevant, practical, and motivating education where learners can actively participate in shaping their own education. The best way to contextualize education is to relate what students are learning to their lives. According to Boyer (2006), children cannot enter the world with confidence and understanding without having a sense of identity and a sense of place. "Place-based education strengthens communities, is inherently interdisciplinary and project-based, it builds on local resources and expertise without great cost" (Boyer 2006, 114–15).

It is also important to note that preparation for teachers is essential since a teacher may or may not have the background in the use of new technologies and equipment to address the community-based science question(s). The teacher professional development in the NSF program occurred on the UK campus, in the local schools, in the field, and at statewide conferences. These sessions involved firsthand experiences for the teachers with water-quality assessment techniques, ecosystem studies, and water-supply issues. The teachers were trained in the use of water test kits and learned how to interpret the results of the chemical analyses and observed biological indicators. Going further and armed with an awareness of and introduction to this technology, they were able to instruct their students in the use of maps to illustrate watersheds and to show where water sampling and quality assessments were done so that the community geography was the local context.

A second example of community-based learning involved the Lexington-Fayette Urban County Government (LFUCG) stormwater education project. During the eighteen-month project, teachers from central Kentucky schools were engaged in a series of professional development workshops in which they broadened their understanding of issues related to stormwater runoff near the schools. In particular, the project offered teachers opportunities to focus on the special challenges that urban water bodies present.

During the LFUCG stormwater project, teachers took part in learning activities set in a classroom-like environment and in outdoor settings. Workshop and field activities modeled the learner-centered, project-based approach. For instance, teacher input was used to guide the selection of topics and the scope and sequence of instruction, as well as the nature of many of the activities

chosen for field experiences. Teachers were consistently engaged in sharing information with one another about how the information they were learning could be applied in their local learning environments. Critically, they discussed ideas about how to involve their students in activities similar to the ones they experienced through the approach modeled in this program. During a trip to a Lexington wastewater-treatment plant, participating teachers frequently expressed interest in providing the same learning opportunity for their students. The concepts teachers learned from these sessions not only affected how they delivered instruction related to stormwater but also influenced how they taught more encompassing topics, such as ecosystems, natural resources, and environmental policy.

A second LFUCG project, which was started in 2012, focused on fourth-grade teacher professional development student projects at another central Kentucky elementary school. Teachers participated in professional development events throughout the school year. They then worked with their students to create multimedia presentations illustrating their best conceptualizations of the school-side stream restoration. As a culminating event, the students presented their creations to the community at a school family night.

A further example of a community-based project involved stream restoration at yet another elementary school in central Kentucky, in which the school implemented a Stream Project and Outdoor Learning Center. Starting with a small grant from the National Fish and Wildlife Foundation to develop a science lab to improve test scores, the project evolved into a larger collaborative effort to reshape the natural environment. Many entities were involved in the project, including federal, state, local, and private agencies and universities. By the end of the project, the students, the community, and the natural environment all benefited from it. The project also reshaped the way students learned and the way teachers taught.

The local championing, which was central to this change in the way to approach science education, was provided by the principal. She dreamed of educating students beyond the school walls. She knew that teaching in the environmental context was good for her students, especially male, special-education, and minority students. As a teacher, she took every professional development class in science that she could find to make the science more meaningful to her students. When she was faced with the opportunity to make the school's campus more beautiful and turn minimally used land into a haven for science learning, she seized the opportunity. She was confident that this would result in increased student learning, and not just in science. However, she was surprised at just how much impact the project had on the school, the community, and the environment.

Our team at UK worked intensively with the science teachers at the school for two school years, 2008–9 and 2009–10. One teacher from each of the six grade levels at the elementary school, along with the science lab teacher, attended professional development sessions throughout the two-year period. This demonstrates a commitment by the local school leader and the classroom teachers to engage students in real and local science education. The professional development program occurred in conjunction with a stream restoration funded by the Kentucky Department of Fish and Wildlife Resources and the National Fish and Wildlife Foundation during the summer of 2009. Much of the teachers' professional development was centered on the restoration project and how to monitor water quality before and after completion of the stream-rehabilitation project. Teachers explored questions regarding watersheds and aquatic ecology once a month during a three-hour professional development session. The teachers incorporated the activities and concepts they explored into their classroom curricula. During May 2009, we took students on a field trip to the creek. Every student in the school made monthly visits to the creek to study and document biological and chemical changes through digital imagery and videography. In addition, the students assisted in the construction of a wetland adjacent to the restored creek by helping spread soil on top of the wetland liner, placing large pieces of woody debris into the wetland, planting native species along the wetland edge, and spreading straw on seeded areas to keep erosion to a minimum.

Today, the school sits alongside a natural Kentucky valley with an abundant array of flowers and grasses. There are seven different habitats that illustrate the variety one would find in a non-human-altered site. The creek now has a natural meander, and wildlife abounds within its waters. Students are learning science within a natural setting and are able to observe and study the impact of humans on the environment. Students and families from the working-class neighborhood surrounding the school come to walk the waters of the stream, looking for the organisms they now know exist there. For some children, the possibility of becoming scientists or wildlife specialists is now a reality because of these new experiences.

The principal believes that her job as the instructional leader is to develop people and change the school's culture; she feels that she has accomplished both. The students learn differently and the teachers teach differently because they use a familiar place, a local context, as a focal point of learning. Most of all, she has developed leaders among her staff and students.

Our most recent project, begun in 2012, expanded the study of water issues to include more emphasis on this sense of place that motivates students. Under a grant provided by the Environmental Protection Agency, staff and faculty from CAFE worked with elementary, middle, and high school students in the Cane Run Watershed of Fayette County, Kentucky, to investigate environmental issues and human impacts related to the Cane Run Creek watershed. Through a combination of classroom and field experiences, the students engaged with the essential question: How do we live sustainably in our watershed? For many of the students, investigating this question meant beginning with the basic issue of defining what constitutes a watershed and how all watersheds are connected in a network.

A primary goal of the project was to construct a map of the Cane Run Creek watershed, which drains portions of northern Fayette and southern Scott Counties. Students were given instruction in the use of GIS mapping technologies so that they could identify sources of pollution in the watershed and locate those sources on the map. As students collected qualitative and quantitative data about such influences as impervious surfaces, agricultural runoff, fertilizer usage, and point sources of pollution, they had the opportunity to learn and develop proficiency in a wide range of science process skills and content knowledge.

Once again, in this project, students' attitudes toward their local environment were affected in observable and measurable ways. Assessments given before and after the instructional activities showed that there were not only shifts in the understanding of science content but also other changes in their willingness to take personal actions that could help improve the health of their local streams.

In conclusion, each of the water-related projects discussed has offered evidence to support the use of project-based learning and place-based approaches to teach science in the schools. The overwhelmingly positive responses from teachers and students over the years and numerous activities demonstrate that substituting a project-based or place-based pedagogy makes teaching more interactive, student centered, and contextually relevant and thus leads to success in motivating more young people to enter and remain in science courses and science-related careers.

Research suggests that placing science content in an everyday context and in relation to students' daily lives (Woolnough 1994; Maltese and Tai 2011) is an important strategy for getting students interested in science lessons; it is a strategy used extensively in our community-based projects to interest students in their local water resources. Active learning is a pedagogical strategy that has positive impacts on attitudes toward science for all students, but especially for females and ethnic-minority students (Oakes 1990). Research shows that student-centered teaching and learning (Baker 2013) and authentic learning experiences (Tyler-Wood et al. 2012) are good for our students. We advocate that teachers make science personal, local, and relevant. Rather than using classic chemical equations or discussing sea-level rise in distant locations, we recommend that they use the chemistry related to their local stream or discuss water-quality issues in their communities and across Kentucky. We are in the company of educators and researchers (Santelmann et al. 2000; Wither 2001; Null 2002; Gunckel et al. 2012) who believe that using local water resources to teach science not only helps raise science achievement but, more important, influences students to take an interest in science as a career and as a lifelong passion.

Acknowledgment

The authors acknowledge editing assistance by Esther Edwards.

References

ACT. 2013. 2012 ACT National and State Scores: College Readiness Benchmark Attainment by State. http://www.act.org/newsroom/data/2012/benchmarks.html (accessed October 18, 2015).

Adderley, Kenneth. 1976. "Project Methods in Higher Education." *Improving College & University Teaching, 24,* 126.

Aud, Susan, Sidney Wilkinson-Flicker, Paul Kristapovich, Amy Rathbun, Xiaoki Wang, and Jijun Zhang. 2013. *The Condition of Education, 2013.* NCES 2013-037. Washington, DC: U.S. Department of Education, National Center for Education Statistics. http://nces.ed.gov/pubsearch (accessed October 18, 2015).

Baker, Dale. 2013. "What Works: Using Curriculum and Pedagogy to Increase Girls' Interest and Participation in Science." *Theory into Practice* 52 (1): 14–20.

Barab, Sasha, Kenneth Hay, and Thomas Duffy. 1998. "Grounded Constructions and How Technology Can Help." *Tech Trends* 43 (2): 15–23.

Blumenfeld, Phyllis C., Elliott Soloway, Ronald Marx, Joseph Krajcik, Marh Guzdial, and Annemarie Palinscar. 1991. "Motivating Project-Based Learning: Sustaining the Doing, Supporting the Learning." *Educational Psychologist* 26 (3 & 4): 369–98.

Borgelt, Ida, Kym Brooks, Jane Innes, Amy Seelander, and Kathryn Paige. 2009. "Using Digital Narratives to Communicate about Place-Based Experiences in Science." *Teaching Science* 55 (1): 41–45.

Boyer, Paul. 2006. *Building Community: Reforming Math and Science Education in Rural Schools.* Alaska Native Knowledge Network Center for Cross-Cultural Studies, University of Alaska, Fairbanks. http://www.ankn.uaf.edu/publications/building_community.pdf (accessed October 18, 2015).

Cleaves, Anna. 2005. "The Formation of Science Choices in Secondary School." *International Journal of Science Education* 27 (4): 471–86. doi:10.1080/0950069042000323746 (accessed October 18, 2015).

Duncan, Arne. 2010. Secretary Arne Duncan's Remarks at OECD's Release of the Program for International Student Assessment (PISA) 2009 Results. December 7. http://www.ed.gov/news/speeches/secretary-arne-duncans-remarks-oecds-release-program-international-student-assessment- (accessed October 18, 2015).

Editorial Projects in Education Research Center. 2013. *Kentucky—State Graduation Brief.* Bethesda, MD. http://www.edweek.org/products/dc/sgb/2013/34sgb.ky.h32.pdf (accessed October 18, 2015).

Glod, Maria. 2008. "Scores on Science Test Causing Concern in U.S." *Washington Post,* December 10. http://articles.washingtonpost.com/2008-12-10/news/36809010_1_international-mathematics-science-education-graders (accessed October 18, 2015).

Grant, Michael. 2011. "Learning, Beliefs, and Products: Students' Perspectives with Project-Based Learning." *Interdisciplinary Journal of Project-Based Learning* 5 (2): 37–69. http://dx.doi.org/10.7771/1541-5015.1254 (accessed October 18, 2015).

Gunckel, Kristen L., Beth A. Covitt, Ivan Salinas, and Charles W. Anderson. 2012. "A Learning Progression for Water in Socio-ecological Systems." *Journal of Research in Science Teaching* 49 (7): 843–68.

Harel, Idit, and Seymour Papert, eds. 1991. *Constructionism.* Norwood, NJ: Ablex.

Harris, Judy H., and Lilia Katz. 2001. *Young Investigators: The Project Approach in the Early Years.* Early Childhood Education Series. New York: Teachers College Press.

Kafai, Yasmin, and Mitchell Resnick, eds. 1996. *Constructionism in Practice: Designing, Thinking and Learning in a Digital World.* Mahwah, NJ: Lawrence Erlbaum Associates.

Kentucky Department of Education. 2006. *Kentucky Testing Reports—IIS7, Kentucky Performance Reports, State Interim Report.* https://applications.education.ky.gov/ktr/default.aspx (accessed October 18, 2015).

Kuwahara, Jennifer. 2013. "Impacts of a Place-Based Science Curriculum on Student Place Attachment in Hawaiian and Western Cultural Institutions at an Urban High School in Hawai'i." *International Journal of Science and Mathematics Education* 11: 191–212.

Lane, Ruth, Daiman Lucas, Frank Vanclay, Sophie Henry, and Ian Coates. 2005. "'Committing to Place' at the Local Scale: The Potential of Youth Education Programs for Promoting Community Participation in Regional Natural Resource Management." *Australian Geographer* 36 (3): 351–67.

Littledyke, Michael. 2008. "Science Education for Environmental Awareness: Approaches to Integrating Cognitive and Affective Domains." *Environmental Education Research* 14 (1): 1–17.

Maltese, Adam, and Robert Tai. 2011. "Pipeline Persistence: Examining the Association of Educational Experiences with Earned Degrees in STEM among US Students." *Science Education* 95 (5): 877–907. http://dx.doi.org/10.1002/sce.20441 (accessed October 18, 2015).

Morgan, Alun. 2009. "Learning Communities, Cities and Regions for Sustainable Development and Global Citizenship." *Local Environment* 14 (5): 443–59. http://dx.doi.org/10.1080/13549830902903773 (accessed October 18, 2015).

Null, Elizabeth N. 2002. *East Feliciana Parish Schools Embrace Place-Based Education as a Way to Lift Scores on Louisiana's High-Stakes Tests.* Rural Trust Featured Project. http://www.eric.ed.gov/contentdelivery/servlet/ERICServlet?accno=ED463136 (accessed October 18, 2015).

Oakes, Jeannie. 1990. *Lost Talent: The Underparticipation of Women, Minorities and Disabled Persons in Science.* http://www.eric.ed.gov/contentdelivery/servlet/ERICServlet?accno=ED318640 (accessed October 18, 2015).

OECD (Organization for Economic Cooperation and Development). 2010. *PISA 2009 Results: What Students Know and Can Do—Student Performance in Reading, Mathematics and Science.* Vol. 1. http://dx.doi.org/10.1787/9789264091450-en (accessed October 18, 2015).

Osborne, Jonathan, Shirley Simon, and Sue Collins. 2003. "Attitudes towards Science: A Review of the Literature and Its Implications." *International Journal of Science Education* 25 (9): 1049–79. http://dx.doi.org/10.1080/0950069032000032199 (accessed October 18, 2015).

Prince, Michael J., and Richard M. Felder. 2006. "Inductive Teaching and Learning Models: Definitions, Comparisons and Research Bases." *Journal of Engineering Education* 95 (2): 123–38.

Reyhner, Jon. 2010. "Place-Based Education." *NABE News,* June/July. http://www2.nau.edu/~jar/NABE/Jun-Jul2010Place.pdf (accessed October 18, 2015).

Rieber, L. P. 2004. "Microworlds." In *Handbook of Research on Educational Communications and Technology,* 4th ed., edited by D. H. Jonassen (Mahwah, NJ: Lawrence Erlbaum Associates), 583–603.

Rumberger, Russell, and Brenda Arellano. 2007. *Student and School Predictors of High School Graduation in California.* California Dropout Research Project. Policy Brief 5. UC Santa Barbara, Gervitz Graduate School of Education. http://cdrp.ucsb.edu/pubs_reports.htm (accessed January 23, 2012).

Rumberger, Russell, and Sun Ah Lim. 2008. *Why Students Drop out of School: 25 Years of Research.* Policy Brief 15. UC Santa Barbara, Gervitz Graduate School of Education. http://cdrp.ucsb.edu/pubs_reports.htm.

Sanchez, Claudio. 2012. "Higher Dropout Age May Not Lead to More Diplomas." http://www.npr.org/2012/01/27/145984943/higher-drop-out-age-may-not-lead-to-more-diplomas (accessed October 18, 2015).

Santelmann, Mary, Hannah Gosnell, and Mark Meyers. 2011. "Connecting Children to the Land: Place-Based Education in the Muddy Creek Watershed, Oregon." *Journal of Geography* 110 (30): 91–106.

Savery, John, and Thomas Duffy. 1995. "Problem-Based Learning: An Instructional Model and Its Constructivist Framework." *Educational Technology* 35 (5): 31–38.

Schiavelli, Mel. 2011. "STEM Jobs Outlook Strong, but Collaboration Needed to Fill Jobs." *US News and World Report,* November 3. http://www.usnews.com/news/blogs/stem-education/2011/11/03/stem-jobs-outlook-strong-but-collaboration-needed-to-fill-jobs (accessed October 18, 2015).

Sobel, David. 2004. *Place-Based Education: Connecting Classrooms and Communities.* Great Barrington, MA: Orion Society.

Sparks, Sarah D. 2012. "Most 8th Graders Fall Short on NAEP Science Test." *Education Week,* May 10. http://www.edweek.org/ew/articles/2012/05/10/31naep_ep.h31.html?tkn=VP XFO3wzO2s%2Bbex2WwFqNNnCfYtzrpCNzSmA&cmp =ENL-EU-NEWS1&print=1 (accessed October 18, 2015).

Trent, Don, and Jon Reyhner. 2002. *Preparing Teachers to Support American Indian and Alaska Native Student Success and Cultural Heritage.* http://www.eric.ed.gov/contentdelivery/servlet/ERICServlet?accno=ED459990 (accessed October 18, 2015).

Tyler-Wood, Tandra, Amber Ellison, Okyoung Lim, and Sita Periathiruvadi. 2012. "Bringing Up Girls in Science (BUGS): The Effectiveness of an Afterschool Environmental Science Program for Increasing Female Students' Interest in Science Careers." *Journal of Science Education and Technology* 21 (1): 46–55.

Van Rooij, and Shahron Williams. 2009. "Scaffolding Project-Based Learning with the Project-Management Body of Knowledge (PMBOK®)." *Computers & Education* 52: 210–19.

Wither, Sarah E. 2001. "Local Curriculum Development: A Case Study." Paper presented at the Annual Meeting of the American Educational Research Association, Seattle, WA, April 10–14, 2001. http://www.eric.ed.gov/contentdelivery/servlet/ERICServlet?accno=ED456022 (accessed October 18, 2015).

Woodhouse, Janice. 2001. "Over the River and through the Hood: Re-viewing 'Place' as Focus of Pedagogy." *Thresholds in Education* 27 (3): 1–5.

Woolnough, Brian. 1994. *Effective Science Teaching: Developing Science and Technology Education.* Bristol, PA: Open University Press.

Young, Kenneth R. 2002. "Minding the Children: Knowledge Transfer and the Future of Sustainable Agriculture." *Conservation Biology* 16: 855–56.

CHAPTER TWENTY-TWO

Using Market-Based Tools to Protect and Improve Water Quality in Kentucky

Wuyang Hu

Sometimes the way to address water resources is at the source of the problem, while other times the geographic source is only one part of the solution. Geography and economics are showing promise as a way to address and, it is hoped, improve water-resource conditions in the future. The key is that people and communities need to fully understand their options and make decisions that are the best for the environment.

Market-based tools were first suggested in the 1960s by economists considering how society could achieve long-term reductions in pollution without causing an undue burden on the economy (U.S. EPA 2004). These economists recognized that instead of direct government orders to polluting entities regarding how much waste or pollutant they would be allowed to discharge, market-based incentives governed by the invisible hand could be used to guide individual behaviors, thereby reducing overall pollution at a cheaper overall cost with less of a negative impact on an economy. One strategy is to let polluters reallocate the pollution they generate among themselves; in other words, they decide who actually does the pollution abatement. Entities with high abatement costs may pollute more (abate less), and those with low abatement costs may pollute less (abate more). If polluting entities are allowed to trade part of their pollution discharges, the units that can be traded are commonly referred to as water-quality credits. A close counterpart to the water-quality trading idea is the carbon-emission trading that resulted from the Kyoto Protocol (Kyoto Protocol 1998).

In regard to water, pollution sources are commonly classified as point sources or nonpoint sources. A point source is a single identifiable polluting source, such as a

wastewater pipe discharging to a river or a creek. Industrial water users or municipal water-treatment plants are examples of point sources. A nonpoint source is a polluter who does not generate a specific identifiable source of pollution. With the exception of concentrated animal-feeding operations, many farms and urban residences are nonpoint sources. Since nonpoint-source pollution is difficult to monitor, the U.S. Environmental Protection Agency (EPA) currently implements total maximum daily loads (TMDLs) only for point sources. Lowering TMDLs for all point sources in a watershed means improved water quality if nonpoint sources maintain their level.

This presents an opportunity for parties involved. If point sources and nonpoint sources are allowed to trade water-quality credits, economic and water-quality-improvement benefits may result. Reallocating water-quality credits through trading could generate significant societal cost savings. Facing ever-stringent TMDLs, point-source entities can find themselves in need of purchasing additional water-treatment equipment or upgrading existing equipment to meet standards. The infrastructure options are typically not cheap, and violating the TMDL regulation may result in a large fine. A solution is needed to comply with pollution regulations while also keeping costs to a minimum. One solution is for point sources to sponsor pollution reductions by nonpoint-source entities because in some situations it is financially advantageous to abate the nonpoint-source pollution. This reduction strategy can be accomplished by purchasing water-quality credits from nonpoint-source entities. If the total cost of purchasing water-quality credits is less than the cost of purchasing or updating equipment, point-source entities will be able to address the pollution requirements while reducing costs. On the other side of the trading scheme, nonpoint-source entities will be financially compensated by point-source entities for maintaining good land-management practices or implementing additional land-management measures to reduce pollution. These measures are known as best management practices (BMPs). Example of BMPs include soil management, pest management, and nutrient management.

As an important player in water quality, agriculture generates particular interest in the public. According to the annual *National Water Quality Inventory,* one of the main documents updating Congress and the public about water-quality conditions, agricultural nonpoint sources account for more than half of the pollutants discharged to the water systems in the United States (U.S. EPA 2004). To address this issue, there exist multiple federally funded programs from the U.S. Department of Agriculture (such as the Conservation Reserve Program and the Conservation Reserve Enhancement Program) and the EPA (such as the Environmental Quality Incentives Program) to help agricultural nonpoint sources reduce pollution. These programs are often referred to as cost-share programs because they cover up to 75 percent of the cost of implementing BMPs on farms. A water-quality trading mechanism will incentivize nonpoint-source entities to reduce pollution because they will receive payments from point-source entities for doing so. Practices of nonpoint sources to protect the environment will still be compensated, but this time, by point sources instead of by government funds. In fact, in some circumstances, the compensation nonpoint sources receive from point sources could be larger than what they would receive from government sources. This will reduce or eliminate the publicly funded programs and save taxpayers money.

The existence of a trading system does not prevent either point- or nonpoint-source entities from investing in technological innovation to reduce pollution. More cost-efficient technology will bring down point sources' demand for water-quality credits and thus push nonpoint sources to lower the unit price for these credits. More effective and cost-saving BMPs that nonpoint sources can adopt will lead to more financial gain from water-quality trading, which may serve as an incentive

for nonpoint-source innovation. In an efficient market, in the end, water-quality-credit prices will respond to the innovations and adjust downward. For society as a whole, technological advancement from either side will lead to lower cost of maintaining and improving water quality. Another benefit associated with market-based tools is that different types of market mechanisms can be tailor designed to fit the specific needs of a community. There is no one trading mechanism superior to other options in every situation, nor does a one-size-fits-all mechanism exist. The best approach is the one that fits the geographic and social conditions of the watershed in question. Two of the longest-existing and most prominent water-quality trading markets are the Lake Dillon watershed in Colorado and the Chesapeake Bay region involving Maryland, Virginia, West Virginia, and Pennsylvania. The Lake Dillon trading market is based primarily on one-on-one negotiation since the region has clearly defined two sides of trading partners. The much larger Chesapeake Bay area features a variety of trading mechanisms because of the complexity of the region (Pinchot Institute 2010).

Using market-based tools such as trading to protect water quality may also encounter a number of issues worth noting. First, not all watersheds are suitable for water-quality trading. Regions with a small number of point and nonpoint sources may not be good candidates. The Kentucky River watershed consists of more than five hundred major point sources, more than two thousand farm nonpoint sources, and many more residential nonpoint sources and is thus a potential case for exploring water-quality trading in the future. Second, nutrients, such as nitrogen and phosphorous, are typically better targets for trading because they are often transferable from one location to another. Both nitrogen and phosphorous are major pollutants in the Kentucky River watershed. Pollutants such as sediments are more challenging to trade because they do not travel well once they are in the water, and their impacts are often limited to a specific area. Third, during a trading process, trades must address the issue of trading ratios and avoid the occurrence of geographic nutrient hot spots. Because of differences in geology and chemical compounds, a reduction in nutrients by one entity in the watershed will need to be converted to an increase in nutrient discharge in another location of the watershed. Trading ratios are used to reflect these differences. Avoiding nutrient hot spots refers to cutting the likelihood of a high concentration of nutrients in any specific region of the watershed due to trading that may generate unacceptable water quality for that region or downstream.

Finally, like any other programs involving multiple participants, a water-quality trading market must consider inputs from various stakeholders. A multidisciplinary research team of economists and engineers at the University of Kentucky is conducting a comprehensive analysis of the feasibility of establishing a water-quality trading market in the Kentucky River watershed to maintain and improve water quality, as well as reduce societal costs, with the goal of a better quality of life for flora and fauna, as well as people and communities.

References

Kyoto Protocol to the United Nations Framework Convention on Climate Change. 1998. http://unfccc.int/resource/docs/convkp/kpeng.pdf (accessed October 18, 2015).

Pinchot Institute for Conservation. 2010. "Nutrient Trading in the Chesapeake Bay Region: An Analysis of Supply and Demand." http://www.pinchot.org/uploads/download?fileId=659 (accessed October 18, 2015).

U.S. EPA (Environmental Protection Agency). 2004. *Water Quality Trading Assessment Handbook: Can Water Quality Trading Advance Your Watershed's Goals?* EPA 841-B-04-001. Washington, DC. http://water.epa.gov/type/watersheds/trading/upload/2004_11_08_watershed_trading handbook national-wqt-handbook-2004.pdf (accessed October 18, 2015).

Further Reading

Childress, R. 2012. "Water Quality Trading Markets for the Kentucky River Basin: A Point Source Profile." Theses and Dissertations—Agricultural Economics, Paper 8. http://uknowledge.uky.edu/agecon_etds/8 (accessed October 18, 2015).

Fernandes da Costa, P. 2011. "Participation in Agricultural Governmental Costs Share Programs in the Kentucky River Watershed." University of Kentucky Master's Theses, Paper 124. http://uknowledge.uky.edu/gradschool_theses/124 (accessed October 18, 2015).

Chapter Twenty-three

Water and People at the Confluence

Amanda Abnee Gumbert

Rivers and streams hold many stories, but in our busy lives we often forget to notice them. This is likely true of many central Kentucky streams—many people see them every day, but they are often taken for granted. As much as I hate to admit it, I am a perfect example of this phenomenon. Growing up in northeastern Kentucky, I saw water every day. We crossed the Licking River on our way to school each morning, and I would watch it flow under the school bus as we crossed over an old steel truss bridge. This bridge was once the main conveyor of traffic along the old Limestone/Maysville Road; a fancier new bridge was put into service in the late 1960s (my father worked on the construction crew), but my classmates lived on the old road, so we took the old route to school. Our bus driver was skeptical of that rickety bridge, and when the river was up, she would make us walk across the bridge instead of riding across for fear that it would collapse and wash us all downstream. Somehow, walking across made it safer for everyone (figure 23.1).

This was the same river my grandfather fished on, the same river his father floated logs down, and the same river that divided sides during the Battle of Blue Licks (thought to be the last battle of the Revolutionary War) (Kentucky Department of Parks). Although I saw this water every day, and it had historical importance in my family, I gave it little thought. I did not realize that my drinking water was drawn from it, nor did I consider its significance in the settling of the area. I noticed when it left its banks and flooded my friend's home, and how it turned chocolaty after heavy rains, but I certainly never dreamed that my daily actions had environmental consequences that could be reflected in that water. The fact is that we all live downstream from something or someone, and how we treat our land and the decisions we make often affect our common

bond: water. We can explore the interface of water and people throughout history, as well as today, and an example I know well is in central Kentucky.

The waters of the Inner Bluegrass have without a doubt influenced and been affected by land use throughout Kentucky's history. Lexington was settled around a natural spring (O'Malley 2007), and although different from the rivers on which most major eastern U.S. cities were established, it provided a water source needed by the settlers and subsequently determined the location of what is currently Kentucky's second-largest city. As the woodland and prairie frontier of the Bluegrass became more agrarian, the streams provided water for livestock and were sometimes straightened to accommodate agriculture and address flooding concerns. These waters were vulnerable to sediment from plowed and grazed fields and pathogens from human and livestock sources.

Prime agricultural land and open spaces lasted into the middle of the twentieth century (United States Department of Agriculture 2007). In 1950, an average of 91 percent of the land area in Fayette County and its surrounding counties was farmland. As the region moved into the modern age, the waters began to be impacted by human activity as a result of land-use choice. The urbanization of the Inner Bluegrass expanded into the countryside like tentacles, as evidenced by a drive along any of the major arteries connecting Lexington and its surrounding county seats. As of 2007, farmland had dropped to 85 percent of land use in these counties (United States Department of Agriculture 2007). Urbanization is known to create more impervious surfaces than open spaces and to affect water resources in many ways (Leopold 1968), including increased runoff and less groundwater recharge. The way we as humans use the landscape is a reflection of

23.1. Looking upstream at the old bridge over Licking River at Blue Licks, Kentucky, with the new bridge in the background, July 2013. The Blue Licks springs and historic resort development were located in the extensive floodplain to the left. The accumulated logs behind the bridge pier indicate the load of sediment and woody debris this river is capable of carrying during high-water events.

our needs, values, and desires, and these land-use choices directly impact the water.

Waterways tell stories of the land, and the history of a waterway is often told by an examination of its watershed. The watershed, or the land that drains to a specific water body, sets the stage for the quality of its receiving stream. A watershed of particular interest in central Kentucky is Cane Run because it recharges the Royal Spring Aquifer, a primary drinking water source for the city of Georgetown. The Cane Run watershed is an example of a mixed-use, karst, Inner Bluegrass watershed. The urban headwaters of the watershed drain the impervious surfaces of northwestern Lexington. The fact that the Cane Run watershed contributes to the drinking water source of a community of approximately 30,000 residents makes it a priority for water quality protection and restoration (University of Kentucky 2011). Concern for water quality in the Cane Run watershed dates back to the 1960s, and it has been the subject of much sampling and monitoring for a multitude of theses and dissertations.

In some places in Kentucky, you can stand right beside the origin of a creek or river, right where the water gurgles out of the ground to meet the daylight and begin its storytelling journey all the way to the Gulf of Mex-

ico. This is true for the headwaters of the Licking River, that river I used to see every day of my childhood, but not for Cane Run Creek; its headwaters are somewhat ambiguous, having been drained into pipes under concrete and asphalt for years, near the intersection of Loudon Avenue and North Limestone Street in Lexington. The current residents of the nearby Loudoun House in Castlewood Park say that the basement used to flood and that springs existed on the lawn, so one would surmise that these springs and seeps were part of the Cane Run headwaters. Today, there is little evidence of a watercourse until you travel approximately one mile from where the headwaters would or should be, to the intersection of New Circle Road (Kentucky 4) and Old Paris Road. At this location, two large stormwater culverts present to the world the first evidence of a surface stream that is Cane Run Creek (figure 23.2).

At this point, the surfaced stream has drained neighborhoods built in the 1930s and later, shopping centers, and parking lots and now flows alongside a busy four-lane road. The stream has been pushed aside to make way for pavement and buildings, so it moves through a narrow passage with retaining walls as stream banks and invasive bush honeysuckle as the urban jungle canopy. The

23.2. First visible evidence of Cane Run Creek in its urban headwaters at the intersection of Old Paris Road and New Circle Road (Kentucky 4) in Lexington, Kentucky, April 2014. Litter and debris clog the stream, a common sight in the urban areas of the Cane Run watershed. The numbers "1974" are stamped on the stormwater infrastructure, indicating the year in which the culverts were installed.

stream no longer has much of a floodplain; now there are parking lots and apartment buildings where flood-waters would have settled, so water runs fast and deep after a rainstorm. The stream has been reduced to just a conduit for stormwater, pollutants, and litter; storm sewers draining surrounding neighborhoods empty into the main channel of Cane Run and its tributaries, carrying the first flush of pollution to the stream. The first flush is the stormwater flow from the first half inch of rain that flushes out nearly 80 percent of accumulated contaminants since the previous rainfall (DeBarry 2004), and it is generally more polluted than the extended stormwater flow. The pollution (e.g., automotive fluids, pet waste, lawn fertilizers and pesticides, and cigarette butts) runs off lawns and paved areas and flows untreated into the stream. In addition, aging and undersized sewer infrastructure results in periodic sanitary sewer overflows during high rainfall. Sanitary sewer overflows and cross connections with storm sewers result in untreated sewage reaching Cane Run Creek.

Farther downstream is the property of Lexmark International, near the intersection of New Circle Road and Newtown Pike. Because of the previously mentioned pollutants, the quality of the water at this location of the stream is questionable at best. Posted signs warn of raw sewage and dangerous pathogens, a result of Lexmark's monitoring, which is supported by data collected by the University of Kentucky's water monitoring station situated on an adjacent stream bank. At this point is the confluence of the main channel and two unnamed tributaries. A confluence is the location where smaller creeks or tributaries come together to form a larger stream or river, mimicking in a way the confluences in human lives. Water is a common thread for all of us, and at this particular confluence, residential urban land use meets industrial land use. Up one tributary is a community park where children play in the creek without regard to water quality or worry over where the water is going or where it has been. Up the other tributary are corporate offices and industrial buildings, inside which, one hopes, employees ponder their corporate responsibility.

The urban waters of Cane Run Creek begin to interact with the groundwater as they flow in a northwesterly direction toward rural land uses: parks, open fields, pastures, research stations, and working farms. The flowing water is visible on the surface during and after rainfalls, but the stream channel may often appear as a dry ditch during low-flow times. In-stream sinkholes, or swallets, can be seen in the main channel of Cane Run Creek and are direct conduits to the groundwater. The groundwater replenishes the Royal Spring Aquifer, and the surface water empties into Elkhorn Creek. It is important to realize that in a karst landscape, surface water can become groundwater in seconds and in the next moment can come back to the surface, all within a stone's throw.

Beyond the confluence on Lexmark's property, Cane Run skirts the edge of Lexington, making its way to the urban fringe and meeting up with the Legacy Trail, a multiuse recreational trail connecting downtown Lexington with the Kentucky Horse Park and eventually beyond. Interestingly, the trail crosses Cane Run Creek eight times on its way out of downtown, presenting the opportunity for a firsthand visitor's experience with the creek. The creek passes through a farm turned research park and enters a city park, where it was the focal point of the first Reforest the Bluegrass in 1999. As a result of this volunteer tree-planting event, the stream banks and floodplain are anchored and shaded by native trees (although a few nonnative trees have found a home here as well). Leaving the park, the stream enters the University of Kentucky (UK) Agricultural Experiment Station. This land has a long history of agriculture (Smith 1981), and as a result, Cane Run Creek and its tributaries have often been straightened, channelized, and denuded of vegetation. Mowing to the edges of stream banks was historically a common practice, not only on this farm but also in places across the commonwealth, in an effort to maintain a tidy, well-groomed appearance. This practice has consequences for the stream. Clipped vegetation along the stream banks exposes the water to sunlight; warmer water temperatures and light energy allow algae to grow in thick mats and choke out beneficial aquatic life when the summertime conditions are right. Frequent mowing also reduces the number of deeply rooted trees, shrubs, and grasses that stabilize and strengthen stream banks, leaving them vulnerable to the magnified erosive forces of runoff flowing downstream from the urban headwaters previously described. This erosive action transfers and deposits sediments in the stream, resulting in poor aquatic habitat and stream banks that retreat into pastures and crop fields as highly productive agricultural soils make their way downstream to the Missis-

sippi Delta. Despite a history of disregarded stream buffers (also called riparian zones), Cane Run Creek and its tributaries now have many areas of no-mow zones and floodplain revegetation projects on the UK Agricultural Experiment Station. Thousands of trees have been planted, vegetation has been left on stream banks, and livestock access to the creek has been restricted in an effort to protect water quality in the Cane Run watershed. This effort was coordinated by a small group of concerned scientists in the UK College of Agriculture, Food and Environment who recognized that the traditional treatment of stream buffers was not achieving healthy streams. A management shift was needed, and although their efforts did not gain them widespread popularity, the team continued to rally for naturalized buffers to benefit downstream water quality.

Beyond the UK Agricultural Experiment Station are multiple institutional entities, including another research park, more farms, and then the Kentucky Horse Park. The entrance of the park provides an up-close and personal view of the main channel of Cane Run Creek. During dry periods, the channel appears as no more than a grassy ditch, while during wet weather the stream flows rapidly. Eroded stream banks are evidence of a mowing history here, too, as park management strives to present a manicured horse-farm appearance expected by park visitors. A conflict emerges between preconceived notions of aesthetic beauty and ecological sustainability. Beyond the entrance to the park, a large swallet is visible in the stream channel, swallowing water from the stream until the aquifer is full, after which the water bypasses the swallet and flows on down the channel toward Elkhorn Creek. This karst feature is an amazing illustration of surface and groundwater interaction and how inextricably the two are intertwined. Polluted surface water equals polluted groundwater.

Around the Kentucky Horse Park are several springs creating tributaries of Cane Run Creek. Like the surrounding properties, these small streams have historically been maintained by mowing right to the water's edge, creating streambank erosion issues (figure 23.3). With technical and financial assistance from watershed project coordinators, this practice is slowly changing, and a more sustainable approach is being taken. An

23.3. Historically mowed stream banks at the Kentucky Horse Park, April 2010. Mowing contributed excess organic matter to the stream in the form of clippings and left stream banks vulnerable to erosion.

example of this is a 2010 stream revegetation project in which approximately five hundred feet of stream bank were planted with native trees, shrubs, grasses, and wildflowers in an effort to slow streambank erosion and filter nutrients from runoff. The planting effort doubled as an educational opportunity; a woodchip walking path allows visitors close access to the stream, and interpretive signage explains the effort to protect water quality. Volunteers helped install the vegetation in time for the 2010 Alltech Fédération Équestre Internationale (FEI) World Equestrian Games to see the wildflowers in their first bloom. This project was a leap of faith on the part of the Kentucky Horse Park as its operators explored a new way of managing stream buffers and began a long-term collaboration with watershed partners to introduce alternative management strategies to a park traditionally managed as a well-manicured landscape.

Beyond the Kentucky Horse Park, Cane Run Creek flows through more farmland, and near the Fayette/Scott County line the Royal Spring Aquifer recharge area and the Cane Run surface watershed diverge. Near this location are mobile-home parks and homes far enough out of town to be beyond the municipal sanitary sewer line, so wastewater is treated with wastewater package plants or on-lot septic tanks. Like other types of infrastructure,

over time, wastewater package plants can fail to operate properly because of age of the facility, inappropriate sizing, or inadequate maintenance, resulting in discharge of improperly treated wastewater into streams. Beyond the Scott County line, the stream flows on toward its confluence with North Elkhorn Creek, which will join its sister branch, the South Elkhorn, and together they will reach the Kentucky River just north of Frankfort. The Royal Spring Aquifer eventually surfaces in downtown Georgetown, where the water is treated and distributed to residents of Georgetown as their drinking water by the Georgetown Municipal Water and Sewer Service.

The current story of the Cane Run watershed presents a somewhat dim outlook for its future. However, despite the present impaired status of Cane Run Creek, there is indeed reason to believe that this stream has a glimmer of hope for the future. The Environmental Protection Agency (EPA), through a consent decree (as described by Schieffer in chapter 16 of this book), has mandated that the Lexington-Fayette Urban County Government (LFUCG) address sanitary and storm-sewer infrastructure, which should reduce sanitary sewer overflows and improve water quality. Further, the Cane Run watershed has been the focus of a scientific team from the University of Kentucky since 2006, when it began the critical task of writing the Cane Run watershed-based plan. A watershed-based plan is a dynamic document that is designed to be a snapshot in time of current watershed conditions, including potential watershed stakeholders, and a plan to protect existing water quality and overall hydrology and subsequently improve it.

The process of writing a watershed-based plan is a very methodical effort: research plans are written, data are collected and analyzed, graphs and charts are developed, technical reports are compiled, and progress reports are submitted. But there is a softer, juicier side to this scientific effort to address water quality issues. Although watershed-based plans might look like technical jargon and a scientific synopsis of the environment (both of which they are), beneath those layers are the efforts of individuals and organizations that exemplify the human dimension to which we can relate. The watershed project led by the University of Kentucky included the development of a watershed council to gather stakeholder buy-in during the development of the watershed-based plan. The council has served as a gathering place of federal, state, and local government representatives, nonprofit organizations, watershed residents, universities, and citizen activists and advocates. Meetings involve information exchange, idea sharing, colorful discussions, and volunteer recruitment for localized projects. Since 2006, the council members have influenced their respective organizations and engaged in a multitude of projects to improve conditions in the Cane Run watershed. Examples of council members' influence include but are certainly not limited to the following:

- In the urban headwaters, neighborhood associations have willingly participated in and assisted with watershed festivals to educate residents about stormwater and human impacts on water quality. Students from UK and local public schools have picked up litter, stenciled storm drains, and visited residents to tell them the story of the watershed.
- In the industrial areas, corporations have embraced their local environment and made Cane Run Creek a corporate priority. The impact that a network of nature-loving employees can bring to bear on an issue as part of an international corporation is amazing. For example, Lexmark International (which is in the urban headwaters) has made Cane Run Creek the focus of an annual clean-up, and its employees have restored and protected stream banks by removing nonnative, invasive vegetation and replacing it with native trees, shrubs, and wildflowers (figure 23.4). This is a corporation striving to improve its local environment on its own property but also seeing the responsibility it has to the larger community.
- On the rural edge, the UK Agricultural Experiment Station has gotten on board with riparian areas, agricultural best management practices, and an educational effort to shift years of historical practices toward more sustainable approaches of agricultural land use. At the Kentucky Horse Park, managers are open minded to innovative ways to reduce stormwater runoff and protect stream banks. Park visitors are appreciating the natural beauty of birds, butterflies, and other pollinators using restored riparian areas.

The Cane Run watershed is an example of how a group of passionate but also diverse individuals have worked together for the common purpose of making a difference in a central Kentucky watershed. Margaret Mead captures it best: "Never doubt that a small group of thoughtful, committed citizens can change the world. Indeed, it is the only thing that ever has" (Mead no date). A small group of people has begun the work that will last decades in the Cane Run watershed to change the water condition for future generations. This work must be carried out by more small groups of committed people not only in the Cane Run watershed but also in many watersheds across Kentucky. It is indeed the commitment of individuals, businesses, organizations, and institutions in the community that is every one of us, making clean water a priority of our lives, that will change the condition of water in Kentucky.

I return to my childhood home from time to time and cross over that same Licking River, only now I do so on the aforementioned new bridge. As I cross, I crane my head to get a good look at the water below, churning its way downstream. It still carries the story of the land and all the ways we use it. It is my hope that more of us will notice and respect our local waterways and become part

23.4. The same stream area depicted in figure 23.3 one year after the revegetation project at the Kentucky Horse Park, July 2011. Native trees, shrubs, and wildflowers reduce erosion, filter contaminants in runoff, and provide shading to the stream.

of a small group of committed people to protect them. So take a look the next time you cross a stream or river, think about its story and its future, and tie it back to your home. How does this waterway affect your daily life? How do your actions affect water quality? Will future generations enjoy this waterway? I challenge you to ask yourself these questions. Embrace the answers, and get involved. What part will you play in shaping water in Kentucky?

References

DeBarry, Paul A. 2004. *Watersheds—Processes, Assessment, and Management.* Hoboken, NJ: John Wiley and Sons.

Kentucky Department of Parks. Blue Licks Battlefield State Resort Park Historic Brochure. http://parks.ky.gov/parks/resortparks/blue_licks/History.aspx (accessed October 18, 2015).

Leopold, Luna. 1968. *Hydrology for Urban Land Planning—A Guidebook on the Hydrologic Effects of Urban Land Use.* U.S. Geological Survey Circular 554. Washington, DC: U.S. Geological Survey.

Mead, Margaret. Institute for Intercultural Studies. http://www.interculturalstudies.org/main.html (accessed October 18, 2015).

O'Malley, Nancy. 2007. *McConnell Springs in Historical Perspective.* Lexington: Friends of McConnell Springs, Inc.

Department of Anthropology, University of Kentucky. http://www.mcconnellsprings.org/images/McConnell Springs_in_Historical_Perspective.pdf (accessed September 18, 2016).

Smith, J. Allan. 1981. *The College of Agriculture of the University of Kentucky: Early and Middle Years, 1865–1951.* Lexington: Kentucky Agricultural Experiment Station.

United States Department of Agriculture. National Agricultural Statistics Service. 2007. *2007 Census of Agriculture.* Washington, DC. http://www.agcensus.usda.gov/Publications/2007/Full_Report/Census_by_State/Kentucky/index.asp (accessed October 18, 2015).

University of Kentucky. 2012. *Cane Run and Royal Spring Watershed-Based Plan.* EPA Project Number C9994861-06. Lexington: University of Kentucky, College of Agriculture, Biosystems and Agricultural Engineering. https://www.bae.uky.edu/CaneRun/PDFs/Cane_Run_WBP_2011.pdf (accessed October 18, 2015).

Further Resource

Kentucky Waterways Alliance and Kentucky Division of Water. 2010. *Kentucky Watershed Planning Guidebook for Kentucky Communities.* http://water.ky.gov/watershed/Documents/guidebook/KY%20Watershed%20Planning%20Guidebook%20-%20Entire.pdf (accessed October 18, 2015).

Contributors

A seventh-generation native of Letcher County, SAM ADAMS grew up steeped in local history and a love of the people and place. As a newspaper reporter, freelance writer, and the author of two books, he has covered politics, crime, features, and environmental stories all over Kentucky. While serving in Volunteers in Service to America (VISTA), he worked on mining, water quality, and forestry issues in Kentucky, Virginia, Tennessee, and West Virginia. He is intimately familiar with the environmental, social, and health problems of Appalachia. Adams holds a BA from the University of Kentucky with a major in telecommunications and a concentration in political newswriting.

CARMEN T. AGOURIDIS is an extension associate professor in the Biosystems and Agricultural Engineering Department and is the director of the Stream and Watershed Science Graduate Certificate at the University of Kentucky. A licensed professional engineer in Kentucky and West Virginia, Dr. Agouridis has expertise in stream restoration and assessment, riparian-zone management, hydrology and water quality of surface waters, and low-impact development. Having received training in Rosgen Levels I–IV along with courses at the North Carolina Stream Restoration Institute and various other workshops, she teaches Introduction to Stream Restoration, a senior-level and graduate-level course at the University of Kentucky.

CHRISTOPHER D. BARTON is a professor of forest hydrology and watershed management in the Department of Forestry and the director of the Appalachian Center at the University of Kentucky. His research focuses on the areas of ecosystem restoration, reforestation, and remediation in stream and wetland habitats and mined lands. In addition, he is examining improved methods for preventing water quality degradation from logging and mining activities. He currently serves as a

coleader of the Appalachian Regional Reforestation Initiative's Science Team and is a cofounder of Green Forests Work.

DAVID R. BROWN is an Associate Professor in the Department of Biological Sciences at Eastern Kentucky University in Richmond. His primary area of research is wildlife population ecology, and much of his work takes place in wetlands habitats. His teaching interests include Wetland Wildlife Management and Biostatistics. Along with Dr. Stephen Richter and their graduate students, he is developing methods to intensively assess the biological condition of wetlands in Kentucky. He received a BA from the University of Colorado, Boulder, an MS from Southeastern Louisiana University, and a PhD from Tulane University. He also held a postdoc position as an Ecology instructor at Louisiana State University. He moved from the Bayou to the Bluegrass in 2008 and has since become involved in Kentucky wetland conservation and research.

JOHN R. BURCH JR. presently serves as dean of library services at Campbellsville University. He received his doctorate in history from the University of Kentucky. His professional activities include serving as a book reviewer for *Library Journal, CHOICE: Current Reviews for Academic Libraries,* and *American Reference Books Annual (ARBA).* In addition to writing numerous journal and encyclopedia articles, he is the author or coauthor of six books. He coedited *The Encyclopedia of Water Politics and Policy in the United States* (Washington, DC: Congressional Quarterly Press, 2011), which was named a 2012 Outstanding Academic Title by *CHOICE: Current Reviews for Academic Libraries.*

DANIEL I. CAREY retired as a hydrologist with the Kentucky Geological Survey (KGS) at the University of Kentucky (UK). He has studied water in Kentucky from a variety of perspectives. As a graduate student at UK, he evaluated water supplies and the use of conservational water pricing. As a consulting engineer, he studied flooding on streams in central and western Kentucky and sinkholes in the Pennyroyal region. He also developed nonstructural flood-damage-reduction measures in eastern and western Kentucky. As a researcher at KGS, he has looked at groundwater, water quality, and water supplies in the Kentucky River basin, the quality of rural domestic water wells, and hydrologic impacts of mining. He has developed maps of groundwater quality in Kentucky. He led the production of strategic plans for water and wastewater development in Kentucky, 2000–2020, for the Water Resources Development Commission. He assisted the Kentucky Infrastructure Authority in its planning efforts, participated in the establishment of Kentucky's Water Resources Information System, and produced groundwater resource reports for every Kentucky county. His geologic maps for land-use planning in every Kentucky county received national recognition. He has also created maps and posters of Kentucky's river basins that illustrate physical characteristics, water quality, water use, and recreational activities. Carey taught graduate courses in geographic information systems and environmental systems at UK. His primary interest was raising earth science awareness for Kentuckians in general and K–12 students in particular with maps, posters, and curricula about the places where we live, work, study, and play.

TRICIA COAKLEY completed her BS in biology and chemistry at Western Kentucky University, where she was also employed as a lab technician at the Ogden Environmental Lab. She then completed a master of geology at the University of Kentucky, where she is employed as the manager of the microbial and metals divisions of the Environmental Research and Training Lab (ERTL). Tricia was raised in Versailles, Kentucky, where she began a lifelong love of Kentucky streams and karst features. Returning to the Bluegrass region for employment at the University of Kentucky created opportunities to become directly involved in the protection and remediation of local waters. Coakley has participated for over fifteen years in citizen action groups focused on water quality

and has implemented innovative molecular fecal source tracking techniques at ERTL to assist these groups, as well as area city governments, in locating the sources of fecal bacteria polluting streams. Her research mission is to empower the citizens of Kentucky to remove pollution from waterways by providing them with data that allow the identification of the contributing sources, including residential, commercial, and agriculture activities.

ANGELA S. CRAIN has been a hydrologist with the U.S. Geological Survey (USGS) Indiana-Kentucky Water Science Center in Louisville, Kentucky, for more than twenty years. Since 2004, she has served as the water quality specialist for the Kentucky Water Science Center and has worked on a variety of water quality projects throughout Kentucky. From 2010 to 2012, she served as the acting USGS Water Science Field Team water quality specialist, providing technical assistance to water quality projects throughout the southeastern United States. She earned a BS in environmental science from Purdue University and an MS in aquatic biology from the University of Louisville.

JAMES C. CURRENS received his BS in geology, with an emphasis on karst hydrogeology, in 1973 from the University of Kentucky. He completed an MS program in fluvial sedimentology at Eastern Kentucky University in 1978. He is employed by the Kentucky Geological Survey as a hydrogeologist specializing in karst aquifers. He is a fellow of the National Speleological Society.

Currens has studied karst geomorphology in the Mammoth Cave area of Kentucky, which resulted in contributions to the understanding of the hydrogeology of the area and a major cave discovery in 1976 that was connected to Mammoth Cave in 1983. His first project as a hydrogeologist was to delineate wellhead-protection areas for two springs used for emergency water supply in Stanford, Kentucky, in the early 1980s. He studied flooding of Sinking Creek, a large karst valley in northwestern Jessamine County, in 1989. He completed a nonpoint-source pollution study of Pleasant Grove Spring in Logan County that involved monitoring a karst spring for agriculturally derived pollutants such as nitrates, herbicides, and bacteria during both base-flow and high-flow conditions in 1991. The reconnaissance phase of the project required delineation of the groundwater basin of Pleasant Grove and several surrounding karst groundwater basins.

Currens was the principal investigator and project manager for a major karst groundwater dye-tracing project in the Inner Bluegrass region in 1998. He conducted karst hydrogeology studies in Monroe County for the Kentucky Department of Transportation in 2005; two studies for the City of Radcliff, one on flooding and another on basin delineation, in 2006; and groundwater tracing for the planned I-66 corridor through southern Kentucky, also in 2006. He conducted another karst hydrogeology study for the Kentucky Department of Transportation in 2009. All these studies were in support of the karst groundwater basin maps, of which eight are presently available. Work continues on these maps with coauthors Joe Ray and Rob Blair, and three more basin maps are planned. The Kentucky Geological Survey has published the majority of Currens's work.

MICHELLE GUIDUGLI-COOK is a biologist in the Water Quality Branch of the Kentucky Division of Water (KDOW). Her research interests pertain to wildlife and habitat conservation of aquatic and wetland systems, especially for amphibians. She assisted in completing an intensification of the National Wetland Condition Assessment for Kentucky. Before joining the KDOW, she worked with Eastern Kentucky University, the KDOW, and other state and federal agencies to develop and test a rapid assessment method for Kentucky's wetlands. She received a BS in Biology from Northern Kentucky University and an MS in Biology (Applied Ecology) from Eastern Kentucky University. Michelle is a Kentucky native and grew up along the Ohio River, which influenced her love of aquatic ecosystems. She currently lives in Georgetown, Kentucky.

AMANDA ABNEE GUMBERT is an extension specialist for water quality at the University of Kentucky. Her area of emphasis is agricultural water-quality issues, with special interests in riparian areas and watershed management. Her work focuses on education and outreach efforts to protect Kentucky's water resources. She received BS and MS degrees in Plant and Soil Science and a PhD in Soil Science from the University of Kentucky. Gumbert grew up on a farm in rural northeastern Kentucky, to which she attributes her appreciation of the land and natural resources. She currently lives in Lexington, Kentucky.

JASON HALE is a registered landscape architect who graduated with a BS in landscape architecture from the University of Kentucky and has over fifteen years of experience in the profession, starting before his graduation in 2001. He has a diverse knowledge of the design and construction industries, which allowed for efficient management of the Landscape Architecture Group at CDP Engineers. He has also served as a part-time instructor since 2006 for the advanced design implementation course offered through the University of Kentucky's Department of Landscape Architecture. His experience ranges from assisting with multimodal transportation and trail studies and small area plan developments to the preparation of construction and contract documents for streetscape improvement projects, shared-use trail networks, and a variety of mixed-use plan developments and amenities in both Tennessee and Kentucky. Most recently he has been employed by Vision Engineering in Lexington, providing his experience with green infrastructure, storm water, and civil/site services, as well as expanding professional services to include landscape architectural design.

CAROL HANLEY has worked as a science educator for thirty years. After receiving her EdD in curriculum and instruction from the University of Kentucky (UK), she joined the Kentucky Department of Education, where she helped develop Kentucky's science content standards. In 2001, she came to UK's College of Agriculture, Food and Environment, where she has been conducting professional development programs for teachers, environmental and STEM (science, technology, engineering, and mathematics) programs for students, and outreach for community members. She has been awarded over $3 million in grant funding. In 2012, Hanley was awarded the Kentucky Science Teachers Association Distinguished Educator of the Year Award.

SUSAN P. HENDRICKS earned her PhD in aquatic ecology at the University of Michigan's School of Natural Resources and Environment. She has been a research scientist in limnology at the Hancock Biological Station since 2004. She is affiliated professionally with the Association of Ecological Research Centers, the Global Lake Ecological Observatory Network (GLEON), the International Association for Great Lakes Research, and the American Society of Limnology and Oceanography.

WUYANG HU earned his PhD in agricultural and resource economics in 2004 from the University of Alberta in Canada, after which he worked as a faculty member at the University of Nevada, Reno. In 2007, he joined the Department of Agricultural Economics at the University of Kentucky as an assistant professor and was subsequently promoted to associate professor in 2010, professor in 2013, and H. B. Price Professor in 2016. After tackling challenges resulting from water shortages in the western dry plains, he now focuses on water quality issues, which are more prominent in the eastern United States. He has published extensively in the popular press and academic outlets on a broad range of issues related to agriculture and the environment. He has been featured on multiple TV interviews with viewers topping 85 million in over 120 countries. He regularly supervises master's and PhD students engaging in a range of relevant research topics. He also teaches both undergraduate- and graduate-level courses in the fields of agricultural and environmental economics and quantitative methods.

BRAD D. LEE is a water-quality extension specialist in the Department of Plant and Soil Sciences at the University of Kentucky. Before joining the faculty at the University of Kentucky in 2009, he was an associate professor and land-use extension specialist at Purdue University. Throughout his academic career, Lee has focused his applied research and education programs on land-use impacts on water quality in agricultural and urban environments. He earned BS and MS degrees in agronomy with a minor in geology at Oklahoma State University and a PhD in soil and water sciences at the University of California at Riverside.

In his research and teaching endeavors, BRIAN D. LEE uses geospatially based analyses and visualization for community decision-making processes during land-use planning, primarily at the watershed/landscape scale. His formal education includes a BS in landscape architecture from Pennsylvania State University and a master of regional planning and a master of liberal arts (landscape ecology) from the University of Pennsylvania. He returned to Penn State to complete a PhD in the School of Forest Resources–Center for Watershed Stewardship while working as a research and teaching assistant. Since 2003, he has been a faculty member at the University of Kentucky, teaching upper-division undergraduate studios and lecture courses.

STEPHANIE M. MCSPIRIT is a professor of sociology and faculty associate of the Center for Appalachian Regional Engagement and Stewardship at Eastern Kentucky University. She is the coeditor of *Confronting Ecological Crisis in Appalachia and the South: University and Community Partnerships,* published in 2012 by the University Press of Kentucky. Her work has been published in the *Journal of Appalachian Studies, Society and Natural Resources, Organization and the Environment,* and the *Bulletin of Environmental Contamination and Toxicology,* among others.

ZINA MERKIN has a BA in visual and environmental studies from Harvard University and an MLA from the University of California at Berkeley. Since 1993, she has worked as a research specialist in the Department of Landscape Architecture at the University of Kentucky, doing land-use planning research, and is currently a part-time graduate student in the UK Department of Geography. She is vice president of the nonprofit trail advocacy group Town Branch Trail, Inc., and has been an active volunteer for over a decade with various efforts by Bluegrass Wildwater Association and American Whitewater for river conservation and river recreation.

GARY A. O'DELL received his PhD from the University of Kentucky and is a professor of geography in the Department of History, Philosophy, Politics, International Studies and Legal Studies at Morehead State University, where he has taught since 2001. Before Morehead, O'Dell spent nearly a dozen years in the employ of the

Groundwater Branch of the Kentucky Division of Water. He has had a lifelong interest in the karst landscape of Kentucky, having been a member of the National Speleological Society for more than four decades, and has personally inspected and documented more than one thousand springs in the state. He was appointed by the governor as state geographer of the Commonwealth of Kentucky for 2014.

ROGER RECKTENWALD has academic degrees in theology and social work. He was employed as a community planner/organizer and subsequently as executive director of the Big Sandy Area Development District, a five-county community development agency in eastern Kentucky; was employed by the Commonwealth of Kentucky as executive director of the Kentucky Infrastructure Authority in 2000–2004; and served for sixteen months as contract manager of the Pikeville (Kentucky) Public Works Department. Since 2006, he has served as director of research and planning for the Kentucky Association of Counties. He assists counties and cities and their respective agencies and special-purpose governmental entities with community planning, local and regional organizing, and utility-system merger and consolidation, as well as drafting of ordinances and interlocal agreements, identification of project funding sources, and other local government and community support activities.

STEPHEN C. RICHTER is a professor in the Department of Biological Sciences and associate director of the Division of Natural Areas at Eastern Kentucky University. His research focuses primarily on amphibian and wetland conservation from ecological and evolutionary perspectives, and all of his studies have applied goals of producing recommendations for land planners and wildlife managers. Currently, he is working with Dr. David Brown to validate a rapid assessment method for wetlands of Kentucky while simultaneously studying the biological integrity of wetlands. Originally from Tennessee, he received a BS in Biology from Berry College in Georgia and an MS in Biology from Southeastern Louisiana University. After receiving his PhD in Zoology from the University of Oklahoma, he was happy to return to southeastern U.S. habitats, amphibians, and wetland systems.

JACK SCHIEFFER, JD, PHD, is on the faculty in the Department of Agricultural Economics, the University of Kentucky. His research and teaching focus on the economics and policy of environmental and natural resource issues. In the area of water resource management, he has studied policies to reduce nonpoint-source pollution from agriculture, the costs of stream degradation as reflected in home values, water quality trading programs, and the problems of water pollution in urban areas.

ANGELA SCHÖRGENDORFER received an MS in computer science from the Vienna University of Technology in Austria, specializing in statistics and intelligent systems. After completing a PhD in statistics at the University of Kentucky, she was a postdoctoral researcher in statistical analysis and forecasting at the IBM T. J. Watson Research Center in Yorktown Heights, New York. She currently works for Google, Inc., as a quantitative analyst for Google Maps and Geo products.

SHAUNNA L. SCOTT is an associate professor of sociology and director of the Appalachian Studies Program at the University of Kentucky. She is also the editor of the *Journal of Appalachian Studies* and coeditor of *Studying Appalachian Studies: Making the Path by Walking.* Her work has been published in such journals as *American Ethnologist, Organization and Environment, Qualitative Sociology, Rural Sociology,* the *Journal of Appalachian Studies,* and *Appalachian Journal.* She is the past president of the Appalachian Studies Association.

JEFFREY W. STRINGER is an extension professor of hardwood silviculture and forest operations in the Department of Forestry at the University of Kentucky. He conducts research and develops science and technology delivery programs associated with silviculture and timber harvesting in eastern hardwoods. He is the author of the Silvicultural Best Management Practices guidelines used in Kentucky and has conducted numerous studies associated with the use of best management practices. He is currently chair of the Kentucky Forestry Best

Management Practices Board and director of the Kentucky Master Logger program.

KELLY TAYLOR has been a science educator since 1979, when he began work as a classroom teacher for Fayette County Public Schools. During a twenty-seven-year career with FCPS, Taylor served as Kentucky Science Teachers Association president and has conducted professional development activities for teachers. Since retiring from classroom teaching in 2006, he has worked in the University of Kentucky College of Education as a supervisor of student teachers and adjunct instructor. He has also worked in the UK College of Agriculture, Food and Environment as a program instructor for several projects involving environmental science and water conservation.

JAMEY WIGLESWORTH has spent most of his life on the banks of the Kentucky River. He has a two-year-old son named Sawyer and two dogs, Satchmo and Pearl. He is an avid birdwatcher and chess player. He hosts bar trivia in the Lexington area and travels the country to see live music, but he is mostly on his way back to the river.

TANJA N. WILLIAMSON is a research hydrologist at the U.S. Geological Survey (USGS) Indiana-Kentucky Water Science Center. She joined the USGS in 2006 as a hydrologist; before that, she was an assistant professor at University of the Pacific, as well as a collaborative consultant for the Ministry of the Environment in Ontario, Canada. Currently, much of her time is spent researching the potential impacts of land use and climate change on stream flow and water availability for human and ecosystem needs. Additionally, she is working on projects that examine the interaction of agriculture and mining with stream systems. She earned a BS in geosciences from Pennsylvania State University and an MS in earth science and a PhD in soil and water science from the University of California at Riverside.

COREY L. WILSON received a BS in landscape architecture from the University of Kentucky in 2008. Upon entering the program in 2004, she quickly identified an interest in water quality and watershed-based landscape research, which she developed throughout her time as an undergraduate through a variety of venues, including projects inside and outside the classroom. She was involved in a number of water quality and watershed-based research and community education/outreach projects in association with the Environmental Protection Agency, the Kentucky Division of Water, and the Kentucky Waterways Alliance and worked as a full-time research assistant to Brian Lee when the Watershed Atlas Project was completed.

EMMA L. WITT received her PhD in soil science from the University of Kentucky in 2012, where her research focused on the Robinson Forest SMZ study. Before her PhD, she completed an MS at the University of Minnesota, also in soil science, where her work focused on forest-fire impacts on mercury in the southern boreal forest. She is a 2004 graduate of the Natural Resources and Conservation Management undergraduate program at UK. She is originally from Lexington, Kentucky, and is currently an assistant professor of environmental studies at Stockton University in Galloway, New Jersey.

DEMETRIO P. ZOURARAKIS received his PhD from the University of Kentucky and has been with the Commonwealth of Kentucky since 1994, working for several state government agencies in various scientific and technologic capacities. His career path has focused on projects and activities related to geospatial analysis for the evaluation, assessment, and monitoring of natural resources, best management practices, and environmental topics. Currently a certified GISP and CMS, he is a GIS and remote sensing analyst with the Division of Geographic Information within the Commonwealth Office of Technology, he has served as technical lead and principal investigator in projects funded by the National Aeronautics and Space Administration and the Environmental Protection Agency; project emphasis was for mapping and quantifying statewide land cover and land use, and their spatio-temporal changes, and geospatial data provisioning for watershed modeling in Kentucky. Demetrio makes his home in Versailles, Kentucky.

Index